Further Advances in the Forest Products Industries

Peter W. Hart, Volume Editor

Brian N. Brogdon, Kenneth M.Nichols and Brian P. Roy
Volume Co-Editors

S.B. Agarwal
Phillip J. Allen
S. Aziz
Sujit Banerjee
P.D. Bennett
J.R. Boerner
Brian N. Brogdon
B.J.W. Cole
Jeffrey Colwell
D.C. Dayton
R. deJong
Donald R. Dimmel
S.T. DiNovo
H.S. Dougal
H. Jeff Empie
Matthew Estes
W.J. Frederick
J.M. Genco
Michael E. Hittmeier
S. Jansson
S.S. Kelly
Gopal Krishnagopalan
David C.S. Kuhn
H.B. Kwon
J. Li
Steven Lien
B.K. Lönnberg
Ted Mao
Thomas J. McDonough
W. Miller
T.L. Moon
R. Nagarajan
Kenneth M. Nichols
D.I. Orloff
Lawrence P. Otwell
V.R. Parthasarathy
J.C. Ransdell
J. Rogers
David Roy
Alan Rudie
Douglas B. Samuels
Victor M. Saucedo
Wei Su
Honghi Tran
V. Turt
Tamara S. Van Til
Vipin K. Varma
K.J. Wåg

AIChE Staff
Maura N. Mullen, Managing Editor
Cover Design: Armand Veneziano

AIChE Symposium Series

Number 315 1997 Volume 93

American Institute of Chemical Engineers

345 East 47 Street New York, N.Y. 10017

American Institute of Chemical Engineers
345 East 47 Street, New York, N.Y. 10017

AIChE will not be responsible for statements or opinions advanced in papers or presented in its publications.

Library of Congress Cataloging-in-Publication Data

Further advances in the forest products industries / Peter W. Hart, volume editor ; Brian N. Brogdon, Kenneth M. Nichols, and Brian P. Roy, volume co-editors.
p. cm. -- (AIChE symposium series. ISSN 0065-8812 ; no. 315, v. 93)
Includes index.
ISBN 0-8169-0740-4
1. Forest products. I. Hart, Peter W. II. American Institute of Chemical Engineers. III. Series: AIChE symposium series ; no. 315.
TS820. F87 1997 97-20716
676--dc21 CIP

FOREWORD

The 1996 Forest Products Symposium represents a compilation of original research papers covering the various interests of the Forest Products Industries. These papers were obtained from the ten conference sessions, sponsored by the Forest Products Division of AIChE during 1996. Additional contributors were obtained through original manuscripts submitted by several academic and industrial researchers for possible inclusion in this volume. A total of 31 papers were submitted for peer review. After careful consideration and review, a total of 17 papers were accepted for publication. Every effort was made to select top quality papers representative of all aspects of the Forest Products Industries. The range of studies reported within this volume vary from theoretical, fundamental understanding to practical application and mill-based experiences. This Forest Products Symposium volume should provide a valuable resource to future academic and industrial engineers working in the Forest Products Industries. The assistance of the authors regarding manuscript preparation in a timely manner and of the volume co-editors is acknowledged and greatly appreciated. The assistance of session chairs in obtaining speaker manuscripts is also welcome. Without the efforts of these people this volume could not have been prepared. The aid of Maura Mullen, AIChE Publications Production Manager, is gratefully acknowledged and she is thanked for producing this volume.

Peter W. Hart, Volume Editor
Westvaco Corporation
Pulping and Bleaching Group
Charleston Research Center
Charleston, SC 29423-8005

CONTENTS

Foreword iii

Advances in Wood Delignification and Bleaching

Application of Alkali Profiling and Real-Time Liquor Analysis to Modified Kraft Pulping
Vipin K. Varma, Victor M. Saucedo and Gopal Krishnagopalan 1

Delignification of Shredded Spruce and Fir Wood Using Sodium Carbonate and Oxygen
J.C. Ransdell, J.M. Genco, S.T. DiNovo and B.J.W. Cole 7

Optimization of Oxygen and Alkali Charges for Mill Scale Oxygen Delignification
David Roy 20

Oxygen Delignification of Southern Hardwoods
S.B. Agarwal, H.B. Kwon, J.M. Genco, W. Miller and B.J.W. Cole 26

A Fundamental Study of a Novel Extraction Stage for an Organoslov-Based Bleaching Sequence
Brian N. Brogdon, Donald R. Dimmel and Thomas J. McDonough. 42

Physico-Chemical Processes Limiting Lignin Removal During Bleaching Extraction
Brian N. Brogdon, Donald R. Dimmel and Thomas J. Mc Donough 52

Advances in Black Liquor Combustion

Distribution of Mass Flows in Black Liquor Sprays
H. Jeff Empie, Steven Lien and Douglas B. Samuels. 58

Characterization of Black Liquor Char Gassification Using Thermogravimetry and Molecular Beam Mass Spectrometry
K.J. Wåg, W.J. Frederick, D.C. Dayton and S.S. Kelley 67

A Fundamental Study of Carryover Deposition in Kraft Recovery Boilers
Ted Mao, David C.S. Kuhn and Honghi Tran 77

Advances in Recycled Fiber Processing

Novel Approaches to Deinking of Xerographic and Laser-Printed Paper
T.L. Moon, J. Li and R. Nagarajan 87

Bleaching of Colored Broke by Elemental Chlorine Free (ECF) and Totally Chlorine Free (TCF) Processes
V.R. Parthasarathy 96

Repulping of Wet-strength Poly-coated Cartons Using Non-chlorine Chemicals
R. deJong, J. Rogers, S. Aziz and H.S. Dugal 104

Advances in Corrosion, Environmental, and Pulp Refining Areas

Corrosion of Composite Tube Air Ports in Kraft Recovery Boiler - Cr_2O_3 Solubility in NaOH
Matthew J. Estes, Alan W. Rudie and Jeffrey A. Colwell 110

Acoustic Enhancement of Dry HCI Scrubbing With Calcium Oxide
J.R. Boerner and D.I. Orloff 116

Operating Experience With a Full Scale Biofilter at a Hardboard Mill
Phillip J. Allen and Tamara S. Van Til 124

Reducing VOCs From Wood Drying By Wet Line Extension
Sujit Banerjee, Wei Su, Lawrence P. Otwell, Michael E. Hittmeier and Kenneth M. Nichols 130

Effect of Specific Edge Loading on Freeness Drop Coefficient
P.D. Bennett, V. Turt, J.M. Genco, S. Jansson and B.K. Lönnberg 134

Index 146

Application of Alkali Profiling and Real-Time Liquor Analysis to Modified Kraft Pulping

Vipin K. Varma, Victor M. Saucedo and Gopal Krishnagopalan
Department of Chemical Engineering, Auburn University, Auburn, AL 36849

The effects of alkali profiled modified kraft cooking and the addition of Anthraquinone (AQ) on pulp properties are discussed in this work. The laboratory system used for the study allows for on-line measurement and control of the effective alkali (EA) concentration in the liquor and helps maintain the EA concentration as even as possible at the desired level during the bulk delignification phase. Anthraquinone (AQ), viewed as a pulping catalyst has also been used in conjunction with alkali profiling. The advantages of using AQ along with liquor profiling are also discussed.

INTRODUCTION

The emphasis of the pulp and paper industry has been on extended kraft cooking ever since the need for reducing the amount of residual lignin entering the bleaching process was felt. The goal of extended kraft cooking is to enhance the delignification, without detrimental effects on the pulp strength properties. In the production of bleached pulp, this would be translated into a more economical and environmentally friendly approach than traditional kraft cooking. In a traditional kraft cook, the period of cook cannot be extended without significant losses to pulp strength and yield. Studies done in the past have shown that pulp properties vary unpredictably with increase in cook time in a normal kraft cook [1]. The techniques proposed to circumvent these problems apply the principle of modifying the cooking chemistry. A commonly used approach is to use additives like hydrogen sulfide, polysulfide, and anthraquinone (AQ) [2]. Another approach is to adjust the concentration of liquor components by liquor profiling [3].

Modified kraft cooking using profiling techniques can be classified into front-end modifications and back-end modifications. In the former case, the entire course of the cook is affected, while in the latter the effect is observed towards the end of the cook. In alkali (OH^- and HS^-) profiling, the hydroxide ion concentration is lower in the initial delignification stage when compared to a normal cook and is maintained at the desired level during the bulk delignification stage. The hydrosulfide ion concentration is kept at high levels during the initial stages of the cook [1]. This essentially falls under the category of front-end modifications. In back-end modified systems, the concentrations of lignin and sodium are kept as low as possible in the residual delignification phase [4].

Several commercial implementations of modified kraft cooking are available. Modified Continuous Cooking (MCC), Extended Modified Continuous Cooking (EMCC), Rapid Displacement Heating (RDH), SuperBatch, and Cold Blow cooking are some of them. MCC and EMCC, implemented in continuous digesters, involve back-end modifications by splitting the alkali addition into several portions. Counter current cooking stages lower the dissolved solids concentration in the back-end [4]. Displacement batch cooking (RDH/SuperBatch) involves impregnation with warm black liquor, followed by hot black liquor and finally cooking with white liquor. This black liquor is mildly alkaline and has a high concentration of hydrosulfide ions. The RDH and SuperBatch processes are front-end modifications and claim enhanced lignin removal and energy savings due to the reduced heating requirements [2,5].

In a normal kraft cook, the initial phase of delignification is accompanied by rapid degradation of the hemicelluloses (galactoglucomannans in softwood) by the peeling reaction. Hydroxy carboxylic acids are the main degradation products. In the case of pine, Glucoisosaccharinic acids are formed due to the degradation of the glucomannans [6]. When the principles of profiling

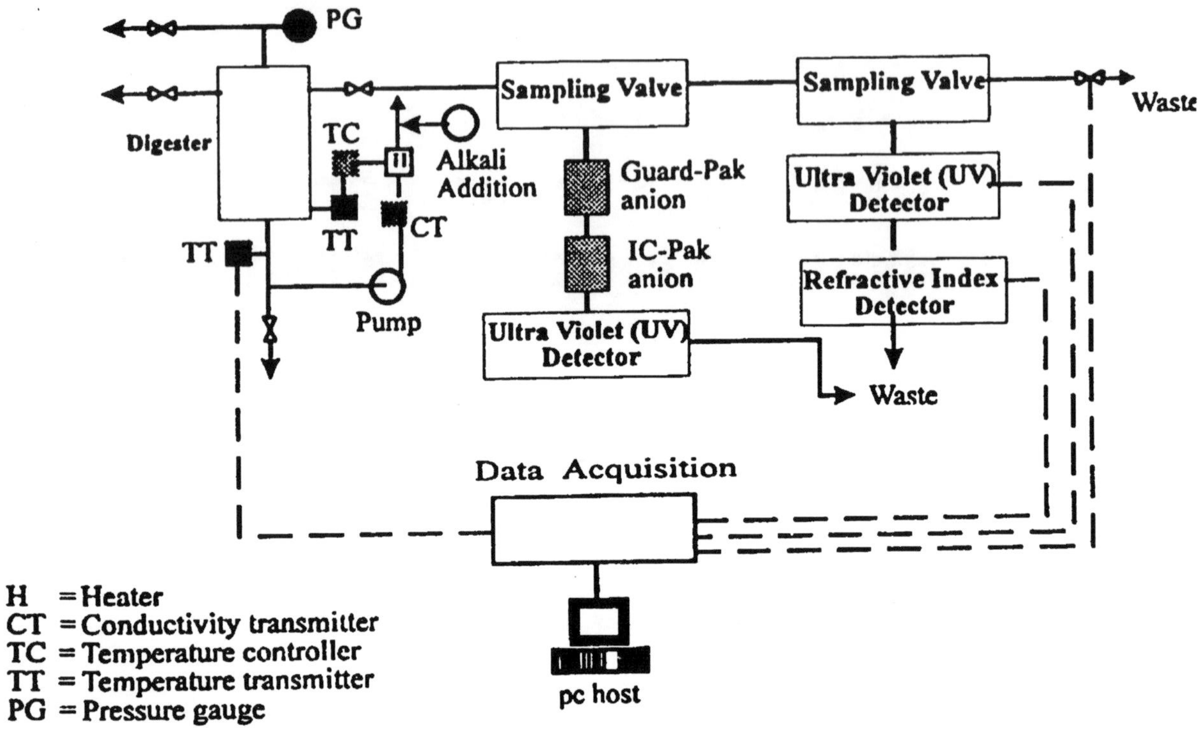

Figure 1. Batch Digester and liquor analysis system

are used in a cook, the process becomes more selective towards lignin removal. Carbohydrate removal during the initial phase is suppressed due to the combined effect of high initial sulfidity and low OH^- ion concentration [1]. Anthraquinone (AQ) is useful for yield and pulping rate improvements, and certain other quinone additives also show similar characteristics [2,7,8]. The dual action of AQ is because it takes part in a redox cycle causing a stabilization of the reducing end-groups in polysaccharides, and also accelerating the cleavage of the lignin polymer [2].

In the present study, liquor profiling is done in a laboratory batch digester, and the EA concentration is monitored using an *in-situ* conductivity sensor. The focus of this work is to investigate the use of alkali profiling as a means of enhancing lignin removal. Secondly, the possibility of using Anthraquinone (AQ) in tandem with profiling, towards achieving similar goals, has also been investigated. Pulp properties being considered are the final kappa number, total pulp yield, and pulp viscosity.

EXPERIMENTAL

Pulping

Figure 1 shows the digester and the associated liquor analysis system used for data collection. The basic digester, supplied by M/K systems, is 6.5 liters in capacity and is equipped with a liquor recirculation pump. An independent controller is used to maintain the temperature during pulping. The liquor heat-up rate can be varied by adjusting the controller settings. In this study, the rate of heat-up to reaction temperature has been kept constant at about 3.75°C/min. The recirculation line is equipped with a Rosemount series 78S platinum RTD temperature sensor, a Micro Motion model D25 density/mass flowmeter, and a Rosemount 222 Toroidal flow-through electrodeless conductivity sensor. A positive displacement pump is connected to the recirculation line near the point of entry into the digester and is used for alkali addition during the bulk phase of delignification.

The liquor analysis system has two sample injection valves with internal loops of 1ml each. Samples are extracted and injected into the various analytical equipment every 5 minutes. A Waters IC-pak anion column, connected to a Waters model 440 dual channel UV absorbance detector, constitutes the on-line sulfide content analysis system. The other sample loop is connected to a Waters Lambda-Max 481 UV spectrophotometer and a Waters 410 Differential Refractometer (RI). UV absorbance at 280nm gives the real-time lignin concentration and the RI detector is calibrated to give the dissolved solids concentration. The on-line liquor analysis system which allows estimation of pulp lignin content and pulp yield, from measurable liquor properties, has been used to model the delignification dur-

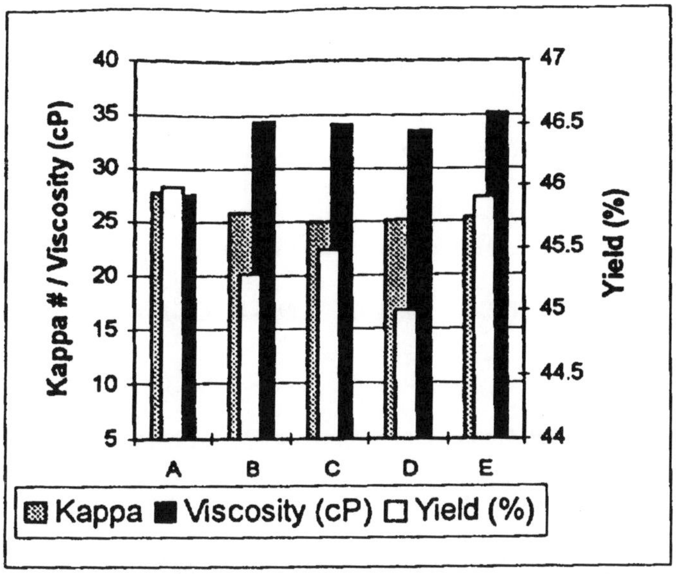

A= Regular Kraft
B= Profile 30 / 12
C= Profile 24.5 / 13
D= Profile 24.5 / 13.5
E= Profile 23 / 14

Figure 2. Effect of Profiling on pulp properties

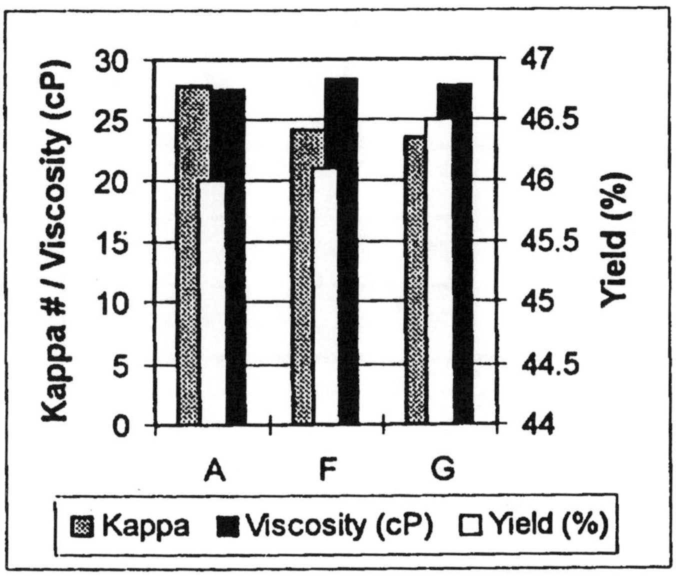

A= Regular Kraft
F= Reg.+ 0.05%AQ
G= Reg. + 0.1%AQ

Figure 3. Effect of AQ on Pulp properties

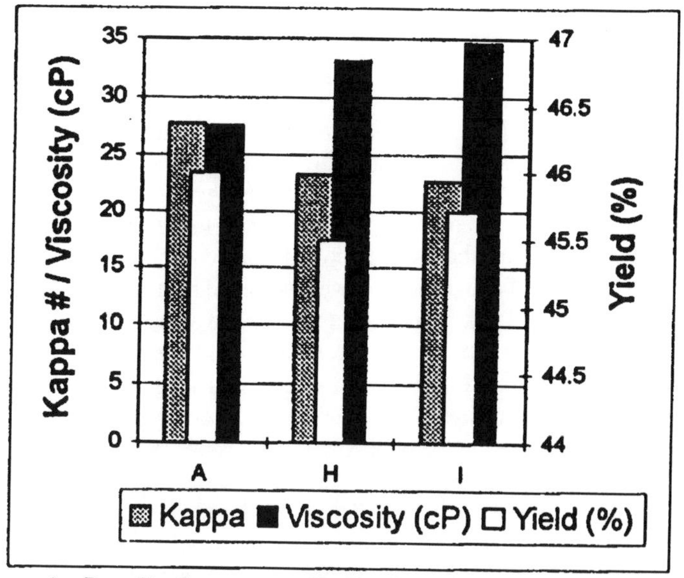

A= Reg. Kraft
H= Profile 30/12 + 0.1%AQ
I= Profile 24.5/13 + 0.1%AQ

Figure 4. Effect of profiling /AQ on pulp properties

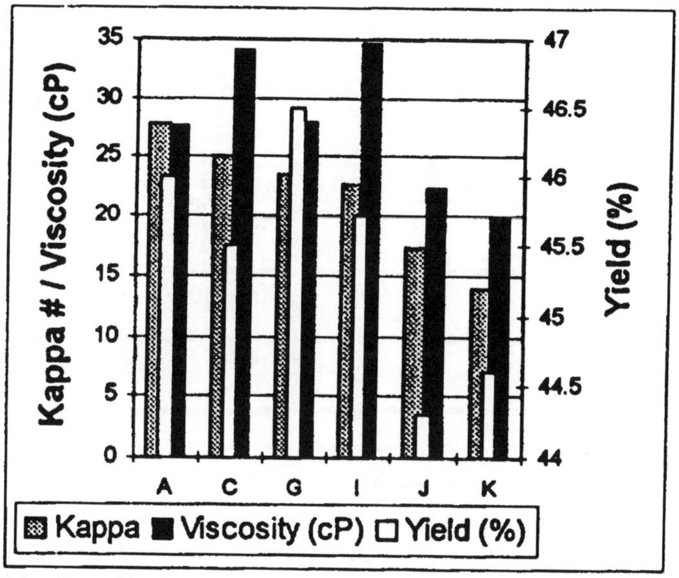

A= Reg. Kraft
C= Profile 24.5/13
G= Reg. + 0.1 %AQ
I= Profile 24.5/13 + 0.1% AQ
J= High EA (25%) profile
K= High EA profile +0.1%AQ

Figure 5. Effect of profiling/AQ on pulp properties

ing the initial and bulk phases in several studies at Auburn [9,10]. Details of the liquor analysis system, and the associated calibration equations are given elsewhere [11].

Hydroxide ion (OH^-) profiling was achieved by injecting sodium hydroxide into the recirculation line of the system using an external positive displacement pump. Batch cooks were performed using 3-5 mm thick southern pine chips, with a liquor to wood ratio of 5.5. In the present study, the lower hydroxide ion concentrations during the initial phase of delignification were achieved by adding all the sodium sulfide and part of the sodium hydroxide at the beginning of the cook. The rest of the sodium hydroxide required for a particular effective alkali charge (EA, % on o.d wood), was added using the external pump to maintain the effective alkali (EA) concentration at the desired level during the bulk phase of the cook. The split addition of alkali resulted in low EA levels and high sulfidities during the initial phase and a desired uniform EA profile during the rest of the cook. Initial sulfidity values in the experiments ranged from 50-80% as opposed to 25-30% encountered in a normal cook. Conductivity measurements and Ion-chromatography measurements were used to obtain the effective alkali (EA) of the liquor at any time. This EA measurement was used for flow-rate adjustment of the external pump, thereby maintaining desired EA level in the liquor.

Liquor Analysis

Calibration equations have been developed, correlating white liquor and black liquor EA to liquor conductivity, incorporating appropriate temperature corrections. In the case of cooking liquor, the dissolved wood substances are also accounted for, as shown in the correlation given below [11]

$$NaOH + 0.5Na_2S = 4.048 \times 10^{-2}\,C + 2.918 \times 10^{-5}C^2 - 0.386Na_2S - 0.283Na_2CO_3 - 0.0694L_{280} \quad (1)$$

where

$NaOH$ = Sodium hydroxide, gpl as Na_2O

Na_2S = Sodium sulfide, gpl as Na_2O

Na_2CO_3 = Sodium carbonate, gpl as Na_2O

L_{280} = Lignin concentration in liquor, gpl

C = Conductivity, mS/cm

Table 1. Enhancing delignification with profiling and AQ

Cook	Initial Chemical Charge (% total)		Initial EA (gpl Na20)	Overall S (%)	EA maintained in Bulk phase (gpl)	Kappa Number	Yield [%b.d wood]	Viscosity cPoise
	NaOH	Na2S						
Kraft 20%EA	100	100	49	30		27.8	46	27.8
Kraft + 0.05%AQ	100	100	49	30		24.2	46.1	28.4
Kraft + 0.1% AQ	100	100	49	30		23.6	46.5	27.9
Profile 1	43	100	29.8	30	12	26	45.3	34.4
Profile 2	32	100	24.5	30	13	25	45.5	34.1
Profile 3	32	100	24.5	30	13.5	25.2	45	33.6
Profile 4	28.5	100	22.8	30	14	25.4	45.9	35.1
Profile 1+ 0.1% AQ	44	100	29.8	30	13	23	45.5	33.1
Profile 3+ 0.1% AQ	32	100	24.5	30	13	22.7	45.7	34.6
High EA Profile	25	100	27.1	30	15.8	17.3	44.5	22.3
High EA Profile +0.1% AQ	25	100	27.1	30	16	14	44.6	19.7

Overall EA 20% (wt.) on wood
High EA 25% (wt.) on wood

Similarly, correlations have been developed for sulfide measurement in white and cooking liquors [12]. Equation 2 gives the relationship for cooking liquor, which accounts for the dissolved lignin.

$$Na_2S = 0.7933*(A_{254} - A_{280}) + 0.0759*EA \quad (2)$$

where

Na_2S = Sodium sulfide, gpl as Na_2O

EA = Effective alkali from Equation 1

$A_{254,280}$ = Absorbances measured at 254 and 280 nm

The two equations are solved iteratively to arrive at EA and Na_2S concentrations at any time.

RESULTS AND DISCUSSION

The purpose of the study was to investigate the effect of split alkali addition, and the addition of Anthraquinone (AQ), on enhancing delignification and on the properties of the resulting pulp. A regular kraft cook performed at 20% EA (on o.d wood) and 30% sulfidity gave a kappa number of 27.8 and was used as a base for comparison. Profiled kraft cooks were performed using the same charge of chemicals. The chemical splitting specifications for the profiles are shown in Table 1. The performances of the different kraft cooks was evaluated by analyzing the pulp samples for kappa number, and determining the yield and viscosity of the unbleached pulp. The kappa numbers and viscosities were obtained in accordance to TAPPI standards. Figures 2-5 illustrate the effects of the cooking conditions on these three pulp properties. In the figures, the initial EA and bulk phase EA for the profiles are indicated as Initial EA Bulk Phase EA. Subsequent discussions also indicate the EA levels in parentheses.

Profiled Cooks

Figure 2 shows the comparison between the regular kraft cook (base) and the profiled cooks carried out with the same charge of chemicals. The profiles being compared for their performance are Profiles (30/12), (24.5/13), (24.5/13.5), and (22.8/14). Profile (24.5/13) resulted in a kappa number drop of 3 units when compared to the base kraft cook and showed a 25% increase in viscosity, at comparable yields. However, Profile (30/12) was not as effective and showed a kappa number drop of 2 units, at comparable viscosities. Earlier studies on the effect of initial phase alkali level have shown that high initial EA results in lower alkali availability during the bulk phase, for the same amount of chemicals, thereby resulting in higher kappa numbers [13]. Insufficient alkali during the initial phase, on the other hand, results in a higher kappa number for the same total alkali charge [13]. The low initial EA in profile (22.8/14) was the reason for the slightly higher kappa number, although viscosity was comparable. This phenomenon can be attributed to insufficient impregnation of alkali during the initial phase, caused by the low EA level. These results serve to substantiate the presence of an optimum initial EA between 23.5-27 gpl, as indicated in earlier studies [13].

Kraft/AQ cooks

AQ levels of 0.05% and 0.1% (on o.d wood) were used in regular kraft cooks. AQ was added and a batch kraft cook was performed at 20% EA and 30% sulfidity. The cooks done with 0.05% and 0.1% AQ charges gave kappa number reductions of 3.6 and 4.2 units respectively. The kappa numbers, yields and viscosities are shown in Figure 3. Due on the slightly enhanced lignin removal at 0.1% AQ, this level of AQ addition was chosen for subsequent profiled cooks.

Profile/AQ cooks

Profiled cooks were carried out with 0.1% AQ charge to determine whether AQ and profiling had synergistic effects. Profile (24.5/13) with 0.1% AQ gave a kappa number drop of 5 units compared to conventional kraft, at comparable yields. A 26% increase in viscosity was also observed, indicating using AQ along with profiling was beneficial in further lowering the kappa number at comparable viscosities. The cook performance was better than that of profile (24.5/13) without AQ. Profile (29.8/13) with 0.1% AQ also performed better than profile (29.8/13) without AQ, in terms of kappa number, viscosity and yield. This indicates the possibility to drive delignification further by supplementing the delignification stage with a combination of alkali profiling, and AQ. The reduced alkali requirement in the presence of AQ, the dual action of AQ, and controlled alkali level seems to work together in enhancing lignin removal, while maintaining pulp properties. The results are shown in Figure 4.

All of the above-mentioned cooks where performed at 20% EA and 30% sulfidity. In order to investigate the possibility of further enhancing lignin removal, cooks were performed at higher EA levels (25% on o.d wood). In the high EA cooks, the initial effective alkali levels were around the optimum level seen earlier and the incremental alkali was used for supplementing the reacting liquor by addition through the external pump. A high EA profile without AQ resulted in kappa number drop of 11 units. The drop in yield was found to correspond to the kappa number drop. However, the viscosity was about 20% lower than that of conventional kraft at 20% EA. Earlier studies have shown that regular cooks at high EA(25%) resulted in an yield loss of 4% and 35%-38% reduction in viscosity compared to 20% EA at the same sulfidity level [13]. This indicates that the incremental EA is more effectively used in profiling. The high EA cook with 0.1% AQ gave a kappa number drop of 14 units compare o the base case. The viscosity in this case was about 25% lower than that of the conventional kraft pulp. This further substantiates the benefit of using AQ in tandem with alkali profiling. Figure 5 shows the comparative performance of these cooks with the best cooks from each category listed above.

The high EA cook with 0.1% AQ shows 50% reduction in kappa number when compared to the base kraft cook. Compared to profile (24.5/13) + 0.1% AQ, there was an additional kappa number drop of 9 units, with corresponding yield loss. However, the associated drop in viscosity is higher and needs to be investigated further. It is expected that EA levels between 20% and 25% on o.d wood, with the incremental EA being used in profiling, might yield pulp with better viscosities. This optimum needs to be identified, and work is in progress to arrive at the alkali level which will give the necessary kappa number drop without reduction in viscosity.

CONCLUSIONS

The results obtained in this study show the efficacy of alkali profiling towards enhancing lignin removal at comparable pulp strength properties. The results also indicate that it is beneficial to use AQ in tandem with profiling, towards achieving the same objective. Further work needs to be done on optimizing the AQ levels that need to be used in combination with profiling, to achieve the same target. Another area needing investigation is the possibility of reducing the overall sulfide charge when carrying out profiled cooks with AQ. Work is in progress to study and optimize these variables.

LITERATURE CITED

1. **Tikka, P. O.**, "Conditions to Extend Kraft Cooking Successfully," *TAPPI Pulping Conference Proceedings* pp 699-706 (Oct. 1992).

2.**Grace, T. M., M. J. Malcolm, eds.**, "Pulp and Paper Manufacture-Alkaline Pulping," 5, *TAPPI Press*, Atlanta, GA, pp 114-123 (1984).

3. **Johansson, B., J. Mjoberg, P. Sandstrom, A. Teder,** "Modified Continuous Kraft Pulping - Now a Reality," *Svensk Papperstidning*, **87**(10), pp 30-35 (Oct. 1984).

4. **Andrews, E. K.**, "RDH Kraft Pulping-liquor Profiling to Extend Delignification, Decrease Effluent Liability and Improve Productivity and Pulp Properties," *TAPPI Pulping Conference Proceedings* pp 607-625 (Oct. 1989).

5. **Pursiainen, S., S. Hiljanen, P. Uusitalo, K. Kovasin, M. Saukkonen**, "Mill-scale Experiences of Extended Delignification with SuperBatch Cooking Method," *TAPPI Journal*, **73**(8), pp 115-121 (Aug. 1990).

6. **Sjostrom, E.** "Wood Chemistry- Fundamentals and Applications," Academic Press, Inc., New York, USA, pp 150-156 (1993).

7. **Holton, H. H., F. L. Chapman**, "Kralt Pulping with Anthraquinone," *TAPPI Journal*, **60**(11), pp 121-125 (Nov. 1977).

8. **Furuya, J.**, "9% Production Gain with Quinone-additive Kraft Pulping in a Batch Digester," *Tappi Journal*, **67** (6), pp 82- 90 (Jun. 1984).

9.**Varma, V. K., A. Krishnagopalan**, "Kinetic Studies of Extended Delignification by Alkali Profiling and Liquor Analysis," AIChE Annual Meeting, 252a, (1994).

10. **Vanchinathan, S., A. Krishnagopalan**, "Kraft Delignification Kinetics Based on Liquor Analysis," *TAPPI Journal* **78** (3), pp 127-132 (Mar. 1995).

11.**Paulonis, M. A., A. Krishnagopaian**, "Kappa Number and Overall Yield Calculation Based on Digester Liquor Analysis," *TAPPI Journal*, **71**(11), pp 185-187 (Nov. 1988).

12. **Liaw, S. J., A. Krishnagopalan**, "On-line Measurement of Sulfide and Alkali Concentrations During Kraft Pulping," *TAPPI Journal*, **75** (6), pp 219- 224 (Jun. 1992).

13. **Varma, V. K., A. Krishnagopalan**, "Selectivity Studies on Extended Delignification Using Alkali Profiling Techniques," *TAPPI Pulping Conference Proceedings,* pp 749- 755 (Oct. 1995).

Delignification of Shredded Spruce and Fir Wood Using Sodium Carbonate and Oxygen

J. C. Ransdell and J.M. Genco
Department of Chemical Engineering, University of Maine, Orono, ME 04469

S.T. DiNovo
Guild Associates, Hilliard, OH 043026

B.J.W. Cole
Department of Chemistry, University of Maine, Orono, ME 04469

Oxygen delignification using sodium carbonate represents a potential alternative pulping method to the kraft process. To minimize effects of low oxygen solubility and oxygen transport from the gas into the solid phase, pulping experiments were conducted in a well-mixed, batch reactor operating at consistencies of 0.5 to 2.0% using shredded feedstock of spruce and fir having an initial freeness of about 700 to 800 ml. Delignification was found to be rapid at 165 C using 40 to 80 psig oxygen over-pressure with an initial carbonate ion concentrations of about 3 to 4 grams per liter. Under these conditions the turbine agitator could supply adequate oxygen to the lignin being de-polymerized. Kappa number values less than 20 could be obtained in 1 hour or less. Strength properties were lower than those obtained from the conventional kraft process although no attempt was made to employ carbohydrate protectors during pulping. Pulp composition, as given by a Ross diagram, was equivalent to that obtained using the kraft process provided the consistency of pulping was about 2% in the well-mixed reactor. The kinetic data could be correlated using an empirical model, based on a reaction rate equation as a function of temperature, Klason lignin content, initial carbonate concentration and oxygen pressure.

$$-dL/dt = k_o \exp(-E_A / RT)\, [CO_3^{-2}]^m\, PO2^n\, L^q \qquad (1)$$

The kinetic parameters ko, EA, m, n and q in the model were determined from experimental data.

INTRODUCTION

The kraft pulping process is the most widely practiced chemical pulping method currently employed in the pulp and paper industry. Approximately 50% of world pulp production (85% of world chemical pulp production) is produced using the kraft pulping process (Biermann, 1993). An environmental concern associated with kraft mills is emission of reduced forms of sulfur and the accompanying odors. Many of the adverse environmental effects could be prevented if an economic sulfur-free chemical pulping process could be developed. One possibility is soda oxygen pulping or its variant the sodium carbonate oxygen pulping process applied to wood; termed the CARBOX process.

Sodium carbonate oxygen delignification represents a promising odor free alternative to the kraft process. One notable advantage of using sodium carbonate as a source of base is its buffering capacity which occurs in the digester resulting from reaction of sodium carbonate with water via the carbonate-bicarbonate equilibria:

$$CO_3^{=} + H_2O = HCO_3^{-} + OH^{-} \qquad (2)$$

Sutton (1987) and Fiala and Nardi (1984) have investigated straw pulping using sodium carbonate and oxygen in a turbo pulper. In application to wood, Marton, Brown and Granzow (1974) report on the oxygen pulping of thermomechanical fiber (shredded wood) in a one stage process using sodium hydroxide and sodium carbonate to eliminate problems with impregnation of chips and diffusion of oxygen.

The sodium carbonate oxygen process differs significantly from the kraft process by the simplification of the chemical recovery process (Kleinert, 1976; and Agarwal, 1993). For this process, shredded wood chips are delignified with a solution of sodium carbonate and oxygen under temperature conditions similar to those used in the kraft process. Deviations from the kraft process include: (1) chip shredding, (2) oxygen production, (3) gas liquid contacting, (4) elimination of the caustization, lime mud washing, calcination, and slaking steps in chemical recovery, and (5) simplification of the recovery furnace by elimination of the need for sulfate to sulfide reduction within the recovery furnace.

Objective and Scope

The development of the soda oxygen process, and its variant sodium carbonate oxygen pulping, has been limited due to three criticisms. These are: (1) mass transfer limitations for oxygen transfer into the liquid phase resulting in long cook times and non-uniform pulping of wood chips, (2) lower pulp strength given by reduced pulp viscosity and wet zero span tensile strength in comparison to kraft pulps (Kleinert, 1976; Abson and Stockman, 1979; and Agarwal, 1993), and (3) reduced heating value of black liquor and lower tall oil recovery (Erickson and Dence, 1975) due to partial oxidation in the digester. The objective of the research reported here was to determine the kinetics governing soda oxygen delignification of shredded wood using sodium carbonate as the base.

RELATED LITERATURE

Ransdell (1996) presents a review of the chemistry of soda oxygen pulping and available mathematical models for quantitative description of the kinetics.

Chemistry of Soda Oxygen Pulping Process

Molecular oxygen reacts with lignin and extractives, and moderately with cellulose and hemicelluloses. The oxidation of wood components initiates the creation of reactive intermediates; organic and hydroxyl radicals, and peroxides (Kleinert, 1976). The radicals form both wood components and dissolved oxygen (Ljunggren and Johansson, 1987; Johansson and Ljunggren, 1994). The protection of cellulose from degradation is essential for the creation of a pulp which is equivalent to kraft pulp in strength (Agarwal, 1993).

Soda Oxygen Reaction Kinetics

The transport of oxygen to the reaction site with lignin in the wood chip has been determined to be a major rate controlling step in oxygen delignification reaction kinetics (Nagano et al., 1974). Mass transfer problems have been overcome by several researchers by introducing new methods of oxygen dispersal and transport. These include the use of continuous reactant flow (Nagano et al., 1974), and rocking or pulsing digesters (Rodden, 1990). The transport of oxygen has also been enhanced through the use of the chip size reduction techniques of shaving or mechanical refining prior to delignification (Agarwal, 1993).

Commercially, oxygen has found success when used as a delignification agent for bleaching of pulp fibers. The lower lignin content of pulp and ease in mass transport to reach the reaction site make oxygen more suitable for bleaching. The major impediments to the use of oxygen as a pulping reagent are: (1) the oxygen concentration near the reaction site with lignin is reduced due to the high reactivity of oxygen with all other wood components, and (2) the low solubility of oxygen, in the high temperature alkaline solution, limits the amount of oxygen available for reaction.

Soda Oxygen Delignification Models

Edwards and Norberg (1973) presented a simple model for oxygen bleaching and kraft pulping delignification kinetics. The reaction rate for oxygen delignification of kraft brownstock pulp is expressed as a function of lignin concentration (as a dimensionless kappa number) alkali concentration, bleaching time, and reaction temperature. The model is empirical, but contains all of the major variables thought to affect the rate of delignification for brownstock pulp.

$$r = k\,(L - L_f)^a\, C^b \qquad (3)$$

where r is the rate of lignin destruction, C is the concentration of hydroxide ion, L_f is the final residual lignin, and the coefficients (a) and (b) are the order of the reaction.

Sarkanen and Johanson (1976) proposed a pseudo-first order reaction model $(-dL/dt = k_L L)$ for the delignification of softwood meal. An activation energy of 30.9 kcal/mole was obtained for delignification conducted at pH 10.0. The effect of pH, given in terms of the hydroxyl concentration and oxygen pressure was determined from appropriate kinetics data. The following empirical kinetics rate equation was presented based on their research.

$$k_L = A\ e^{-30.9/RT}\ [OH^-]^{0.30}\ [P_{O2}]^{0.3} \qquad (4)$$

It was also shown that the percent carbohydrate retained is a function of the pH of pulping. Higher pH (12.9) led to significant carbohydrate losses even in excess of the kraft pulping process. The use of the equation for delignification of wood chips is hypothesized and a set of proposed oxygen concentration gradients within the wood chip are shown.

Olm and Teder (1979) proposed a general kinetics model for oxygen delignification of kraft brownstock with initial kappa number of 30 (bleaching), which has come to be accepted as an industry standard for bleaching reaction kinetics. Softwood brownstock was delignified under constant and well defined conditions to low kappa number. Two delignification regimes exist when the data are plotted as log kappa number versus time. In the first regime, lignin is removed quickly. In the second regime, the delignification rate decreases significantly. A long slow delignification period follows the initial delignification period. The following model equation was presented by Olm and Teder (1979) and assumes two pseudo-first order parallel reactions.

$$-dK/dt = k_1 [OH^-]^{0.1} P_{ox}^{0.1} K_1 + k_2 [OH^-]^{0.3} P_{ox}^{0.2} K_2 \quad (5)$$

where, $K = K_1 + K_2$ = Total Lignin (6)

and K_1 and K_2 are the amount of easily and slowly eliminated lignin respectively.

In the initial stage, the reaction rate is about 20 times higher than in the final stage. The oxygen bleaching experiments were carried out in a laboratory digester with forced circulation. The pulp was constrained in a wire basket at 8% consistency (11.5 g liquid/gram pulp). The ratio of bleaching liquor to brownstock pulp was 200g liquor/gram pulp to keep the alkali concentration constant during the bleaching without alkali addition. Also, a high supply of oxygen going to the reaction was maintained. The Olm and Teder model expresses oxygen delignification as two pseudo-first order reactions. Additionally, the reaction selectivity was shown with a curve of bleaching time versus carbohydrate intrinsic viscosity and number of cellulose chains.

Hsu and Hsieh (1988) reported that many inaccuracies existed for the commercial application of Olm and Teder's model (1979). They went on to prove that the reaction rate model was inaccurate at low pulp densities. Reaction kinetics were much more rapid than given by Olm and Teder when pulp was bleached at low consistency. Hsu and Hsieh (1988) modified Olm and Teder's (1979) model to have a form:

$$-dK/dt = k \exp(-E/RT) [OH^-]^m P_o^n K^o \quad (7)$$

A non-linear least squares Gauss-Newton method was used to estimate the coefficients m, n, and o. This modification was made following experimentation at ultralow pulp consistency in the region where inter-phase mass transfer rates were found to have no effect on the overall delignification rate.

As with Olm and Teder (1979), Hsu and Hsieh (1988) found that a two stage delignification model better represented the data by model fit.

$$-dK/dt = 2.46 \exp(-3.6x10^7/RT) [OH^-]^{0.78} P_o^{0.35} K^{3.07} [u(t) - u(t - 2)] +$$

$$143.49 \exp(-7.1x10^7/RT) [OH^-]^{0.70} P_o^{0.74} K^{3.07} [u(t - 2)] \quad (8)$$

The unity functions, u(t) and u(t-2), were used to imply that in the first two (2) minutes of reaction time, the kinetics equation as described by the first term on the right hand side; during the rest of the time, the reaction is described by the second term.

Myers and Edwards (1989) used a two stage, first order kinetic model to establish general rate equations for oxygen delignification of kraft pulps. The model was similar to the one given by Hsu and Hsieh (1988), in that the delignification rate was broken into two periods, one rapid and one slow. Oxygen was modeled kinetically as dissolved oxygen concentration in the liquid phase, expressed as kg/m^3, as opposed to using the partial pressure of oxygen. The rate of oxygen delignification of eucalyptus kraft pulp was found to be a function of temperature, lignin content, applied oxygen pressure, and alkali concentration by Adam, Vincent, Nguyen, and Mathews (1994). Existing rate equations for oxygen delignification of kraft pulps proposed by Myers and Edwards (1989) were shown to be inadequate for predicting accurate delignification profiles for eucalyptus kraft pulps. Alternative rate equations were presented, based upon oxygen delignification of two eucalyptus kraft pulps of different initial kappa numbers.

The development of a suitable rate expression for describing the complex solid, liquid, and gaseous reactions which comprise the soda oxygen pulping process beginning with wood has not been previously reported in the literature. The oxygen bleaching studies starting with low lignin content pulps (Olm and Teder, 1979; Hsu and Hsieh, 1988) have been expanded in this study to shredded softwood.

EXPERIMENTAL

Details of the experimental methods are summarized in the Appendix. In preliminary experiments, two digesters of different design were used to deduce the effects of mixing. In these experiments, a slow-speed rocking digester and a well-mixed digester were used to study and compare oxygen delignification. In the rocking digester, the pulp slurry was cascaded through the oxygen phase. Pulping in this cascade reactor was then compared to pulping in a well-mixed stirred tank. These reactors represented extremes in movement of the contens of the digester as they relate to transfer of oxygen into the liquid phases.

In the well mixed continuous stirred tank, pulping experiments covered a range of temperatures (300, 330, and 345 °F), oxygen pressures (41 and 82 psig), carbonate concentrations (2.95, 3.94, and 4.92 g/l as $CO_3^=$), and reaction time (0 to 2 hours) for the development of data to fit an experimental model. An effort was made to identify and minimize gas phase and liquid phase mass transfer as these effects will differ as a result of reactor selection. Solid phase mass transfer has been determined in this study to be negligible for a feedstock of shredded wood. A stirred batch tank reactor and a rocked batch tank were used in these experiments. The stirred tank was operated at 800 RPM and 2.0 or 0.5% consistency. The rocked tank was rotated through a 180° arch at 2 and 6 rpm. Three sodium carbonate levels were used; 60% on wood as Na_2O (330°) {5.25 grams of Na_2CO_3 per liter}, 80% on wood as Na_2O (300°, 330°, and 345°) {7.0 grams of Na_2CO_3 per liter}, and 100% on wood as Na_2O (330°) {8.75 grams of Na_2CO_3 per liter}. The reaction time was selected based upon the reaction temperature.

RESULTS AND DISCUSSION

Mixing Effects

The comparison of pulps produced by the soda oxygen process with a well mixed reactor (see Figure 1) as opposed to rocking for mixing reveals significantly increased reaction time for rocking. A typical pulping time of 3.25 hours in the rocking digester could be reduced to 0.75 hour by mechanical mixing. In well mixed experiments the rotational speed was 800 rpm and the consistency was 2%. Hsu and Hsieh found that mass transfer effects were minimal at a consistency of 0.4%. This indicated the great importance of gas to liquid phase mass transfer. A low consistency of 0.5% to 2.0% ensures an adequate supply of oxygen at the reaction site, minimizes mass transfer effects in the liquid phase, and avoids deleterious lignin condensation reactions which inordinately extend the period of exposure to oxygen and thus loss of pulp strength. In Figure 1, the high Kappa Number values were estimated from gravimetric lignin determinations.

The rocking digester facilitates the transfer of oxygen to the liquid phase by creating a large gas/liquid surface area through cascading. The wood undergoes only moderate levels of shear in comparison to the well mixed reactor. This data would suggest that the availability of bulk phase oxygen for reaction is rate controlling as opposed to the degree of bulk phase agitation intensity. High agitation intensity during pulping is thought to be the reason for the lower pulp strength of soda oxygen pulps produced in the well-mixed digester.

Figure 1 also compares pulping of shredded wood at 2 rpm rocking speed versus 6 rpm. At 6 rpm an increased dissolution rate of oxygen into the reaction environment is hypothesized as the reason for the increase in delignification rate. An increase in mixing speed signifies an increase in the number of times that the pulp slurry will cascade through the gas phase. This will promote gas dissolution into the liquid phase as well as more uniform liquid mixing. The rate of delignification is dependent on dissolved oxygen availability in the liquid phase. The higher rate of cascading for the pulp slurry across the gas phase promotes more rapid delignification. This data supports the conclusion of Nagano et al. (1974) in which the most rapid delignification was achieved when oxygen replenishment was highest.

Delignification Kinetics

Thirty-two pulping experiments were performed to develop a data-base for evaluation of a reaction kinetic model. The pulp consistency was chosen to be 0.5% to minimize mass transfer effects as shown by Hsu and Hsieh (1988). The sodium carbonate concentrations used in the 0.5% consistency experiments (2.95, 3.94, and 4.92 g/l as CO_3^{-2}) were chosen to ensure $CO_3^{=}$ ion concentrations equivalent to those present when the consistency values are 2 to 3%; which is more typical of what would be used commercially rather than 0.5%.

Effect of Pulping Temperature. Of the three variables studied, that is the temperature, oxygen pressure and sodium carbonate concentration, temperature influences the delignification rate to the greatest extent. Oxygen delignification is a very strong function of temperature and is illustrated clearly in Figures 2 and 3. Figure 2 compares the Klason lignin versus time data at 300 and 330 °F; while Figure 3 summarizes similar data at 330 and 345 °F. Raising the temperature greatly increases the rate of reaction, decreasing the lignin content of the pulp. Kappa number values less than 20 can be obtained in 30 minutes at 330 F.

Effect of Oxygen Pressure. During the preliminary study, oxygen availability was determined to play a significant role in the soda oxygen pulping process, in the comparison of the use of the rocking and well mixed digesters. However, under well mixed conditions at the low pulp consistency of 0.5%, the increase in the oxygen pressure from 41 psi to 82 psi had only a minimal effect on the delignification rate. This is illustrated in Figure 4. This was interpreted to mean that under conditions of low consistency and high turbulence an abundance of oxygen is available for reaction, limiting the effect created by changes in the oxygen pressure.

Sodium Carbonate Concentration. In preliminary experiments performed at low sodium carbonate dosage (20% as Na_2O on wood at 0.5%) consistency, delignification was slow. These results were interpreted to mean that lignin fragments, which increase in hydrophillicity on reaction with OH^-, were not being effectively dissolved in the aqueous phase because of the low concentrations of sodium carbonate. Higher values of sodium carbonate, expressed as percent on wood, were used to avoid low OH^- concentration in the liquid phase. Sodium carbonate concentration variation (60%, 80%, and 100% as Na_2O on wood at 0.5% consistency) did not result in a significant change in delignification behavior for softwood pulp. This is illustrated in Figure 5.

Rejects and Screened Yield

Typical data for screened and un-screened yield and rejects are plotted as a function of pulping time in Figures 6 and 7. These data resulted from pulping at 0.5% consistency in the well mixed reactor stirring at 800 RPM. Un-screened pulp yield (Figure 6) decreased with pulping to approximately 40 to 50% of the initial wood feedstock depending upon temperature and time. Yield reduction represents the loss of extractives, lignin, and some carbohydrates.

Screened yield increased to a maximum with pulping and then was followed by a decrease with continued delignification. A dip occurs initially in the screened yield as a function of cooking time or kappa number (see Figure 7). This represents the minimum pulping time required to remove a sufficient quantity of lignin to fiberize the coarse wood fiber bundles associated with the shredded wood feedstock. It is assumed that at low lignin levels near the end of the pulping process, carbohydrate loss is accelerated by abrasion from mixing and reaction with sodium hydroxide and oxygen in the pulping environment. The percent rejects decreased with pulping time or kappa number (see Figure 7). Rejects are characterized as un-pulped fiber bundles. Following delignification, the bundles can be easily fiberized reducing the particle size to that of the individual fibers. The decrease in rejects is matched by the increase in screened accepts following the initial dip.

The Ross Diagram

A Ross diagram for the pulping of shredded spruce and fir using sodium carbonate and oxygen are summarized in Figures 8. The Ross diagram is a locus of the lignin to carbohydrate ratio in the pulp $[L_P/(1-L_P)]$ plotted versus the un-screened pulp yield. The Ross diagram gives a rough indication for selectivity of pulping and shows the composition of lignin to carbohydrates that are removed as the pulping continues and the yield drops.

Figure 8 shows the data for the delignification experiments conducted in the well mixed digester at 2% consistency, 20% sodium carbonate calculated as Na_2O, 165.5 C (330 F) and 82 psig oxygen pressure. For purposes of comparison, the kraft pulping data for red spruce (Busayasakul, 1986) are also shown in Figure 8. Busayasakul studied the kraft pulping of red spruce under conditions of liquor to wood ratio of 8, active alkali of 24%, 30% sulfidity, 170 C temperature and variable cooking time. The line for perfect selectivity is shown in Figure 8 assuming that there are no carbohydrates losses during pulping. Comparison of the data in Figure 8 indicates the selectivity of the sodium carbonate oxygen cook and the kraft delignification are approximately the same; that is, the lignin to carbohydrate ratios are similar.

The selectivity is further illustrated in Figure 9 which presents a comparison of the total pulp yield plotted versus kappa number for the sodium carbonate-oxygen and kraft delignification experiments. It is not surprising that the selectivity curves are similar since the temperature (165.6 and versus 170 C) and alkali concentration calculated on an Na_2O basis were about the same; 24% and 170 C for the kraft cook versus 20% and 165.6 C for the Na_2CO_3 oxygen experiments. Peeling reactions would be expected to be a strong function of alkali concentration and time at temperature as shown by Genco and co-workers (1990).

Figure 10 shows the Ross diagrams for the delignification experiments conducted using sodium carbonate and oxygen under conditions of 0.2% and 2% consistency conducted at 165.6 C (330 F) using equal concentrations of sodium carbonate (3.94 g/l as CO_3^{-2}). This illustrates that consistency at which the delignification occurs has an important effect on consistency.

Pulp Strength

Wet Zero Span Tensile Strength. Intrinsic pulp strength was determined by performing the wet zero span tensile test on unbeaten samples of every pulp. The wet zero span test was used in place of the 0.5% CED viscosity test (Cowan, 1986). Physical properties were determine on a fully pulped brown stock pulp following processing the pulp in a PFI mill and preparing Standard Tappi Handsheets.

As a general outcome, wet zero span tensile strength decreased with exposure to oxygen and mixing in the

pulping environment (See Figure 11). As the kappa number and lignin content decreased, there was a corresponding decrease in wet zero span tensile strength.
Commercial kraft Northeast softwood brown-stock pulp was obtained from a local kraft mill that process mixed Northeastern softwood; predominately spruces, balsam fir, and hemlock. The brown-stock sample was obtained after washing and screening and had a wet zero span tensile strength of about 32 psi. There was however a significant effect of consistency when comparing delignification experiments performed at 0.5 and 2.0% consistency; with the lower consistency (0.5%) experiments having a lower wet zero span tensile strength (see Figure 11).

The above results suggest that the agitation level affects pulp strength. A high level of agitation was maintained in the well mixed batch digester for determination of the rate and high oxygen dissolution with minimal mass transfer limitations. Also, the high shear field present in the refiner required for size reduction of the wood chips to provide shredded wood may have led to a decreased fiber strength in comparison to traditional kraft pulps. It may be possible to preserve more intrinsic fiber strength by pulping the softwood in a low agitation environment with carbohydrate protection provided adequate oxygen mass transfer can be maintained.

Pulp Physical Properties. Pulp processing data are summarized in Figure 12 as Canadian Standard Freeness versus revolutions in the PFI mill. For comparison the kraft brown-stock pulp is also shown on the same curve. The soda oxygen pulp had a much higher freeness loss coefficient and indicates that it beats quite easily. Unfortunately, two of the four points on the beater curve were at low freeness values (4,000 and 5,000 revolutions). The tensile index is plotted versus the freeness in Figure 13 and indicates that the soda oxygen pulp had about 80% of the tensile strength of the kraft pulp over the entire freeness range except at high freeness where the tensile index was the same (600 to 650 ml. range). The tear versus density plot is shown in Figure 14. These data show that at equivalent sheet density the soda oxygen pulps have approximately 75 to 80% of the tear value of conventional northeast softwood kraft pulp.

Global Delignification Model

A global kinetic model of similar form to that proposed by Sarkanen and Johanson (1976) was selected for correlating the kinetic data. The Sarkanen and Johanson model is the only model available for describing soda oxygen delignification that is based upon wood as the starting raw material. All other models start with pulp and treat oxygen bleaching in the presence of caustic. Virtually all oxygen delignification models, including the Sarkanen and Johanson model, are empirical and treat the oxygen delignification reaction as a three body problem. Rates are proportional to the lignin content [L], carbonate ion concentration $[CO_3^{-2}]$ and oxygen pressure $[P_{O2}]$ and have the form:

$$-\frac{dL}{dt} = k[CO_3^{-2}]^m \, P_{O2}^n \, L^q \qquad (9)$$

In equation (9) the reaction order is expressed in terms of the empirical constants m, n, and q. The reaction rate coefficient of k is influenced by temperature according to the Arrhenius equation:

$$k = k_o \exp(-E_A / RT) \qquad (10)$$

where k_o is the frequency factor, E_A is the activation energy, R is the universal gas constant, and T is the absolute temperature. (°K).

In the present analysis, Equation (10) was modified according to the Sarkanen and Johanson model to take the form:

$$-r_a = -\frac{dL}{dt} = k\, L^q \qquad (11)$$

Equation (11) indicates that the rate constant is assumed to be of order q in the lignin content given by the Klason lignin and not as the kappa number as given by Sarkanen and Johanson (1976). The rate constant k is taken to be dependent upon temperature (T), and the activation energy (E_A) is given in equation (10) but is expanded to include the effects of the sodium carbonate concentration taken as the initial concentration of carbonate ion $[CO_3^{-2}]$, and upon the initial oxygen pressure (P_O) raised to the powers m and q, respectively. Thus the rate constant k takes the form:

$$k = k_o \exp(-E_A / RT)\left[CO_3^{-2}\right]^m \, P_O^n \qquad (12)$$

In solving for the model parameters shown above, Equation (12) was applied to all of the data to determine the rate coefficient (k) and the empirical constant (q). This was accomplished by plotting the rate data ($-r_a$) on log-log paper and solving for the coefficient k and q as the intercept (log k) and slope (q) of the linear regression line.

$$\log_{10}\left[-r_a\right] = \log_{10}\left[\frac{-dL}{dt}\right] = \log_{10} k + q\log_{10} L \tag{13}$$

Details of this procedure are presented by Ransdell (1996).

The above procedure was applied to all of the curves to determine a set of values for q and k applicable to the various oxygen pressures, temperatures and initial concentrations of sodium carbonate for the entire data set of 32 experiments performed at 0.5% consistency in the well-mixed reactor. An average value of the slope (q) was determined using all of the rate curves. The values of the empirical variables (k_o, E_A, m, n and o) were determined by applying successive regression analyses to the data base for the rate coefficient (k). The final form of the global rate model is:

$$-r_a = \frac{-dL}{dt} = k\ L^{1.28} \tag{14}$$

with the rate constant (k) being:

$$k = 2.27x10^{+15}\ exp\left(-138{,}000 / RT\right)\left[CO_3^{-2}\right]^{0.42} P_O^{0.54} \tag{15}$$

The values for the model parameters and applicable units are summarized in Table 1.

The global model is shown as the curved lines in Figures 2, 3, 4, and 5 and is compared to the experimental data shown as symbols. The global model represents the data to within about 5%. The dependence on Klason lignin is 1.28 and illustrates that there is a greater dependence than the first order dependency normally associated with soda oxygen delignification during bleaching of kraft brown-stock pulp (Olm and Teder, 1979).

Of the three variables studied, that is the temperature, oxygen pressure and sodium carbonate concentration, temperature influences the delignification rate to the greatest extent. Oxygen delignification is a very strong function of temperature and is illustrated clearly in Figures 2 and 3. Figure 2 compares the data at 300 and 330 °F; while Figure 3 summarizes the data at 330 and 345 °F. The strong function of temperature is clearly shown by the activation energy which has a value of 138,000 joules/gram mole (32,959 calories/gram mole). The dependency of the oxygen over-pressure is seen in Figure 4; which is the second most important variable. The empirical constant (n) for the oxygen dependency is 0.54. The data for Klason lignin versus time coincide or over-lay each other for all of the sodium carbonate concentrations evaluated. This indicates that there is virtually no dependency of the delignification rate on the sodium carbonate concentration provided sufficient alkali is available from the carbonate-bicarbonate equilibria given in Equation (1). This is believed to mean that high amounts of hydroxyl ion (OH^-) are present from the decomposition of carbonate ion ($CO_3^=$) to satisfy that required for the delignification reactions. This is seen clearly in Figure 5 and also by the value of the empirical coefficient (m) of 0.42 and shows the buffer capacity of the sodium carbonate.

CONCLUSIONS

The kinetics of oxygen delignification of wood have been found to be mass transfer dependent. The development of new pulping techniques which remove the mass transfer restrictions of the soda oxygen process will determine the future of the process. It is important that the carbohydrate degradation reactions be controlled to avoid loss of strength in the brown-stock pulp. This possibly can be done either by controlling the time the oxygen spends in contact with the wood or by the use of a protector. Also, the present study clearly showed that high degrees of agitation lower the strength of the resultant pulp. Thus, the proper level of agitation is important in maintaining pulp while still providing oxygen to the reaction site.

Oxygen availability in the reaction environment is critical for the delignification of shredded softwood. The condensation of lignin fragments is assumed to play a significant role in the delignification of softwood by the soda oxygen process. In order to minimize the effect of lignin condensation, the reaction was accomplished at low pulp consistency (0.5 to 2%) during the kinetics study. A low pulp consistency maximized the concentration gradient between lignin in the solid phase and dispersed suspended lignin.

At higher pulp consistencies a method of stripping the reaction environment of suspended lignin fragments may be required in order to maintain high delignification rates. Work should be done to develop protectors to increase the intrinsic strength of the pulp obtained in the soda oxygen delignification process which is an odorless alternative to the kraft process.

APPENDIX: EXPERIMENTAL METHODS

The experimental methods are summarized in detail by Ransdell (1996).

Feedstock Preparation

Two samples of spruce and fir softwood chips were obtained from Champion Pulp and Paper Company of Bucksport, Maine.

Preliminary work in the current study and that by Agarwal (1993) has shown the importance of mass transfer effects in the soda oxygen process. Low solubility of oxygen in the liquid phase and poor penetration of the pulping liquor into the chips requires a greater degree of feedstock size reduction than in the kraft process. This has necessitated shredding of the wood into a coarse wood meal prior to exposure to the pulping liquor. For industrial significance the particle size of the shredded wood was characterized using the Canadian Standard Freeness Test. The shredded wood used in these experiments had a drainage rate of approximately 700 to 800 ml CSF. Chip shredding was accomplished using a laboratory scale disk refiner. The softwood sources were shredded from chips in a thermomechanical pulping refiner in a single pass for high freeness values. Prior to shredding, the chips were steamed for 15 minutes at atmospheric pressure.

Liquor Preparation

Pulping liquors were prepared from reagent grade sodium carbonate and tap water in the clean digester vessel after wood addition. The pulping liquor was oxygenated after the mixed suspension of wood and liquor were brought to cooking temperature.

Reactor Selection

Preliminary experiments of soda oxygen pulping were performed on shredded softwood in a rocking digester and also in a continuous stirred tank (Ransdell, 1996). The rocking digester was a simulation of a rotary digester. The effects of cooking time, rocking rate or mechanical mixing, temperature, and pulp consistency were investigated. Measurements made of pulps included: total yield, accepts, rejects, kappa numbers or permanganate number, wet zero span tensile strength, and viscosity.

The rotational speed and consistency for the continuous stirred batch digester were set to minimize mass transfer effects. The cylindrical batch stirred tank reactor and the rocking digester both of approximately 12.5 liters total volume were attached to an oxygen gas cylinder. Temperature was monitored by two K-type thermocouples. and a resistance thermocouple device. Heat was supplied electrically through a heating mantle which surrounded the digester shell. A temperature controller was utilized for the continuous application of heat at a rate set by the operator. Reactor pressure was monitored by pressure gage. Mixing speed for rocking or turbine was set by using a variable speed direct current control. The turbine mixing shaft was sealed by the use of two standard shaft seals between which pressurized water was introduced to provide a barrier between the reaction environment and the atmosphere.

Digestion Procedure

Prior to each pulping, moist shredded wood of the desired weight was added to the clean laboratory digester, along with dry reagent grade sodium carbonate powder, and sufficient water to make 6 liters. The reactor was sealed and mixing initiated. Entrapped air was purged from the digester after the reactor was heated above 100 °C. The reactor was brought to temperature over a time period of approximately forty five minutes. As desired, oxygen was added when the digester reached the desired cooking temperature. During the kinetics study, the reactor was sampled under pressure at the end of the cook at temperature. After cooling, the pulp and liquor obtained were separated. The digester was rinsed clean to obtain any residual wood fibers. The collected pulp was washed to remove residual pulping liquor and stored under refrigeration prior to evaluation.

Analysis Methods

Standard laboratory methods were used to determine lignin content, percent carbohydrate, and reagent concentrations. Established pulp testing methods have been developed for industrial applications and are given in the TAPPI Standard Methods Publications.

Lignin analysis methods included both Klason lignin and kappa or permanganate number titration. The Klason lignin was determined by following the method of Effland (1977) based on a modification of TAPPI Useful Method U-250. The kappa (T-236) or permanganate number (T 214) for pulp is an indirect method of determining the lignin content empirically by titration.

Pulp samples obtained from the laboratory digester were screened in a Sommerville pulsing screen across a 0.006 inch slot shives screen. Rejects were dried, and weighed. The approximate dry weight of the accepts was determined based on the percent moisture and the overall moist sample weight. TAPPI standard hand sheets (T-205) were produced from each pulp sample for wet zero span tensile strength analysis, following the manufacturer's method (Cowan, 1996).

REFERENCES

1. **Abson, D., and V.E. Stockman,** "fundamental properties of soda oxygen and kraft pulps" TAPPI **62**(2):69-72(1979).

2. **Adam, H., D. Vincent, K. Loi Nguyen, and J.F. Mathews,** "Kinetics of oxygen delignification of eucalypt kraft pulp," *Appita*, 47(3):217-220 (1994).

3. **Agarwal, A.**, "Development of an aqueous based sulfur free pulping process using oxygen," *M. S. Thesis*, University of Maine, Orono, Maine (1993).

4. **Busayasakul, N.**, "Total Organic Carbon as an Indicator for Wood Delignification," Ph.D. Thesis, University of Maine, Orono, Maine (1986).

5. **Biermann, C. J.**, "Essentials of Pulping and Papermaking," Academic Press, Inc., pp. 61-62 (1993).

6. **Cowan, W.F.** "Zero short span tensile testing can determine basic paper properties," *Pulp and Paper Magazine*, pp. 84-86 (1986).

7. **Edwards, L., and S. Norberg**, "Alkaline delignification kinetics: a general model applied to oxygen bleaching and kraft pulping, " *TAPPI*, **56**(11):108 (1973).

8. **Effland, M.,** "Modified process to determine acid-insoluble lignin in wood and pulp," *TAPPI* **60**(10):143-144 (1997).

9. **Erickson, M., and C. W. Dence**, "Yield and composition of tall oil from oxygen-alkali pulped thermomechanical fiber," *TAPPI*, **58**(8):190 (1975).

10. **Fiala, W. and F. Nardi**, "The NACO Process," Straw in Papermaking Symposium, Pira Paper & Board Division (October 16, 1984).

11. **Genco, J. M., Busayasakul, N., Medhora, H. K., and W. Robbins**, "Hemicellulose retention during kraft pulping," *Tappi Journal*, **73**(4) 223-233 (1990).

12. **Hsu, C. L., and J. S. Hsieh**, , "Reaction kinetics in oxygen bleaching," *AICHE Journal* **34**(1):116-122 (1988).

13. **Johansson, E and Ljunggren, S.**, *Journal of Wood Chemistry and Technology*, **14**(4): 507-525 (1994).

14. **Kleinert, T. N.**, "Mechanisms of delignification in oxygen-alkali system," *TAPPI*, **59**(9):122-124 (1976).

15. **Ljunggren, S. C. and Johansson, E. C.**, 1987 International Oxygen Delignification Conference, TAPPI Proceedings, p.125 1987.

16. **Marton, R., Brown, A., and Granzow, S.** "Oxygen Pulping of TMP Fiber," 1974 Non-Sulfur Pulping Symposium, Madison, Wisconsin (October 16-18, 1974)

17. **Myers, M. R., and L. L. Edwards**, , "Development and verification of a predictive oxygen delignification model for hardwood and softwood kraft pulp," *TAPPI* **72**(9):215 (1989).

18. **Nagano, T., S. Miyao, and K. Takeda**, "Hopes oxygen pulping process-its basic concept and some aspects for the reaction of oxygen pulping," Non-sulfur Pulping Symposium, October 16-18, Madison, Wisconsin, TAPPI, 53-61, (1974).

19. **Olm, L., and A. Teder**, "The kinetics of oxygen bleaching," *TAPPI*, **62**(12):43-46 (1979).

20. **Ransdell, J.** "Kinetics of Softwood Pulping with Sodium Carbonate and Oxygen," PhD Dissertation, University of Maine, Orono, Maine (1996).

21. **Rodden, G.**, "Soviets introduce oxygen-alkali pulping process," *Pulp and Paper Canada*, **91**(4):31-32 (1990).

22. **Sarkanen, K. V., and L. N. Johanson** "Use of oxygen in pulping," *AICHE Symposium Series, Chemical Engineer in Pulp and Paper Technology*, **72**(157):50-55 (1976).

23. **Sutton, P.**, "No Sulfur, No Chlorine: No Problem," *Pulp and Paper International*, pp 48-50 (April, 1987).

ACKNOWLEDGEMENTS

Funding for this project was provided by the Federal Environmental Protection Agency through Guild Associates, Inc., 4089 North Leap Road, Hilliard, Ohio 43026, under EPA SBIR Contract No. 68D30078. The authors wish to express their appreciation to Proserfina Bennett, Amos Cline, Keith Hodgins, and Joel Preston for help in research reporting and the use, construction, and maintenance of the laboratory apparatus and testing equipment.

Table 1
Summary of Model Parameters
Oxygen Sodium Carbonate Delignification of Shredded Spruce and Fir

---

L = Klason lignin, % on pulp
t = Time at temperature, hours.
E_A = Activation Energy, 138,000 joules/gmole
R = Gas Constant, 8.3144 joules/gmole K
T = Temperature, K
k_o = Empirical rate constant, $2.27 \times 10^{+15}$
CO^{-2} = Initial carbonate concentration (2.95, 3.94 or 4.92) grams/liter
P_O = Initial oxygen pressure (41 or 82) psig
m, n, q = Empirical coefficient
$m = 0.42$
$n = 0.54$
$q = 1.28$

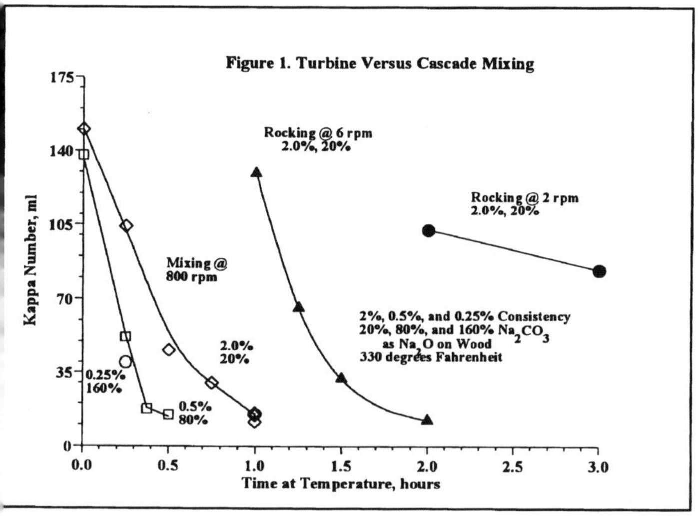

Figure 1. Turbine Versus Cascade Mixing

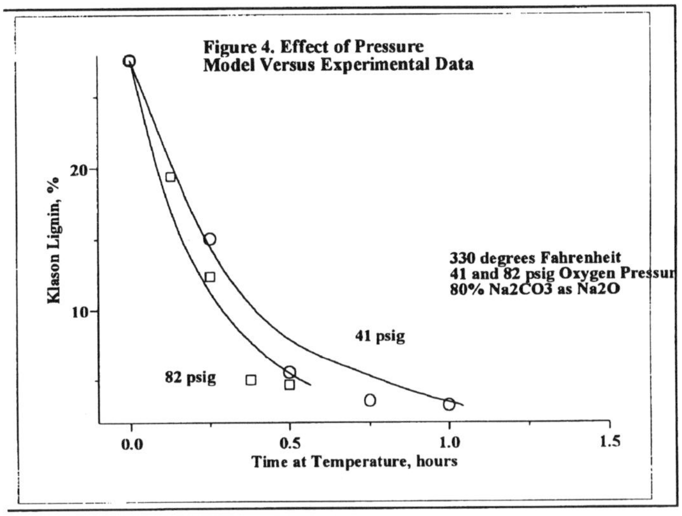

Figure 4. Effect of Pressure
Model Versus Experimental Data

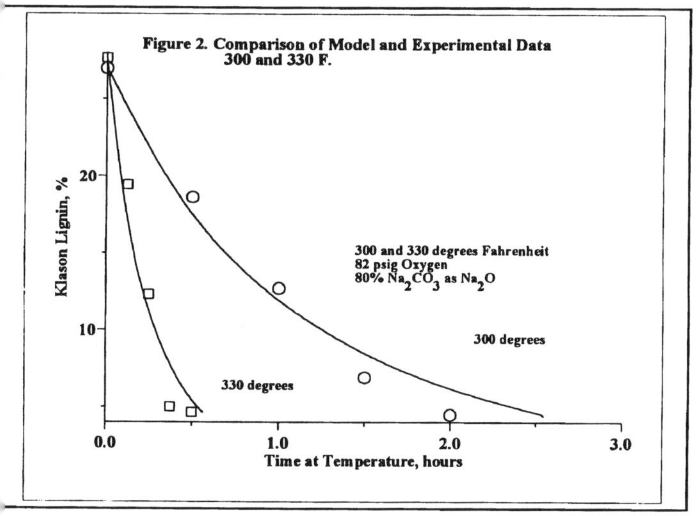

Figure 2. Comparison of Model and Experimental Data
300 and 330 F.

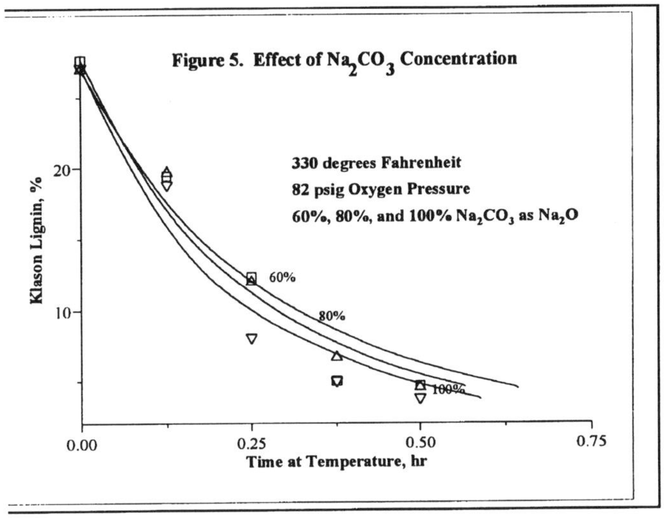

Figure 5. Effect of Na_2CO_3 Concentration

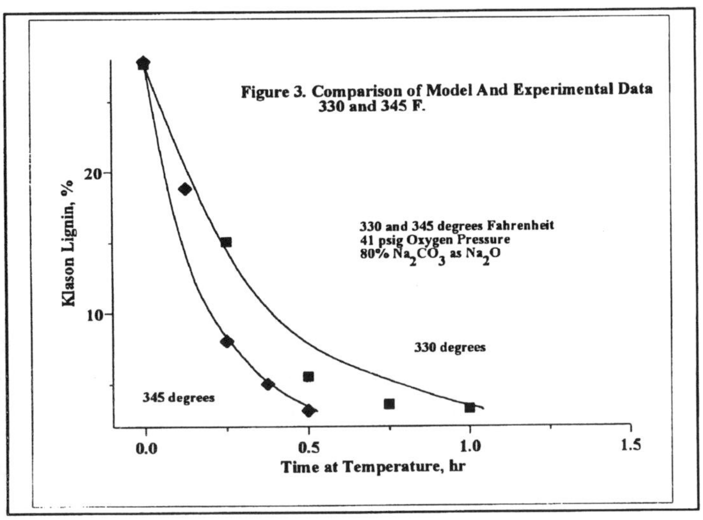

Figure 3. Comparison of Model And Experimental Data
330 and 345 F.

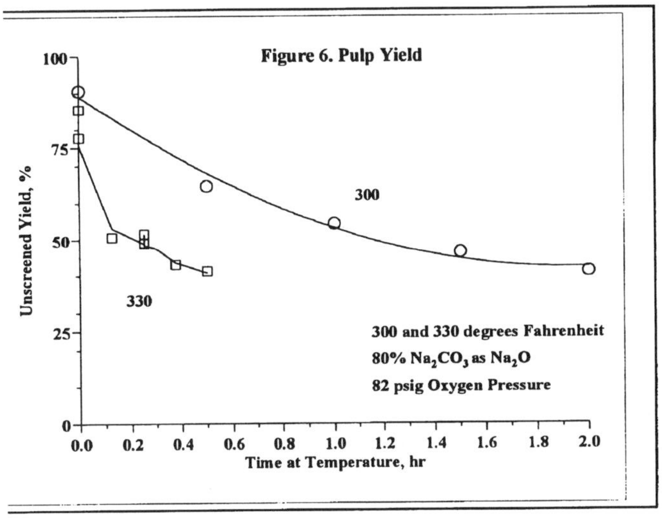

Figure 6. Pulp Yield

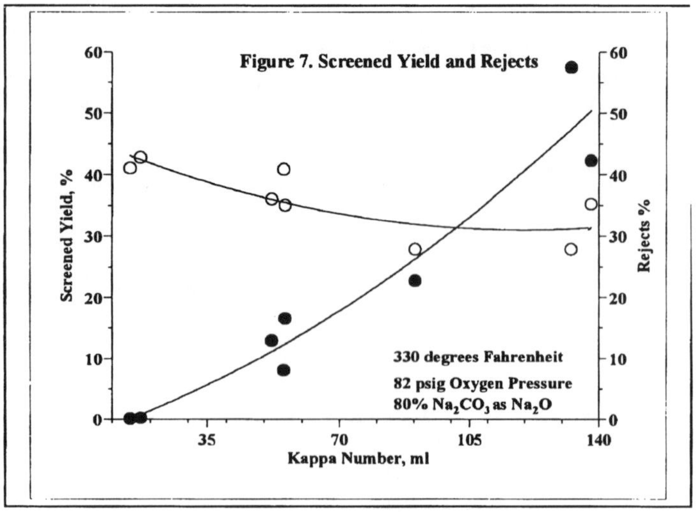

Figure 7. Screened Yield and Rejects

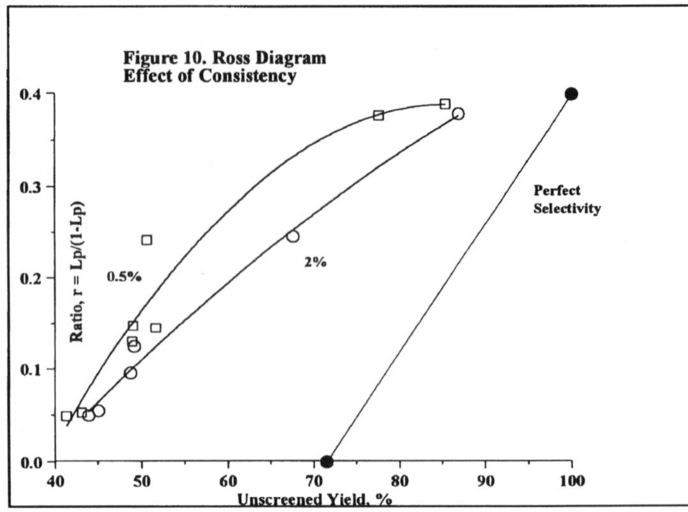

Figure 10. Ross Diagram Effect of Consistency

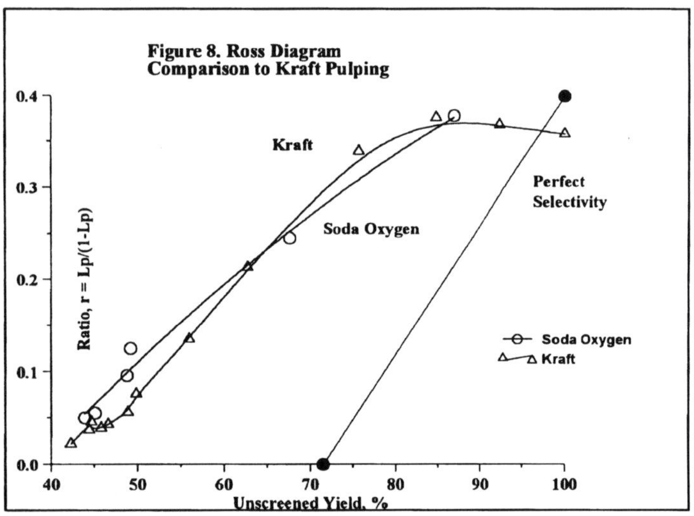

Figure 8. Ross Diagram Comparison to Kraft Pulping

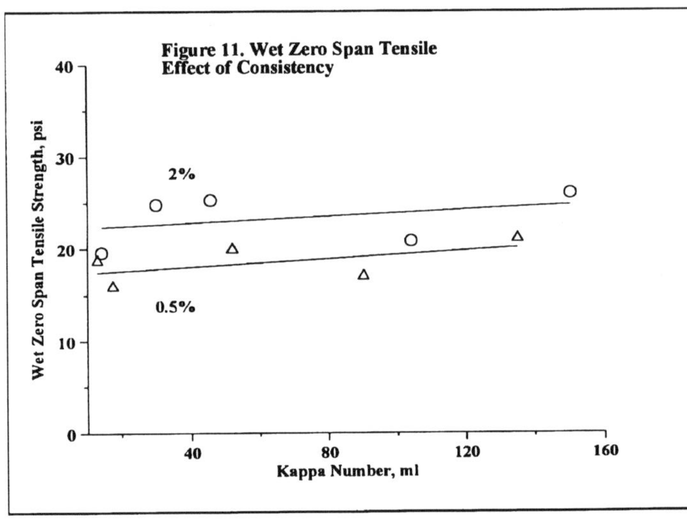

Figure 11. Wet Zero Span Tensile Effect of Consistency

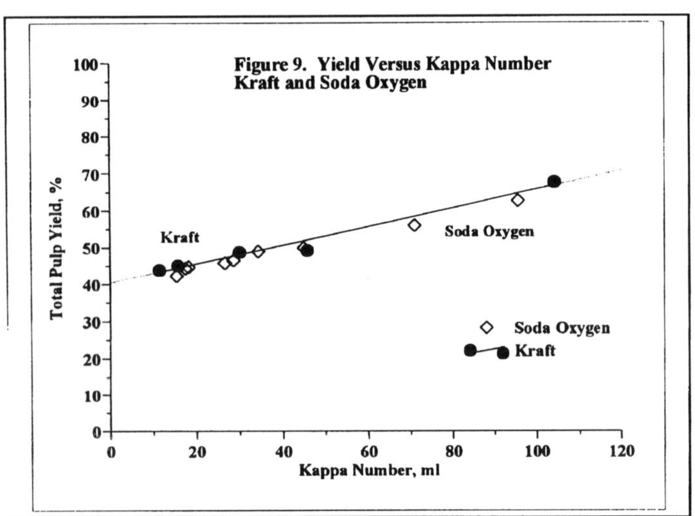

Figure 9. Yield Versus Kappa Number Kraft and Soda Oxygen

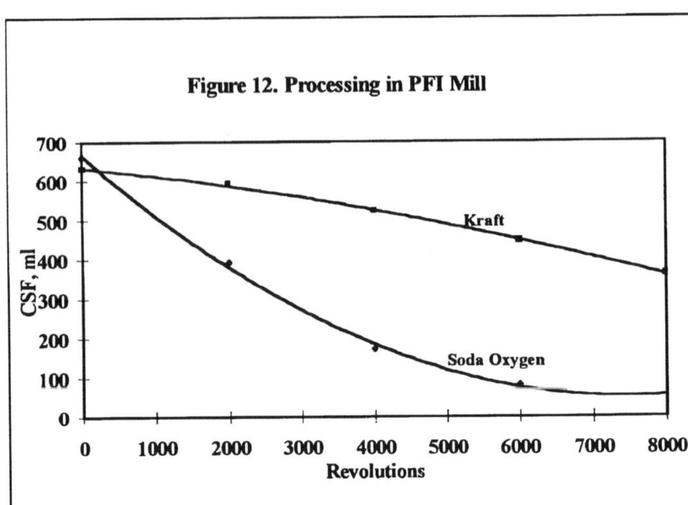

Figure 12. Processing in PFI Mill

Figure 13. Tensile Strength

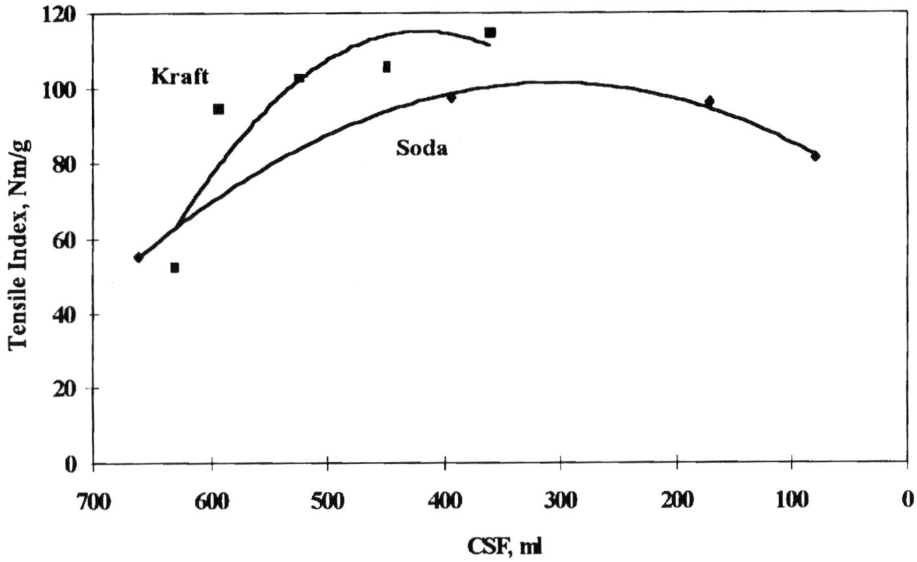

Figure 14. Tear Versus Density

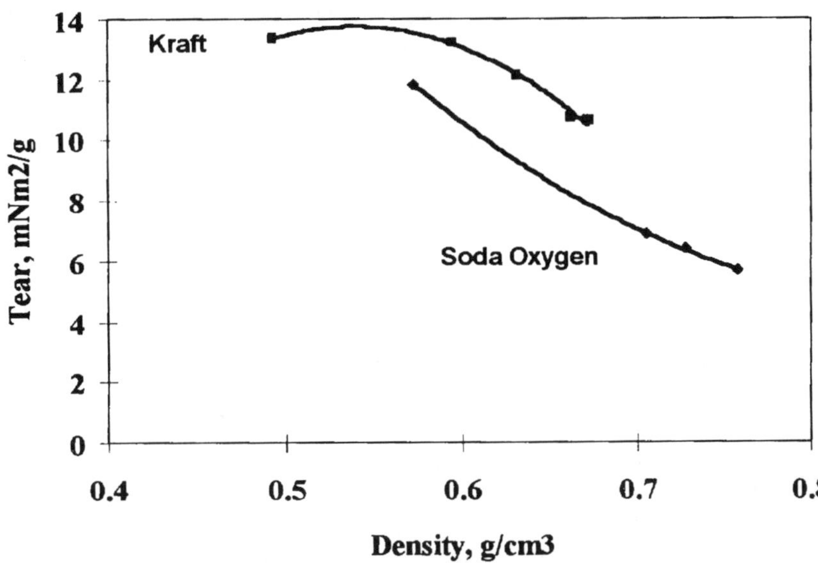

Optimization of Oxygen and Alkali Charges for Mill Scale Oxygen Delignification

David Roy
P. H. Glatfelter Company, Spring Grove, PA 17362

The degree of delignification achieved across medium consistency oxygen reactors is dependent upon any number of variables: oxygen charge, alkali charge, pH, consistency, dissolved solids, temperature, and pressure. Although each variable has some cost associated with its manipulation, oxygen and alkali costs are typically the most significant, and are also the easiest to quantify. In a mill environment, optimization of chemical addition must consider costs, equipment constraints, and final pulp properties. In this study, oxygen addition was evaluated as a function of wash filtrate residual alkali. Cost and pulp strength data were then combined with this information to develop a chemical addition control strategy to be followed by mill operating personnel.

INTRODUCTION

The P.H. Glatfelter Company's Spring Grove, Pennsylvania mill is an integrated pulp and paper facility that produces uncoated and coated fine papers. As part of its Pulp Mill Modernization Project, the mill installed a medium consistency oxygen delignification system on its pine fiberline. The system, which was commissioned in July, 1994, follows a Kamyr continuous digester and was designed for 45 percent delignification at a production rate of approximately 300 ADTPD.

As part of an overall pine fiberline optimization project by the Technical Department, a study was conducted to determine the relative importance of operating variables on delignification. Among the variables initially considered were reactor pressure, temperature, pH, oxygen charge, alkali charge, consistency and dissolved solids. As operating experience increased and system limitations became more clear, operating targets for temperature, pressure and consistency were defined. This shifted the focal point of the study to the effects of washing and chemical charge.

DISCUSSION

System Design

The O_2 system is a typical medium consistency design. The typical brownstock permanganate number range from the digester is 18.0 to 20.0. Following the digester, brownstock is run through a pressure knotter and followed by two stages of washing on Impco compaction baffle (CB) filters. Oxidized white liquor (OWL) and 50 lb. steam are added to the second filter repulper, ahead of the reactor feed pump. Oxygen and 150 lb. steam addition follows the reactor feed pump. The pulp, chemicals, and steam are mixed in a high shear mixer before entering the reactor, which has a residence time of approximately 60 minutes. The reactor discharges into an atmospheric blow tank, and is followed by two stages of washing on conventional drum washers, with screening between the stages. The pulp is then pumped to high density storage before being delivered to the bleach plant. A sketch of the system is shown in Figure 1.

The initial operating guidelines provided for the system called for maintaining a reactor feed temperature of 85°C, and a reactor inlet pressure of 620 kPa. Oxidized white liquor addition was to be adjusted so as to maintain a minimum reactor outlet pH of 10.5. Oxygen addition was to be manipulated for delignification control.

Study History

During the first several months in which the system was operated, the amount of delignification achieved varied widely, seemingly independent of the oxygen charge as shown in Figure 2. In this case, delignification is determined using the difference between reactor inlet and outlet permanganate numbers. Permanganate number control was poor, with operating personnel being hesitant to make adjustments because of the unpredictability of the changes. As the operators' experience increased, delignification control improved greatly. However, the oxygen charge required to meet the permanganate number target still fluctuated greatly from day to day.

Questions arose regarding the significance of process variables other than oxygen on delignification. In a paper by Miller et al. [1], the effects of black liquor carryover and washing on delignification were discussed. This provided a starting point for a comprehensive study of the effects of operating variables on delignification.

PRELIMINARY STUDY

The preliminary study was designed to assess the significance of reactor pressure, feed temperature, inlet and outlet pH, oxygen charge, alkali charge, consistency, and conductivity (dissolved solids). Filtrate residual alkali, both before and after the reactor, were later included as study variables [2,3]. Temperature and

Figure 1. Pine fiberline - P.H. Glatfelter Co.

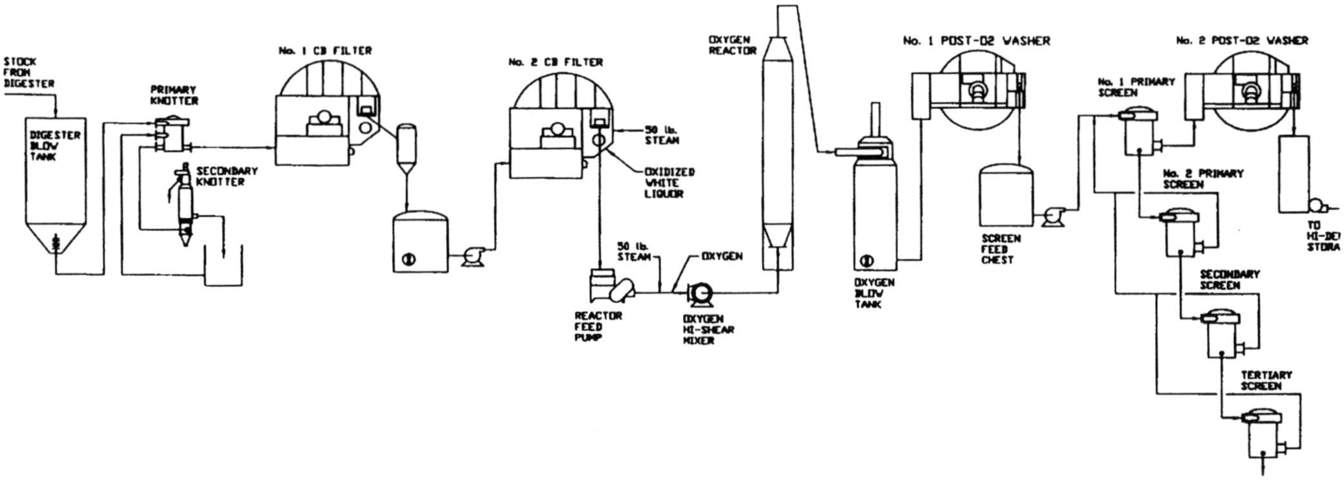

Figure 2

Delignification v. Oxygen Charge

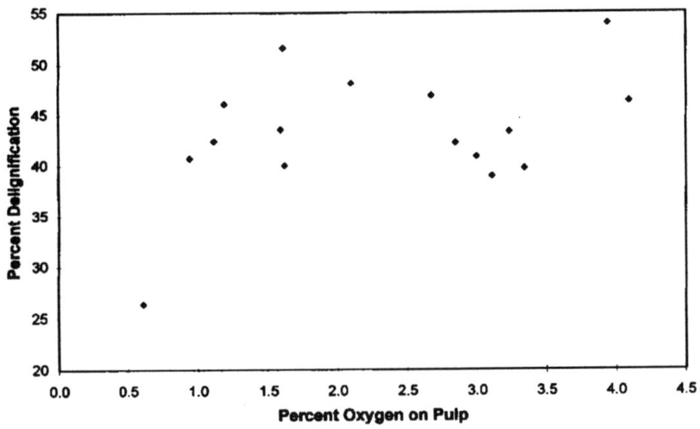

Figure 4

Oxygen Efficiency v. Inlet Residual Alkali

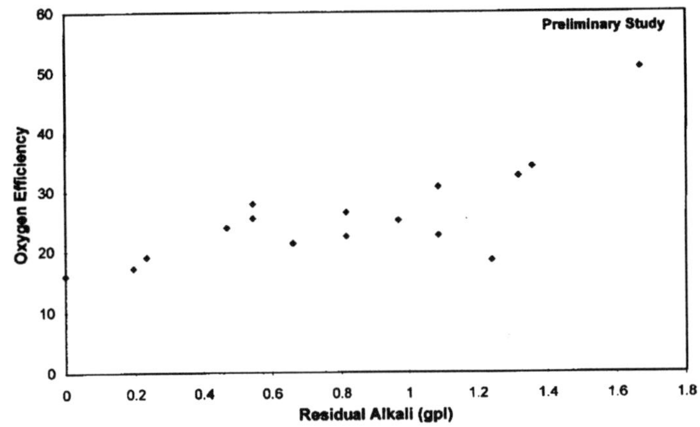

Figure 3

Oxygen Efficiency v. Outlet Residual Alkali

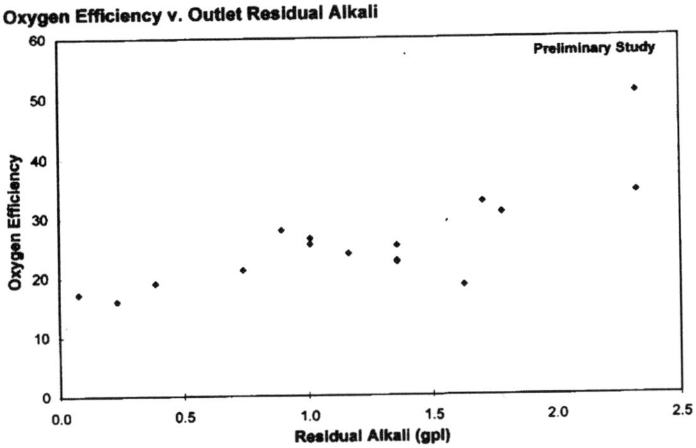

Figure 5

Oxygen Efficiency v. Outlet Residual

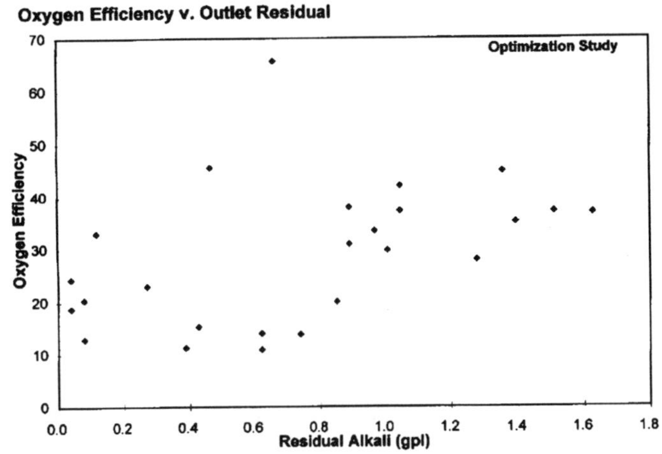

Figure 6

Oxygen Efficiency v. Inlet Residual Alkali

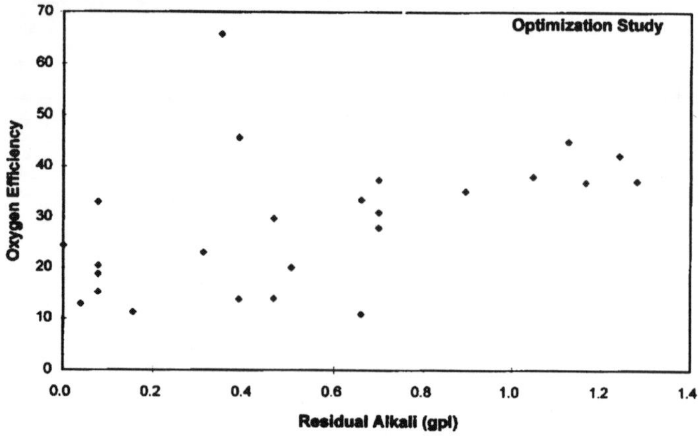

Figure 9

Outlet Residual Alkali v. Alkali Charge

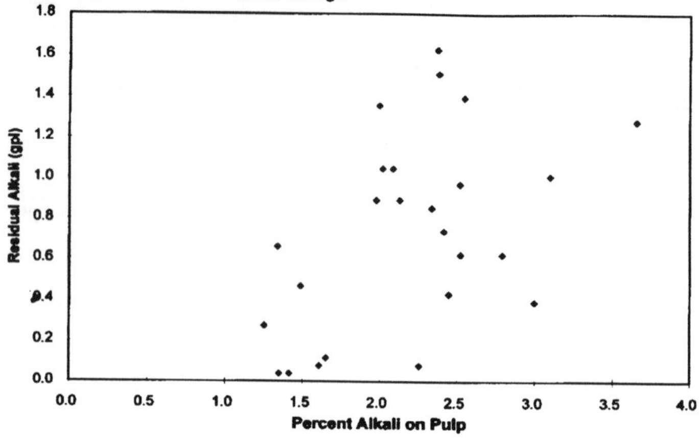

Figure 7

Strength Loss v. Outlet Residual Alkali

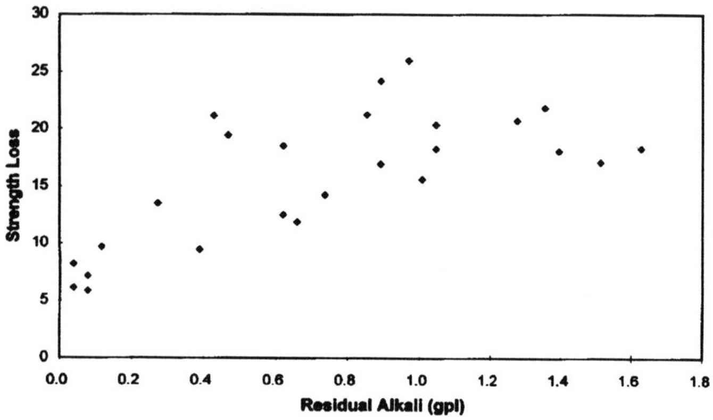

Figure 10

Deviation from Standard Normalized Cost v. Alkali Charge

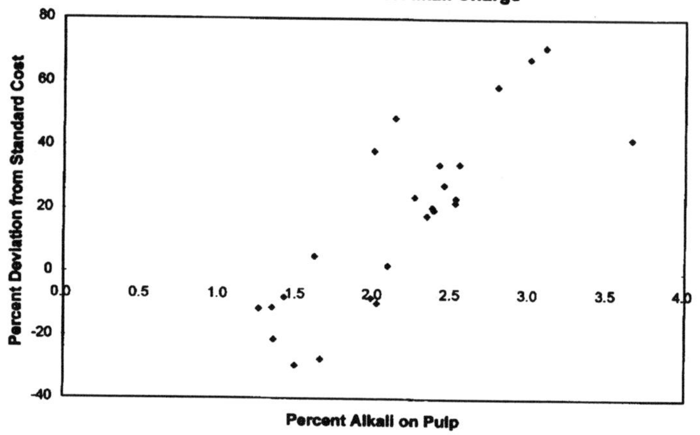

Figure 8

Strength Loss v. Inlet Residual Alkali

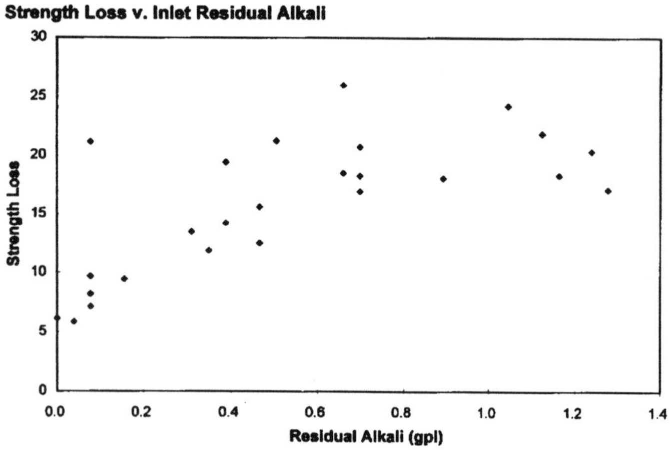

Figure 11

Inlet Conductivity v. Inlet Residual Alkali

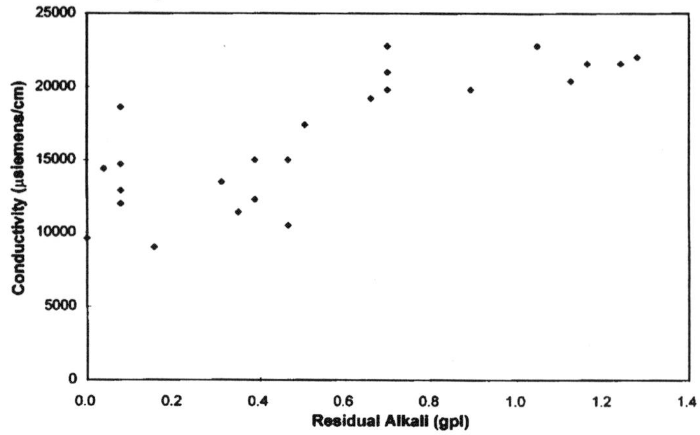

pressure were quickly eliminated from consideration because of system idiosyncrasies limiting their manipulation.

Prior to and during the data collection period of the preliminary study, control of wash water flows and chemical addition was done subjectively by operating personnel. Wash water flow into the second post oxygen washer was adjusted based on the digester operating rate and weak liquor inventory. Oxidized white liquor addition was base-loaded according to the digester operating rate, with corrections made only if the pH strayed outside the 10.5-12.5 target. Oxygen flow was manipulated based on the permanganate numbers around the reactor, with a reactor outlet target of 10.0.

Over a period of approximately six months, filtrate and pulp samples were collected around the oxygen reactor under a variety of operating conditions. The reactor feed temperature was fixed at 86°C, and the inlet pressure set at 620 kPa. Ahead of the reactor, samples were collected both prior to and after OWL addition. The sample collected prior to the OWL addition was used to measure filtrate conductivity (dissolved solids), filtrate residual alkali, and the brownstock permanganate number. The second sample was used for determination of pH and brownstock consistency.

The post O_2 sample was collected from the bottom of the blow tank, following the addition of consistency dilution, and was used to quantify conductivity, residual alkali, pH, and permanganate number. The post O_2 sample was collected approximately 75 minutes after the pre O_2 sample to compensate for reactor and blow tank residence times. These times were determined using temperature bump tests at different production rates, and allowed samples to be collected regardless of rate.

When sufficient data had been collected, its interpretation then became the task. As before, the relationship between delignification and oxygen dosage was weak. There was also no direct relationship between delignification and pH, conductivity, alkali charge, or residual alkali. Delignification was found to be especially poor during periods of low reactor feed consistency, suggesting channeling, but this did not occur often enough to explain the other phenomena observed.

As data were being collected during typical operation, the delignification target was always moving due to the normal fluctuation of incoming digester permanganate numbers. This made it difficult to determine whether something obvious was being overlooked, or if the testing plan was flawed. In trying to find a way to evaluate the data without the effect of the varying delignification rate, it was decided to try to normalize the data.

The variables first chosen for normalization were the percent delignification and the percent oxygen applied. The term *Oxygen efficiency* was coined for this normalized variable whereby:

$$\textit{Oxygen efficiency} = \frac{\%\ \text{Delignification}}{\%\ \text{Oxygen applied}} \qquad [1]$$

All of the study variables were plotted against the *oxygen efficiency* to search for some type of correlation.

The puzzle began to fit together when the plots of *oxygen efficiency* versus residual alkali (Figures 3 and 4) were analyzed. The graphs showed that the *oxygen efficiency* increased linearly with the residual alkali. Encouraged by this discovery, additional samples were tested and found to agree with the previous data, further supporting the finding. Once again, pH, conductivity and alkali charge did not correlate.

For the case of the outlet residual alkali, it was found that the *oxygen efficiency* was approximately 55% greater when a residual alkali of approximately 0.3 gpl or greater remained. As the residual increased, so too did the efficiency.

Following this discovery, the next logical set of questions began to surface. What are the economics of using oxygen or alkali preferentially? How is pulp strength affected? Are bleach plant operations impacted? With these findings in hand, and these questions remaining, a shift in focus to optimizing chemical charges was started.

OPTIMIZATION STUDY

The transition into the optimization phase of the project was not immediate, and allowed enough time to further evaluate the data and prepare a work plan for the next phase. The similarity of the plots of *oxygen efficiency* versus both inlet and outlet residual alkali flagged an earlier concern about the validity of filtrate data from the post O_2 sample. In hindsight, it was decided that the influence of No. 1 post O_2 filtrate on the O_2 blow tank filtrate sample was too great, and was masking the true properties of the filtrate which had carried through the reactor. This resulted in piping modifications which allowed the post O_2 sample to be collected from the reactor blow line before dilution, providing a more representative sample.

A second sampling change was made on the reactor feed sample with the elimination of the second sampling point. Concerns about inadequate mixing for the pre O_2 pH sample cast doubts on its usefulness. Improvements made to the sampler on the CB filter allowed for the collection of the consistency sample at that point, thus eliminating the need for the second sample.

The most significant change to the plan was the inclusion of strength testing. The choice for this study was the use of wet zero span tensile (ZST) using the Pulmac Troubleshooter. The Troubleshooter, a zero span tensile instrument which was relatively new to the mill, had been purchased for this type of application, and was a good fit.

At the start of the optimization sampling period, the operating guidelines were essentially the same as during the preliminary study. The reactor feed pressure remained at 620 kPa, and the feed temperature at 86°C. Oxidized white liquor addition was still manipulated to maintain a target pH range of 10.5-12.5, but the dosages were kept as low as possible due to cost considerations.

After a sufficient number of data points had been generated at the low OWL dosages, the dosage was then gradually increased to create data at higher levels of residual alkali. Several bump tests were conducted whereby the OWL dosage was increased dramatically for approximately two hours in order to increase the outlet residual alkali without affecting the inlet residual. Excessive residual alkali had previously been found to negatively impact bleach plant operations by increasing the brownstock pH. Thus for the optimization study, residual alkali levels were kept down to avoid this problem, and did not reach the levels achieved during the preliminary study.

RESULTS

Once the data from the optimization study were compiled, the challenge became its interpretation. Because of the changes in the sampling procedure, there was some speculation as to what could be expected.

Oxygen efficiency, the basis for the optimization phase of the study, proved to be less useful this time around. For the case of the reactor outlet sample, the relationship between *oxygen efficiency* and residual alkali was not as well defined. As seen in Figure 5, the *oxygen efficiency* for the 0.4 to 0.6 gpl residual range was extremely unpredictable, making it difficult to ascertain anything from the plot. The reactor inlet sample, however, seen in Figure 6, displayed similar behavior as that observed previously. The main difference between the two data sets is the greater number of data points located below 0.5 gpl residual for the second study. It appears that the *oxygen efficiency* reaches a plateau at a residual of approximately 1.2 gpl, suggesting an upper threshold above which additional alkali is wasted.

In considering strength data, the wet ZST values would be expected to decrease with increasing delignification. In order to account for this, a normalized strength parameter was created. This parameter, called *strength loss* is defined as:

$$\text{Strength loss} = 100 * \frac{\text{\% Decrease in ZST}}{\text{\% Delignification}} \qquad [2]$$

As done with *oxygen efficiency* in the preliminary study, *strength loss* was plotted against all of the test variables to check for a relationship. The only obvious correlations were found to be both inlet and outlet residual alkali. For the case of the outlet residual, the *strength loss* increased linearly with the residual alkali until it reached a residual of approximately 1.4 gpl, where it leveled off, as shown in Figure 7. The same behavior was exhibited for the inlet alkali, where the *strength loss* increased linearly with the residual to approximately 0.75 gpl before reaching a plateau, as shown in Figure 8. This decrease in strength with increasing alkali residual was not surprising, as Olm and Teder had previously documented the increase of carbohydrate degradation with increasing alkali concentration (4).

The lack of correlation between alkali charge and outlet residual alkali is shown in Figure 9. For the purpose of maintaining pulp strength, the use of filtrate pH as the control parameter for alkali addition failed. Because OWL is much more expensive than oxygen, any deviation above the standard alkali charge results in a significant increase in operating costs. This is shown in Figure 10, where the deviation from standard operating costs, normalized for the degree of delignification, is plotted versus the alkali charge.

CONTROL STRATEGY PLAN

For both strength and economic reasons, the recommended course of action is to delignify using oxygen preferably, adding only enough alkali to maintain the oxygen dosage just below the maximum. Although the simplest theoretical solution to this problem would be to use residual alkali as the control parameter, the time involved in running the residual alkali titration makes it prohibitive for the operators to perform this test on an hourly basis. Although there are currently on line alkali analyzers available for black liquor, none of the units researched were capable of measuring at the low levels involved for this work. This required some type of compromise in developing a control strategy.

In viewing the plot of the reactor inlet conductivity versus inlet residual alkali, seen in Figure 11, such an opportunity was noticed. At a residual alkali of approximately 0.75 gpl, a level which was determined to be excessive for this application, the relationship between residual and conductivity improved greatly. Because the measurement of conductivity is simple, this provides an easy method of determining whether the level of alkali in the filtrate is excessive. Thus, at a conductivity of 17500 microsiemens/cm or greater, the operators can quickly identify a need to reduce the alkali charge. Residual alkali titrations can then be performed once per shift, with a target range of 0.05-0.20 gpl.

Further improvement to this strategy is the application of OWL on a percentage or ratio basis rather than on the existing flow basis. This would automatically correct for changes to the mass fiber flow caused by consistency and flow variations.

The recommended control strategy is in the midst of being implemented, and has not yet been evaluated.

CONCLUSIONS

The preferential use of oxygen over alkali for delignification results in the minimization of fiber *strength loss* as measured by wet zero span tensile. Maintaining measurable levels of alkali in oxygen system filtrates can improve the *oxygen efficiency* at the expense of strength. Because chemical costs are site specific, the preferential use of oxygen on delignification costs will also be site specific. In this study, the preferential use of oxygen resulted in significant cost savings.

Filtrate residual alkali may be used as a control parameter for the addition of oxidized white liquor. The measurement of filtrate conductivity can be substituted for residual alkali at the levels where alkali is present in excess, but is not satisfactory at the target level. Filtrate pH, the original standard for OWL addition, was not useful in this study. The use of ratio based controls can be used to further improve OWL control, particularly for systems which experience regular swings in production rate.

LITERATURE CITED

1. **Miller, W.J., *et al.*,** "Effect of entrained black liquor carry-over on medium consistency oxygen delignification," *TAPPI Journal*, **74** (2), pp. 117-122 (Feb. 1991).
2. **Genco, J.M.,** University of Maine, Orono, Maine. Personal communication.
3. **Miller, W.J.,** IMPCO, Nashua , New Hampshire. Personal communication.
4. **Olm, T. and Teder, A.,** "The Kinetics of Oxygen Bleaching, " *TAPPI Journal*, **62** (12), pp. 43-46 (Dec. 1979).

Oxygen Delignification of Southern Hardwoods

S.B. Agarwal, H.B. Kwon, J.M. Genco
Department of Chemical Engineering, University of Maine, Orono, ME 04469

W. Miller
IMPCO Pulping Technologies, Division of Beloit Corporation, Nashua, NH 03061

B.J.W. Cole
Department of Chemistry, University of Maine, Orono, ME 04469

This paper describes the medium consistency oxygen delignification of southern hardwood. A mixed southern hardwood kraft brownstock pulp having an initial kappa number 13 was delignified in a medium-consistency (12%), oxygen reactor under industrial conditions of alkali concentration (1.5, 2.5 and 3.5%), variable time (5, 10, 20, 40, and 60 minutes), and temperature (90, 100 and 110 C). Data are presented for the change in kappa number, intrinsic viscosity, wet zero span tensile strength, brightness, and residual alkali following exposure to oxygen. The oxygen delignification data for kappa number were fit to an empirical rate equation of order "q".

$$-r_a = -\frac{dK}{dt} = kK^q$$

The reaction rate constant (k) was correlated to the initial alkali concentration [OH⁻, g/liter], temperature (T, K), activation energy (E_A, kJ/mole) and oxygen pressure (P_{02}, psig) by the equation:

$$k = 6.59x10^6 \exp(-107.2 / RT)[OH^-]^{0.92} P_{O2}^{0.53}$$

The model may be used to infer effects of time, temperature, oxygen pressure and concentration on the change in kappa number. Since the rate equation is empirical and includes mass transfer and kinetics effects (q=7.7), it must be used within the range of variables investigated. The rate equation is compared to other models available in the literature for estimating the change in kappa number during oxygen delignification.

INTRODUCTION

Oxygen delignification is becoming increasingly important in fiber lines for the 21st century. The use of an oxygen stage in ECF and TCF bleaching sequences can significantly reduce emissions of chlorinated lignins and COD in wastewater effluent while saving chemical costs through the use of lower amounts of chlorine dioxide, hydrogen peroxide and other oxidizing agents. Compared to Cl_2 and ClO_2 stages, an oxygen delignification stage has both lower reactivity and selectivity (McDonough, 1992). Poor selectivity is related to carbohydrate degradation while lignin reactivity is hindered by mass transfer of oxygen into the fiber wall (Hsu and Hsieh, 1988) and lignin condensation reactions (Ljunggren and Johansson, 1987).

For optimization and control of an oxygen delignification stage, information on delignification kinetics is required. A kinetic model combined with the stoichiometry of the process (amount of bleaching chemicals required to reach a given kappa number) can be used to describe and control the process. Axegard, Moldenius and Olm (1979), taking the example of chlorine dioxide bleaching, showed how kinetic models can be used to achieve optimum control of the process. Kovasin, Uusitalo and Markku (1987) showed how a kinetic equation can be used in the dimensioning of oxygen delignification reactors.

Few systematic studies of hardwood delignification appear in the open literature. The primary objective of the work described in this paper is to present a comprehensive set of data applicable to medium consistency delignification of southern hardwoods. A second objective is to present a mathematical model for describing the kinetics of oxygen delignification of such species. In this study, kinetic data were collected in a medium-consistency (12%), laboratory reactor under varying conditions of alkali concentration, temperature and oxygen pressure. The oxygen delignification curves were fit using an empirical kinetic model suggested by Sarkanen and Johanson (1976) and later used by Ransdell (1996) in describing the oxygen delignification of shredded wood using sodium carbonate with oxygen. Data are also presented for the effects of various process variables on the kappa number reduction, selectivity, brightness, alkali consumption and wet zero span tensile strength.

Difficulties in Developing a Kinetic Model

Several notable difficulties arise in the development of a kinetic model for oxygen delignification applicable to industrially significant situations. Oxygen delignification is a heterogeneous reaction which involves three phases comprising fiber (solid), water (liquid) and oxygen (gas). To explain the kinetics of the process completely, mass transfer effects cannot be neglected. Hsu and Hsieh (1985A, 1985B) showed that the effect of mass transfer on the oxygen delignification process under medium and high consistency can be significant. However, any mathematical model incorporating mass transfer phenomena along with the chemical reaction step would require an extremely complex theoretical treatment. Furthermore, it would be very difficult to simulate the mixing pattern in commercial reactors on a laboratory scale. This problem has been bypassed in most kinetic studies on oxygen delignification by considering "overall" rates wherein the mass transfer and reaction kinetics are lumped into single "global kinetic models".

Another problem is that lignin chemistry and the distribution of lignin within the fibers is not well understood. This inhibits fundamental study of the reaction kinetics. Kinetic studies of the delignification process have been handled empirically, therefore, largely because of the urgent need for data applicable to practical situations.

Oxygen Delignification Chemistry

The chemistry of oxygen delignification is quite complicated due to the complexity of the lignin structure as well as the number of different oxygen species formed during the reaction. The mechanisms of oxygen bleaching have been studied using model compounds (Gierer and Imsgard, 1977; Ljunggren and Johansson, 1987; Ljunggren and Johansson, 1994) and by examining residual and dissolved lignins (Berry and Fleming, 1987). These studies which have been summarized by McDonough (1996), show that free phenolic hydroxyl groups play a major role in oxygen delignification. The phenolic group is ionized under strongly alkaline conditions, generating a site with high electron density which undergoes transfer of a single electron to molecular oxygen or any of the available radical species (Figure 1.a).

The resulting phenoxy radicals have several resonance structures which then undergo coupling reactions or reactions with various radical species such as molecular oxygen, hydroxyl radical, hydroperoxy radical, or the superoxide anion radical. The products formed by these reactions include muconic and other organic acids, carbonyl structures and carbon dioxide (Figure 1.b).

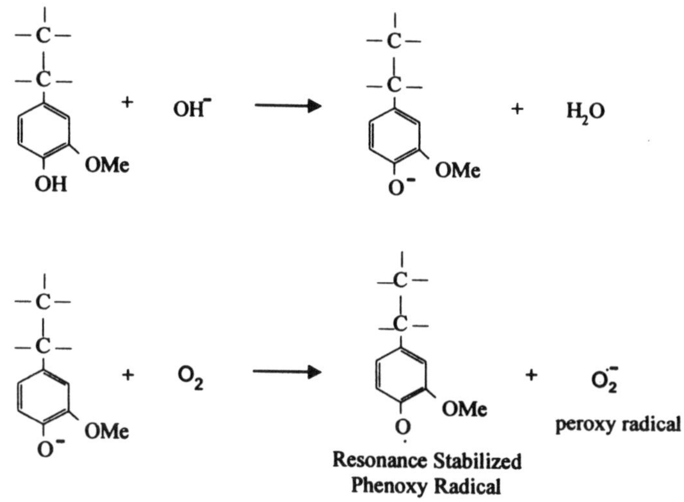

Figure 1.a. Initial attack of oxygen on phenolic nuclei (Singh and Dillner, 1979)

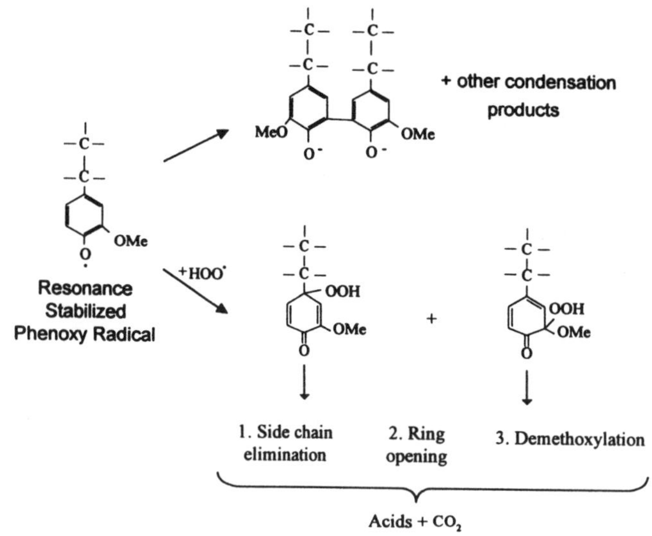

Figure 1.b. Possible reactions of lignin via phenoxy radicals (Ljunggren, et al., 1987)

Side chain elimination, ring opening and demethoxylation reactions are all possible which break up the polymeric structure of the lignin molecule and make the resulting fragments more water soluble (Gierer and Imsgard, 1977). The coupling products formed increase the molecular weight of the lignin fragments making them less soluble (San Clemente, 1981).

PREVIOUS LITERATURE

The kinetics of oxygen delignification have been reviewed by Perng and Oloman (1994) and Guven et al. (1995). Virtually all the oxygen delignification models are empirical in which the rate of delignification is considered to be proportional to the lignin content expressed as the kappa number [K], hydroxide ion concentration [OH^-] and oxygen pressure [P_{O2}] in the form of a "power law". This takes the form given by equation (1) :

$$-\frac{dK}{dt} = k_1[OH^-]^m[P_{O2}]^n K^q \qquad (1)$$

The constants m, n and q are determined empirically from experimental data. The reaction rate coefficient k_1 depends on the temperature and is given by the Arrhenius equation :

$$k_1 = A\exp(-\frac{E_A}{RT}) \qquad (2)$$

The model described by equations (1) and (2) involves five constants that must be determined experimentally (A, E_A, m, n, and q). The reaction order (q) with respect to the lignin content is sometimes higher than three; see for example the work of Hsu and Hsieh, (1987, 1988). This is interpreted to mean that the delignification process is hindered; most likely by diffusion or mass transfer effects and/or by lignin condensation reactions similar to the ones which occur during kraft cooking.

Olm and Teder Model

Several investigators have treated oxygen delignification by considering that the drop in the kappa number takes place over two distinct time periods, a rapid initial step followed by a long period over which the kappa number drops very slowly. The reason for the two periods is thought to be related to the various lignin linkages comprising its structure which have been shown by Johansson and Ljunggren (1994) to have widely different half lives when exposed to oxygen. The disadvantage of this approach is that a two region model expands the number of empirical coefficients to eight (8) from five coefficients required by equations (1) and (2).

The most widely accepted kinetic model for softwood is that of Olm and Teder (1979) who defined the initial kappa number (K_0) as the sum of the quickly (K_{01}) and slowly (K_{02}) eliminated lignin.

$$K_0 = K_{01} + K_{02} \qquad (3)$$

Equation (1) was applied to the two regions in the form of two first order kinetic models. The corresponding two-region delignification equation takes the form :

$$-\frac{dK}{dt} = k_1[OH^-]^{0.1}[P_{O2}]^{0.1} K_1 + k_2[OH^-]^{0.3}[P_{O2}]^{0.2} K_2 \qquad \text{(Softwood)} \qquad (4)$$

where the constants k_1 and k_2 are the rate constants in the initial and final phases respectively and K_1 and K_2 are the "easily and "slowly" eliminated lignin. The data of Olm and Teder (1979) have also been represented using the one region global model (Kovasin, Uusitalo and Viilo, 1987):

$$-\frac{dK}{dt} = k_3\,[P_{O_2}]^{0.5}[OH^-]^{0.6} K^{3.2} \qquad \text{(Softwood)} \qquad (5)$$

By performing oxygen delignification tests varying the reaction time, temperature, oxygen pressure and alkali dosage, the resulting kappa number reduction can be expressed in either form suggested by equations (4) or (5).

Other Available Models

The Olm and Teder model was later corrected for mass transfer by Hsu and Hsieh (1987, 1988). Edwards and Norberg (1973), Myers and Edwards (1989), and Vincent, Nguyen and Matthews (1994) postulated a "floor" or a final equilibrium value for the kappa number. All other workers have assumed a non-equilibrium condition with an asymptotic decrease in kappa number with time. Perng and Oloman (1994) have summarized the available rate models together with the appropriate rate constants applicable to the initial and final reaction rate periods.

Most of the oxygen delignification models have been developed for softwoods (see Table 1). Myers and Edwards (1989) developed a general delignification model applicable to both softwoods and hardwood kraft pulps for initial kappa numbers ranging from 11 to 128 and for all consistencies up to 30%. They used the dissolved oxygen concentration rather than the partial pressure in the kinetic expression. Vincent et al. (1994) collected kinetic data for Eucalyptus kraft pulp and fitted the results to the two-stage first order model of Myers and Edwards (1989). New equation parameters were determined specific for Eucalyptus pulp.

Because oxygen delignification rates depend greatly on the wood raw material and pulping conditions, it is worthwhile to perform laboratory tests to create a kinetic equation for each specific case in question to ensure reliable results.

EXPERIMENTAL

Oxygen delignification was carried out on mixed hardwood kraft pulp samples which were obtained from a mill in the southeastern United States. The initial kappa number of the pulp samples was 13.2. Several runs were carried out at varied conditions of temperature (90, 100 and 110 C), alkali concentration (1.5, 2.5 and 3.5% based on pulp), oxygen pressure (20, 60 and 100 psig), and time (5, 10, 20, 40 and 60 minutes). After each run, measurements were made for the kappa number, viscosity, wet zero span tensile strength, brightness, residual alkali and pH. Details of the experimental procedure are summarized in Appendix A.

RESULTS AND DISCUSSION

Delignification Rate Equation

The delignification data were correlated using the general kinetic model given by equation (6). The dependency of pressure and initial sodium hydroxide concentration can be incorporated into the rate coefficient (k) as suggested by Sarkanen and Johanson (1976) who investigated the use of oxygen for pulping wood meal.

$$-\frac{dK}{dt} = k\,K^{q} \qquad (6)$$

where

$$k = k_1\,[OH]^m[P_{O2}]^n = A\exp[E_A/RT][OH]^m[P_{O2}]^n \qquad (7)$$

The units used in the correlation are:

K = kappa number
[OH] = sodium hydroxide concentration (g/L)
P_{O2} = oxygen partial pressure (psig)
t = time (hr)

The dependence of k_1 on temperature is given by the Arrhenius relationship shown in equation (2). Sarkkanen and Johanson proposed the empirical rate equation of order "q" to express the reaction kinetics for oxygen delignification of wood meal. Integration of equation (6) under conditions of constant oxygen pressure and initial alkali concentration gives the equation for the kappa number as a function of time (Levenspiel, 1972).

$$K^{1-q} - K_0^{1-q} = (q-1)kt, \qquad q \neq 1 \qquad (8)$$

Equation (8) may be rewritten as

$$K = \left[\frac{1}{K_0^{q-1}} + (q-1)\,k\,t\right]^{-\frac{1}{(q-1)}}, \quad q \neq 1 \qquad (9)$$

The parameter q which denotes the rate order with respect to the kappa number cannot be found explicitly from equation (8); so a trial and error solution was used following the differential and integral methods of analysis (Levenspiel, 1972). The other parameters were determined by linear regression of the experimental data. Details of the calculation procedure are presented in Appendix B. The values of the parameters for southern hardwood with the appropriate units are summarized in Table 2. The value of q was found to be 7.7. The rate constant k was found to be:

$$k = 6.59x10^6\exp(-107.2/RT)\left[OH^-\right]^{0.92} P_{O2}^{0.53} \qquad (10)$$

It should be emphasized here that the model represented by equation (6) is purely empirical and does not quantitatively represent the combined phenomena of diffusion and competitive chemical reactions that occur during the delignification process. The rate order with respect to the kappa number (q=7.7) is very high. Nevertheless, the model can be used effectively in technical applications such as reactor sizing.

Effects of Alkali Concentration

Figure 2 shows the effect of alkali level (NaOH =1.5, 2.5 and 3.5%) or initial concentration (NaOH = 2.04, 3.49 and 4.77 grams/liter) on the delignification of southern hardwoods. Data are presented for the kappa number reduction when the temperature (90 C) and the oxygen pressure (100 psig) were held constant and the initial alkali was varied. At a given time, the greatest reduction in kappa number was obtained at the highest level of the alkali concentration (3.5% application) The dependence on the initial alkali concentration was found to vary with the 0.92 power; which is almost a linear dependence upon the initial alkali concentration.

Effect of Temperature

Similarly, Figure 3 illustrates the effect of temperature (90, 100 and 110 C) on delignification. The data are shown for a 2.5% NaOH level and 100 psig oxygen pressure which were held constant. Figure 3 shows that the greatest reduction in kappa number occurs at the highest level of temperature (110 C) when the alkali concentration and the oxygen pressure are set constant. The frequency factor in the Arrhenius relationship is $A=6.89x10^6$ and the "activation energy" parameter E_A is 107.2 kJ/mole (25,620 calories per gram mole). The low activation energy is most likely related to hindrance of the actual reaction by diffusion effects.

Effect of Oxygen Pressure

Figure 4 shows the effect of oxygen pressure (100, 60 and 20 psig) on the delignification curves for 100 C and an alkali level of 2.5%. Again the greatest reduction in kappa number was for the highest value of the oxygen pressure (100 psig) but the effect was much less pronounced than the effect of alkali concentration and temperature, especially for the higher oxygen pressures (60 and 100 psig). The dependency of the rate equation is to the 0.53 power.

Comparing the three primary process variables indicates that the variables listed in order of decreasing importance are the temperature, alkali concentration and, lastly, oxygen pressure.

Kappa Number Dependence

The dependence upon kappa number given by the empirical rate equation (6) is shown in Figure 5. The rate equation predicts a very strong dependence upon kappa number. The kappa number versus time curves flatten out at long values of time and at all values of time at low kappa number values. Since mill pulp was used in this study, it was not possible to determine the effect of kappa number on the rate equation because that would require numerous samples over a wide range of kappa number values; which would be difficult to obtain in the mill situation. Since the reaction rate parameter (q=7.7) was determined from only one sample having kappa number 13.2 mL, it would be prudent to use the reaction equation only within a narrow range about this value, say 10 to 16 mL.

Comparison With Other Models

Most of the models have been developed for softwoods. The only models found in the literature which correlate hardwood data are the generalized model of Myers and Edwards (1989), which was developed for both softwoods and hardwoods, and that of Vincent et al. (1994) which was developed for Eucalyptus Kraft pulp. The Myers and Edwards' model is difficult to use because they correlated delignification data to the dissolved oxygen concentration instead of the oxygen partial pressure in the gas phase which is the parameter known in the plant and design situations.

The model developed in this paper was compared, however, with both the Vincent (1994) and Hsu and Hsieh (1988) models. Guven and co-workers (1995) found that the Hsu and Hsieh model, although developed for softwoods, gave a reasonable fit to southern hardwood oxygen delignification data. The models of Vincent (1994) and Hsu and Hsieh (1988) are both two-region, first order models for kappa number reductions. As such they require eight (8) parameters each to express the dependence on temperature, alkali concentration and temperature dependence; namely four parameters for each region. The results of this comparison are shown summarized in Figure 6 for data taken at 90°C, 3.5% caustic and 100 psig. Both the Vincent et al. and the Hsu and Hsieh model predict a higher initial as well as final rate of delignification than observed experimentally.

Viscosity

During oxygen delignification competitive reactions occur and are reflected in a decrease in the molecular weight of the carbohydrate polymers in the pulp. Figure 7 illustrates a plot of the intrinsic viscosity [η] plotted versus time (t) for a temperature of 110 C and oxygen pressure of 100 psig showing the alkali application level as a parameter (1.5%, 2.5% and 3.5% NaOH). These correspond to initial alkali concentrations of 2.05, 3.40 and 4.77 grams NaOH/liter. The alkali application is a major variable that affects the intrinsic viscosity. The temperature and oxygen pressure used in the oxygen stage had considerably less effect than the initial alkali concentration on the intrinsic viscosity data for the pulp. The data in Figure 8 illustrate the correlation between the 0.5% CED Tappi viscosity and intrinsic viscosity.

Selectivity

For oxygen delignification, selectivity can be defined as the ratio of the delignification rate to the carbohydrate degradation rate. It is an important factor as it determines how far the delignification can proceed without excessive loss of pulp strength. The selectivity for a given pulp is usually expressed as a plot of viscosity versus kappa number. Figure 9 shows the effect of the alkali concentration on the selectivity for oxygen delignification at 100 C and 100 psig oxygen pressure. It can be seen that lower alkali concentration gives better selectivity. Curves similar to Figure 9 were obtained at other temperatures showing the major effect concentration has on the selectivity.

The effect of temperature on selectivity is illustrated in Figure 10 for an alkali application of 2.5% and 100 psig oxygen pressure. Varying the temperature did not show a significant effect on the selectivity compared to the initial alkali concentration (see Figure 9). Similar results were seen for the other oxygen pressures investigated (20 and 60 psig), which also did not show any significant effect.

Residual Alkali
Typical data for residual alkali concentration (grams per liter) are shown in Figure 11. This figure plots residual alkali versus concentration for the delignification conducted at a temperature of 100 C and an oxygen pressure of 100 psig. The data show that after an initial rapid sorption of caustic, the residual alkali drops slowly over time during the second phase of the reaction.

Brightness
Figure 12 shows a plot of brightness versus the kappa number obtained in all experiments conducted in this study. The brightness was found to depend only on the kappa number regardless of the process conditions. Perng and Oloman (1994) suggested a relation between the brightness and the kappa number as given by equation (11).

$$B = B_0 \exp[a(K_0 - K)^b] \tag{11}$$

A similar relationship was assumed and the parameters were determined by performing a non-linear regression analysis.

$$B = 30.31 \exp[0.0693 * (13.18 - K)^{1.06}] \tag{12}$$

A regression coefficient of 0.905 was obtained.

Wet Zero Span Tensile Strength
There was no noticeable effect of the process variables on the wet z-span tensile strength. A graph of wet z-span versus viscosity was determined for all the experiments performed (Figure 13). A linear relationship given by equation (13) was assumed, giving a regression coefficient of 0.74.

$$Z = 0.017[\eta] + 10.2 \tag{13}$$

where $[\eta]$ = Intrinsic Viscosity, cm^3/g

CONCLUSIONS

There are various difficulties associated with the development of a kinetic model for the oxygen delignification process. These difficulties arise because of the heterogeneous reaction system and the complexity of lignin structure. Because of the complexity of the problem, oxygen delignificaition kinetics have been traditionally treated by using homogeneous empirical models which generally have a power law representation. The power law incorporates both mass transfer and chemical reaction effects in one simple empirical equation. It can be used for sizing oxygen delignification reactors.

In the present study, oxygen delignification data have been collected for a mixed southern hardwood kraft brown stock pulp. Data were obtained over a wide range of industrially significant process conditions at 12% consistency. The delignification data for southern hardwood kraft pulp can be correlated using a general global kinetic model having five empirical parameters. The model suggests a high rate order with respect to the kappa number (q=7.7). This high dependence upon the kappa number most likely results from the form of the equation and the fact that it incorporates both mass transfer and chemical reaction affects. The model should not be used for kappa number values significantly different than that used in the present study (13.2).

Temperature and alkali concentration have the strongest effect on the decrease in the kappa number with time while oxygen pressure did not show any significant effect over the range 20 to 100 psig O_2 pressure. The initial alkali concentration was found to have a strong effect on the selectivity whereas temperature and oxygen pressure did not appreciably affect the selectivity. The residual alkali data indicate that caustic is rapidly sorbed into the fiber. The alkali decreases slowly with time during the second phase of the reaction. Brightness was found to depend only on the kappa number. The effect of the process variables on the z-span tensile was not clear. Previous models available in the literature do not accurately describe the delignification data for southern hardwoods reported here.

REFERENCES

1. Axegard, P., Moldenius, S. and Olm, L., "Basic Chemical Kinetics Equations are Useful for an Understanding of Pulping Processes", Svensk papperstid, 82(5): 131(1979).
2. Berry, R. M. and Fleming, B. I., "Why Do Chlorination and Extraction Fail to Delignify Unbleached Kraft Pulp Completely?", Holzforschung, 41(3): 177(1987).
3. Cowan, W.F., "Component Quality Testing Improves Rapport between Pulp-Paper Mills", Pulp and Paper Magazine, 60(12),108 (1986).
4. Croon, I. and Andrews, D.H., "Advances in Oxygen Bleaching: 1. Demonstrations of its Feasibility and Scope", Tappi 54(11): 1893(1971).
5. Edwards, L. and Norberg, S., "Alkaline Delignification Kinetics", Tappi, 56(11): 108(1973).
6. Gierer, J. and Imsgard, F., "The Reaction of Lignin with Oxygen and Hydrogen Peroxide in Alkaline Media", Svensk Papperstid., 80(16): 510(1977).
7. Guven, E., J.M. Genco, , W. Miller, and P. Hart, "Medium Consistency Oxygen Delignification of Southern Hardwoods", Paper presented at the Forest

Products Division Session at the 1995 Tappi Pulping Conference, Chicago, Illinois (November, 1995).
8. Hsu, C.L. and Hsieh, J.S., " Fundamentals of Oxygen Bleaching: Measurement of Oxygen Diffusion Rates in Medium and High Consistency Pulp", Tappi, 68(6): 92(1985A).
9. Hsu, C.L. and Hsieh, J.S., "Effects of Mass Transfer on Medium Consistency Oxygen Bleaching Kinetics", Tappi, 68(11): 126(1985B).
10. Hsu, C.L. and Hsieh, J.S., "Oxygen Bleaching Kinetics at Ultra-Low Consistency", Tappi, 70(12): 107(1987).
11. Hsu, C.L. and Hsieh, J.S., "Reaction Kinetics in Oxygen Bleaching", AIChE, 34(1): 116(1988).
12. Kovasin, K., Uusitalo, P., and Villo, M., "Dimensioning of Oxygen Delignification Reactors", 1987 International Oxygen Delignification Conference, 223-230, Tappi Press, Atlanta, Georgia.
13. Levenspiel, O., Chemical Reaction Engineering, 2nd Ed, John Wiley and Son, New York, Chapter 3, pp. 41-86 (1972).
14. Ljunggren, S.C.H. and Johansson, E.C., "Reaction Kinetics of Lignin Structures during Oxygen Bleaching- Effects of Solvent, Oxygen Pressure and pH", 1987 International Oxygen Delignification Conference, Tappi Proceedings, p.125 (1987).
15. Ljunggren, S. and Johansson, E., "The Kinetics of Lignin Structures during Oxygen Bleaching: Part 4- The Reactivities of Different Lignin Model Compounds and the influence of Metal Ions on the Rate Degradation", Journal of Wood Chemistry and Technology, 14(4): 507 (1994).
16. McDonough, T.J., "Oxygen Bleaching Process", Tappi, 69(6): 46(1986).
17. McDonough, T., J., "Kraft Pulping Bleaching Technology: A Brief Overview of Basic Principles and Current Trends", Institute of Paper Chemistry, Atlanta, GA (1992).
18. McDonough, T. J., contained in "Pulp Bleaching - Principles and Practice", edited by C. W. Dence and D. W. Reeves, Tappi Press, Atlanta (1996), pp 218-223.
19. Myers, M.R. and Edwards, L.L., "Development and Verification of a Predictive Oxygen Delignification Model for Hardwood and Softwood Kraft Pulp", Tappi, 72(9): 215(1989).
20. Olm, L. and Teder, A., "The Kinetics of Oxygen Bleaching", Tappi, 62(12): 43(1979).
21. Perng, Y.S. and Oloman, C.W., "Kinetics of Oxygen Bleaching Mediated by Electrochemically Generated Ferricyanide", Tappi, 77(7): 115(1994).
22. Ransdell, J., "Kinetics of Softwood Pulping using Sodium Carbonate and Oxygen", PhD Dissertation, University of Maine, Orono, Maine (1996).
23. San Clemente, M.R., Sarkanen, K.V., and Sundin, S.E., "Alkaline Autoxidation of 4-ethyl guaiacol", Svensk Papperstid., 84(3): R1(1981).
24. Sarkanen, K.V.,and Johanson, L. N., "Use of Oxygen Pulping", AIChE Symposium Series, Chemical Engineering in Pulp and Paper Technology, 72(157), 50-55 (1976).
25. Singh, R. P., and Dillner, B. C., contained in "The Bleaching of Pulp", edited by R. P. Singh, 3rd edition, Tappi Press, Atlanta (1979), p. 161.
26. Vincent, H.D., Loi Nguyen, K. and Mattews, J.F., " Kinetics of Oxygen Delignification of Eucalypt Kraft Pulp", Appita, 47(3): 217(1994).

ACKNOWLEDGEMENTS

Financial support for Mr. Agarwal was provided by the U.S. Department of Agriculture, Grant number 94-34158-0028. The authors wish to thank Ms. Angela Genco for determining the intrinsic viscosity and Mr. Eugene Abbott for his assistance in performing the experiments.

Table 1

Oxygen Delignification Models Available in Literature

Literature	Type of Pulp
Hartler et al. [1970]	Pine sulfate pulp
Edwards & Norberg [1973]	Kraft pulp
Evans et al. [1979]	Kraft and soda anthraquinone pulps
Olm & Teder [1979]	Kraft softwood
Hsu & Hsieh [1988]	Southern pine Kraft pulp
Myers & Edwards [1989]	Scandinavian pine Kraft pulp
Vincent et al. [1994]	Eucalyptus Kraft pulp

Table 2

Parameters Used in Delignification Model

Parameters	Value
Alkali Dependence (g/liter)	$m = 0.92$
Oxygen Dependence (psig)	$n = 0.53$
Kappa Number Dependence	$q = 7.7$
Frequency Factor	$A = 6.59x10^6$
Temperature (K) Dependency	$E_A = 107.2$ kJ/mole

APPENDIX A

EXPERIMENTAL PROCEDURES

Oxygen Delignification

The oxygen delignification reactor has been described by Guven et al. (1996). Southern hardwood pulp samples were used in the oxygen delignification experiments. Oxygen delignification experiments were carried out in a 20 liter autoclave equipped with a mixer designed to approximate the low level of mixing in an oxygen up-flow tower. The speed of the mixer was set at 23 RPM and was constant for all the experiments during the oxygen delignification reactions. The reactor had previously been shown to simulate kappa number reduction seen in mill trials. All experiments were conducted under constant conditions except for the reaction time, the reaction temperature, the amount of caustic, and the oxygen pressure. Oxygen delignification stage conditions were:

The weight of the pulp : 120 oven dry grams
Pulp consistency : 12 %
Temperature : 90, 100, 110 $^{\circ}C$
Oxygen pressure : 20, 60, and 100 psi
Reaction time : 0, 5, 10, 20, 30, and 60 minute
Amount of caustic used : 1.5, 2.5, 3.5 % on pulp
Amount of $MgSO_4$ used : 0.5 % Mg^{+2} ion on pulp (added as $MgSO_4$)

For each run, $MgSO_4$ was mixed into the pulp and the pulp sample was then placed into a polyethylene mixing bag. After sealing the sample bag, each side of the bag was heated for about 10 minutes before the oxygen delignification experiments. The pulp was placed into the reactor which had been pre-heated to the experimental reaction temperature. Pre-heated NaOH was added into the reactor and thoroughly mixed into the pulp. The reactor was covered very quickly but removing the cover caused a decrease in temperature. When the reactor reached the reaction temperature oxygen was added to the reactor at the desired pressure. At the moment the reactor mass reached the reaction temperature and the oxygen was added, the time was started. The oxygen pressure was held constant by an outlet flow valve located on the top of the reactor.

At the end of the oxygen delignification experiments, the oxygen pressure inside the reactor was released to the atmosphere by completely opening the outlet flow valve. Water was then added into the reactor and the pulp slurry was taken out through a bottom valve. The pulp was washed thoroughly and handsheets were prepared.

Pulp Properties

The kappa number of the pulp was measured according to TAPPI standard method T236 os-76. Fiber strength was estimated by measurement of intrinsic viscosity and zero span tensile strength. The intrinsic viscosity [η] was estimated according to ASTM standard D1795-62 which is similar to SCAN-C 15:62. The wet zero span tensile strength was estimated using a Pulmac tester based upon the work of Cowan (1986). As a check, the 0.5% CED Tappi viscosity was also measured by following Tappi Standard T-230. ISO brightness was determined using a Technibrite Micro TB-1C brightness meter. Residual alkali was determined by titration following Tappi Standard T-625.

APPENDIX B

CALCULATION PROCEDURE

The delignification data were fit to the general model summarized by equation (6). The order (q) cannot be found explicitly from equation (6) so a trial-and-error solution must be made. To accomplish the data fit, a value for the order q must be selected and k calculated. The value of q which minimizes the in k is the desired value of q.

The calculation was performed using both the differential and the integral methods of analysis. The differential method was used to estimate the value of q. Then the integral method was used to see how well the obtained value of q agrees with the data. The integral method gives more accurate results because the data do not have to be differentiated. Hence the value of k was obtained from the integral method.

Differential Method

To use the differential method, equation (6) can be written as :

$$\log(-dK / dT) = \log(k) + q\log(K) \tag{B1}$$

The parameters k and q were then found for each data set by plotting log(-dK/dt) versus log(K). The average value of q obtained for all the data was 7.7.

Integral Method

Equation (9) may be written:

$$\frac{K^{1-q} - K_0^{\ 1-q}}{(q-1)} = k\,t \tag{B2}$$

where K_0 is the initial kappa number.

The left hand side of equation (B2) was plotted against t. The plot should be a straight line passing through the origin with slope equal to k. The q value used in the integral method was the arithmetic mean of the q values obtained from the differential method (q=7.7). The values obtained with oxygen partial pressures of 20 and 60 psi were not included when evaluating the average; but, they served as a check on the value of q=7.7. Other values of q were tried but did not improve the results.

The values for k are summarized in Table B1 using a value of q = 7.7. The results obtained when the oxygen partial pressure was changed at 100 C and 2.5% sodium hydroxide concentration are summarized in Table B2. The integral method appears to give better results, as seen by the value of the regression coefficient r^2.

Table B1
Values of k as a Function of Temperature and Initial Alkali at 100 Psig Oxygen Pressure

Set	k	r^2
(1) 90 C, 3.5% NaOH	9.285E(-8)	0.999
(2) 90 C, 2.5 % NaOH	4.220E(-8)	0.942
(3) 90 C, 1.5% NaOH	3.060E(-8)	0.957
(4) 100 C, 3.5% NaOH	3.210E(-7)	0.877
(5) 100 C, 2.5% NaOH	1.435E(-7)	0.977
(6) 100 C, 1.5% NaOH	1.053E(-7)	0.915
(7) 110 C, 3.5% NaOH	5.587E(-7)	0.996
(8) 110 C, 2.5% NaOH	5.043E(-7)	0.965
(9) 110 C, 1.5% NaOH	1.263E(-7)	0.915
Mean		**0.949**

Table B2
Values of k as a Function of Oxygen Pressure at 100 C Temperature and 2.5% Alkali Concentration

Set	k	r^2
(1) Partial pressure = 20 psi	6.276E(-8)	0.973
(1) Partial pressure = 60 psi	1.223E(-7)	0.976

Estimation of A, E_A, m and n

Knowing the value of k, the values of A, E_A, m and n were estimated by applying equation (7).
For the calculation of E_A, equation (7) can be written as :

$$k = k_2 \exp(\frac{-E_A}{RT}) \qquad \text{(B3)}$$

where $k_2 = A\,[OH]^m\,[P_{O2}]^n$ was assumed to be constant for a particular sodium hydroxide concentration and oxygen partial pressure. Equation (B3) can be then be written as :

$$\ln(k) = \ln(k_2) - \frac{E_A}{RT} \qquad \text{(B4)}$$

Thus, the value of E_A was estimated from a regression analysis for ln(k) versus [-1/(RT)] for the three different sodium hydroxide concentrations. The average of the three values obtained in this manner was **1.07225×10^5 Joules/gmol.**

To get m, equation (B3) can be written as :

$$\log(k) = \log(k_3) + m \log[OH^{-1}] \qquad \text{(B5)}$$

where $k_3 = A \exp[-E_A/(RT)]\,[P_{O2}]^n$ is a constant for a particular temperature and oxygen pressure. The average value of m obtained was **0.92**

Similarly, to get n, equation (B3) can be written as :

$$\log(k) = \log(k_4) + n \log[P_{O2}] \qquad \text{(B6)}$$

where $k_4 = A \exp[-E_A/(RT)]\,[OH]^m$ = constant for a particular temperature and alkali concentration. The average value of n obtained was **0.53**

Knowing E_A, m and n, equation(B3) can be solved for A for different sodium hydroxide concentrations and temperatures. The average value of A obtained was **6.588×10^6**

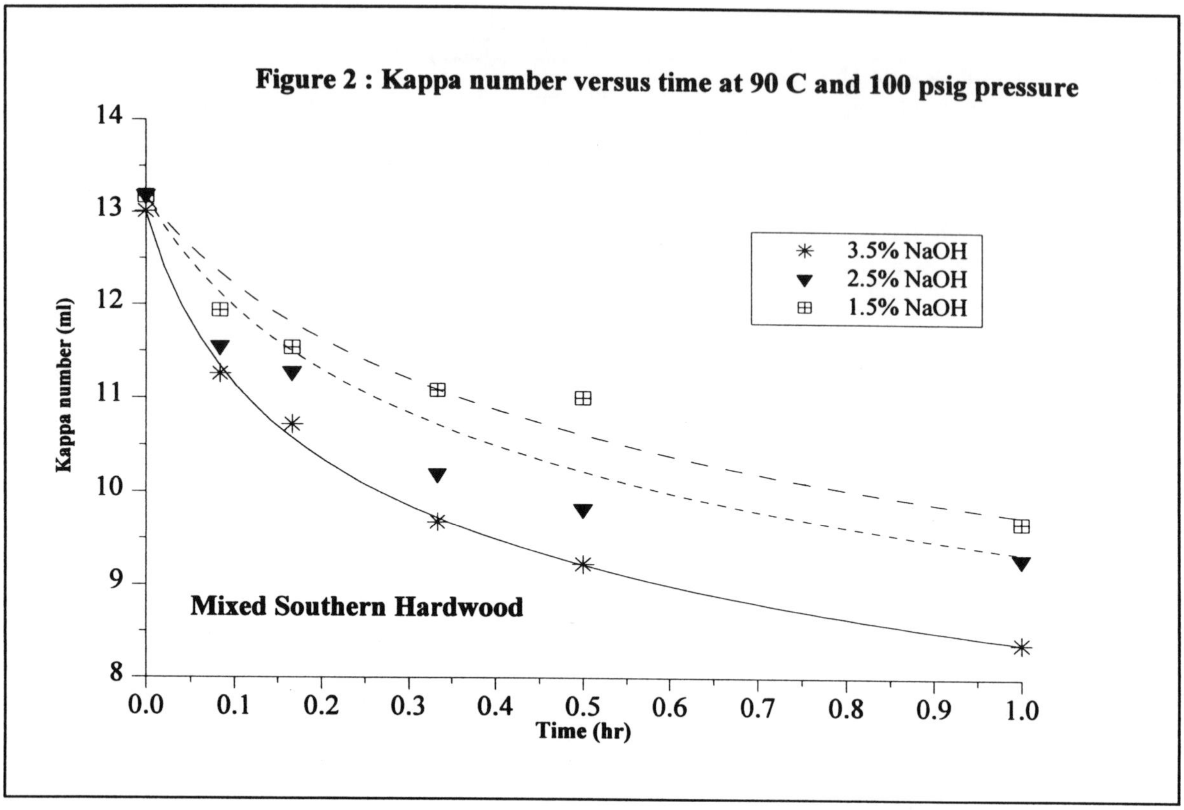
Figure 2 : Kappa number versus time at 90 C and 100 psig pressure
Kappa number (ml)
14
13
12
11
10
9
8
3.5% NaOH
2.5% NaOH
1.5% NaOH
Mixed Southern Hardwood
0.0
0.1
0.2
0.3
0.4
0.5
0.6
0.7
0.8
0.9
1.0
Time (hr)

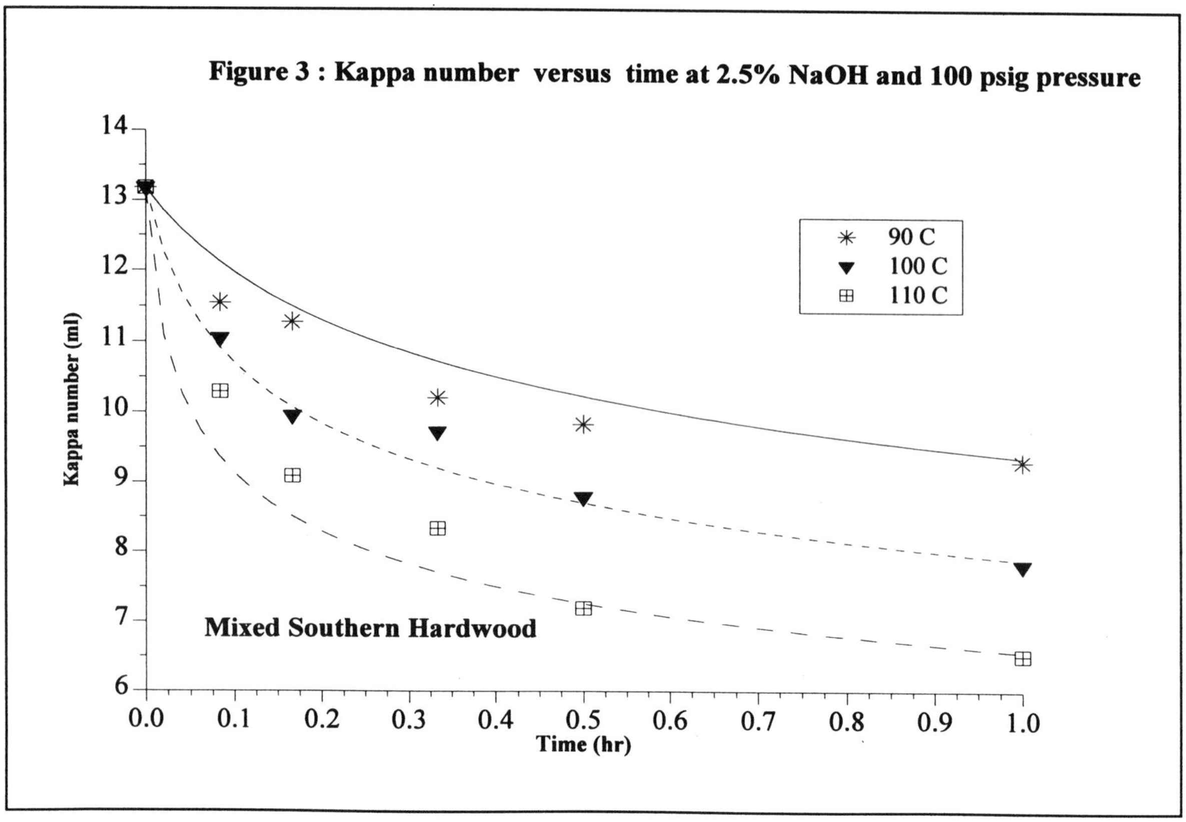
Figure 3 : Kappa number versus time at 2.5% NaOH and 100 psig pressure
Kappa number (ml)
14
13
12
11
10
9
8
7
6
90 C
100 C
110 C
Mixed Southern Hardwood
0.0
0.1
0.2
0.3
0.4
0.5
0.6
0.7
0.8
0.9
1.0
Time (hr)

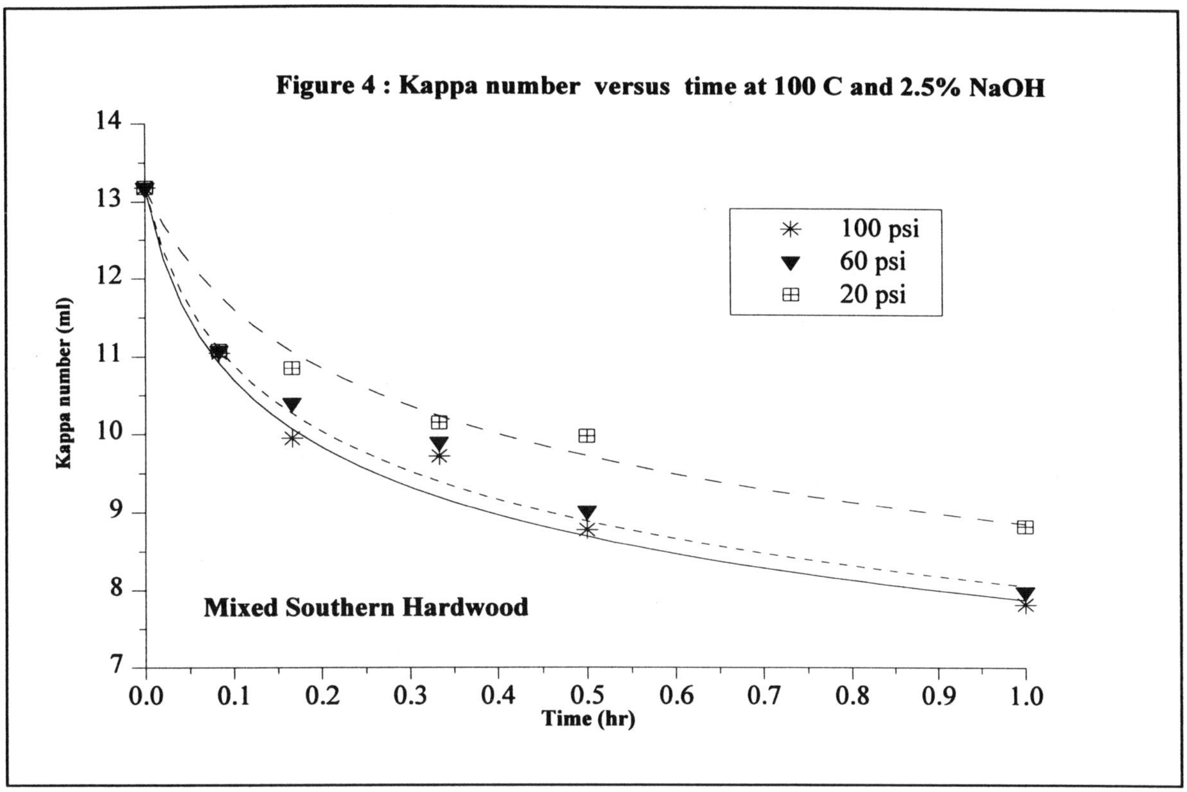
Figure 4 : Kappa number versus time at 100 C and 2.5% NaOH
Kappa number (ml)
Time (hr)
100 psi
60 psi
20 psi
Mixed Southern Hardwood

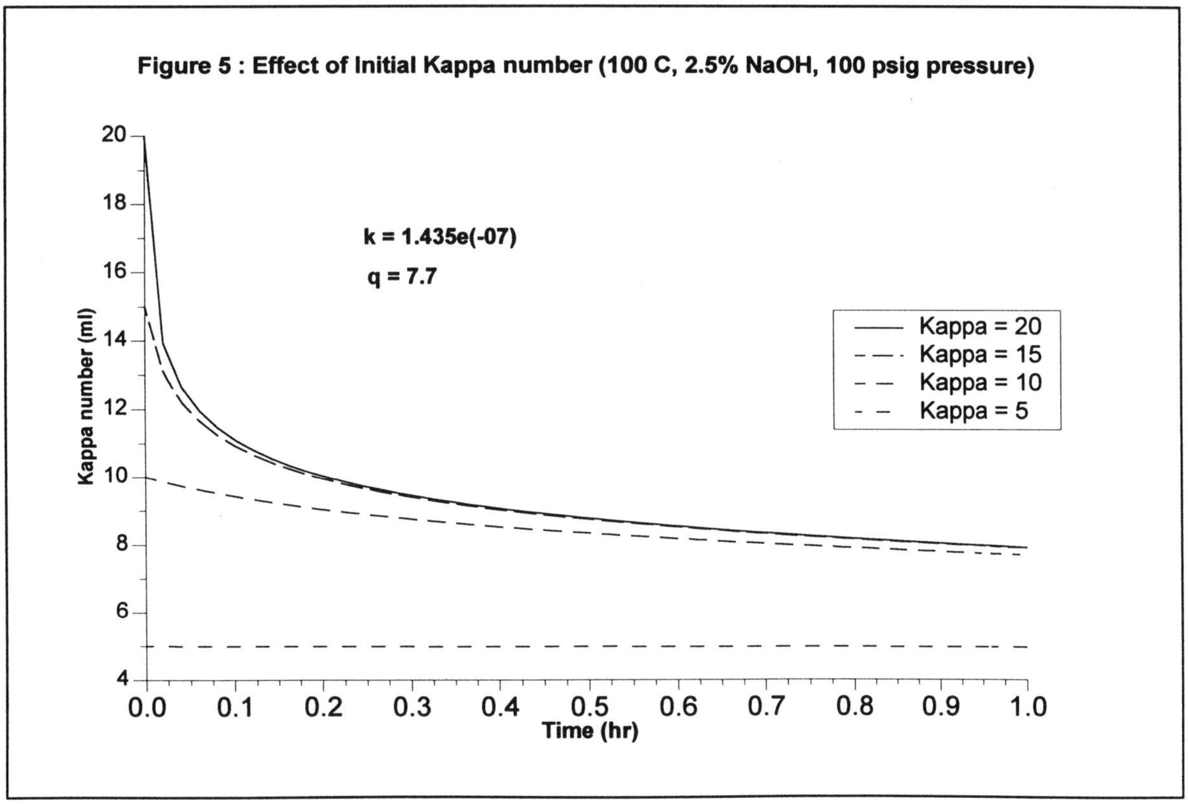
Figure 5 : Effect of Initial Kappa number (100 C, 2.5% NaOH, 100 psig pressure)
k = 1.435e(-07)
q = 7.7
Kappa = 20
Kappa = 15
Kappa = 10
Kappa = 5
Kappa number (ml)
Time (hr)

Figure 6 : Comparison of models (90 C, 3.5% NaOH, 100 psig pressure)

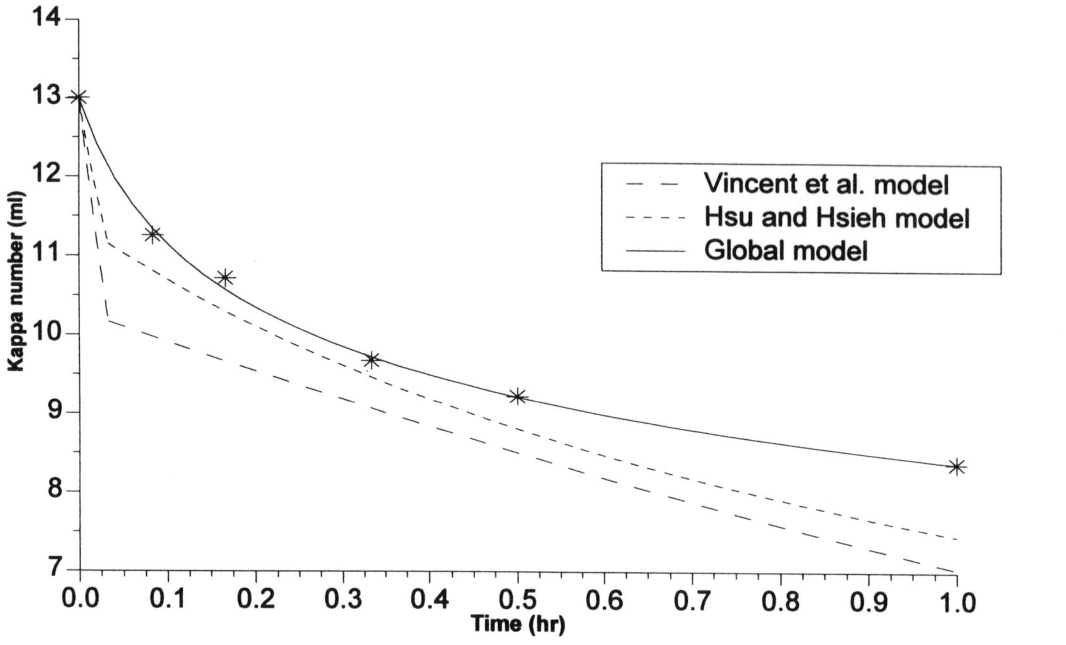

Figure 7 : Intrinsic viscosity versus time at 110 C and 100 psig pressure

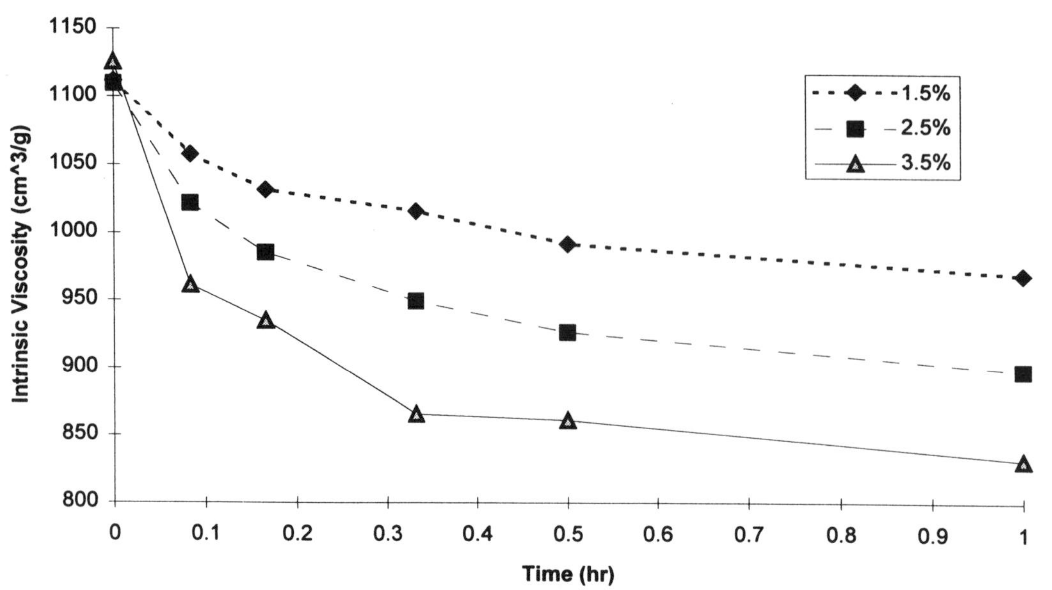

Figure 8 : 0.5% Tappi viscosity vs intrinsic viscosity

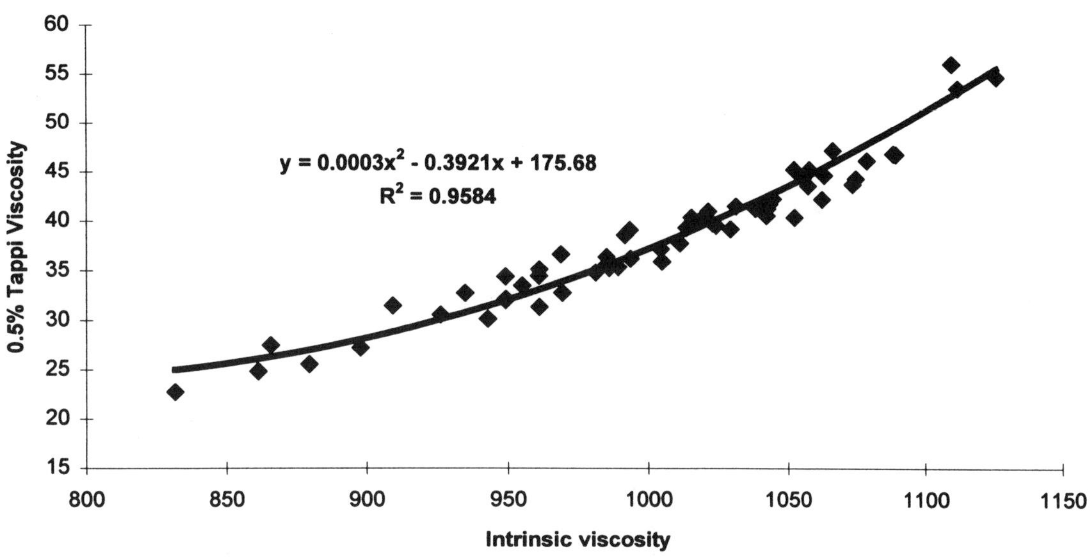

Figure 9 : Selectivity at 100 C and 100 psig pressure

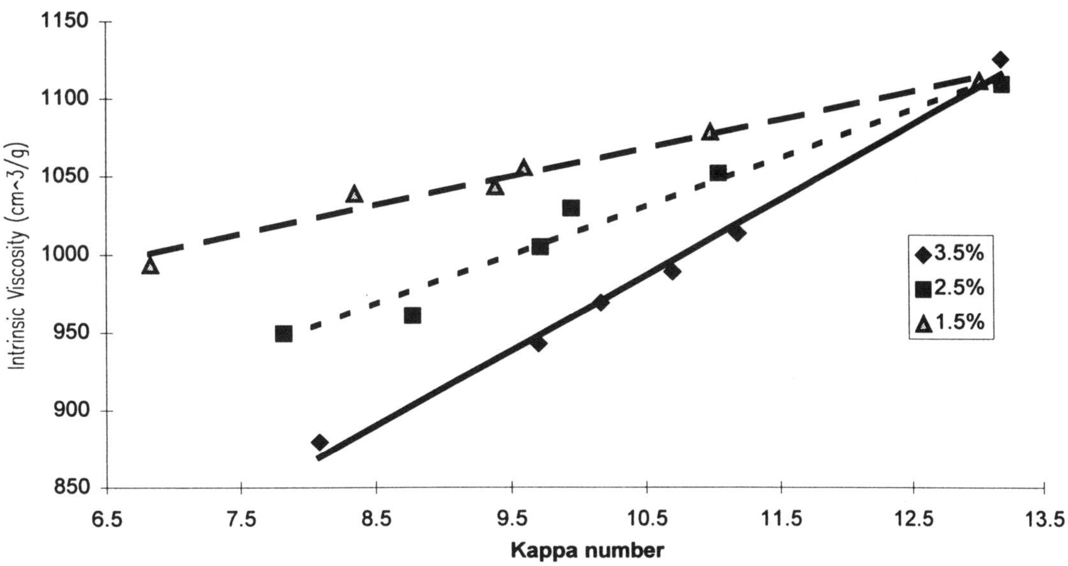

Figure 10 : Selectivity at 2.5% NaOH and 100 psig pressure

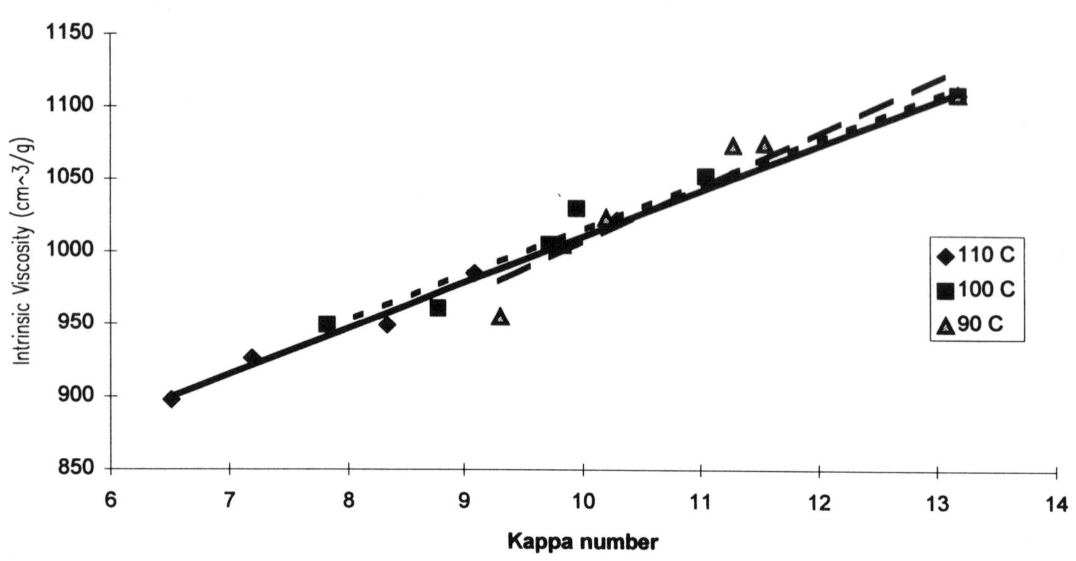

Figure 11 : Residual alkali concentration at 100 C and 100 psig pressure

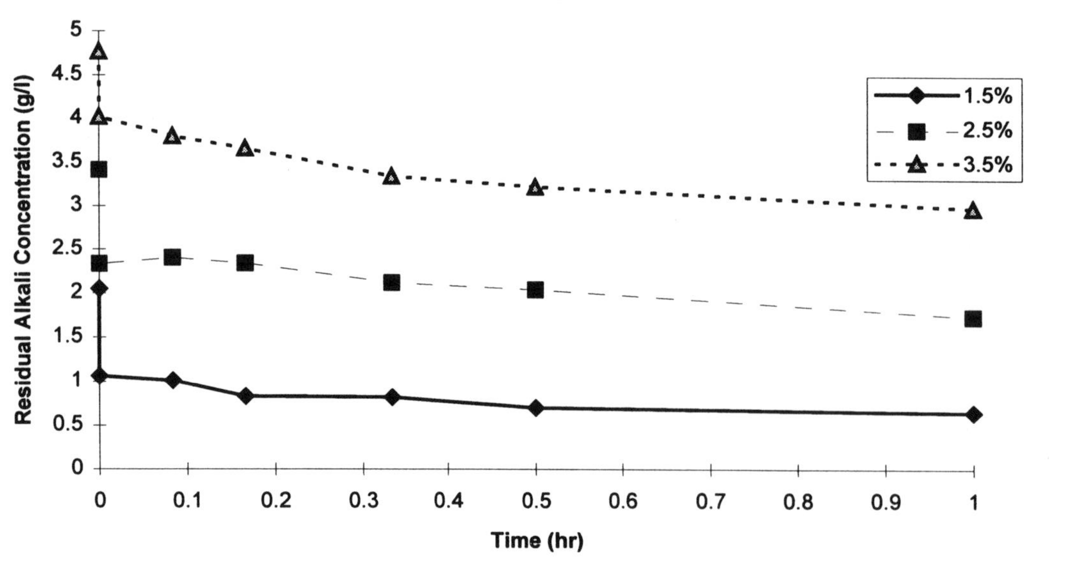

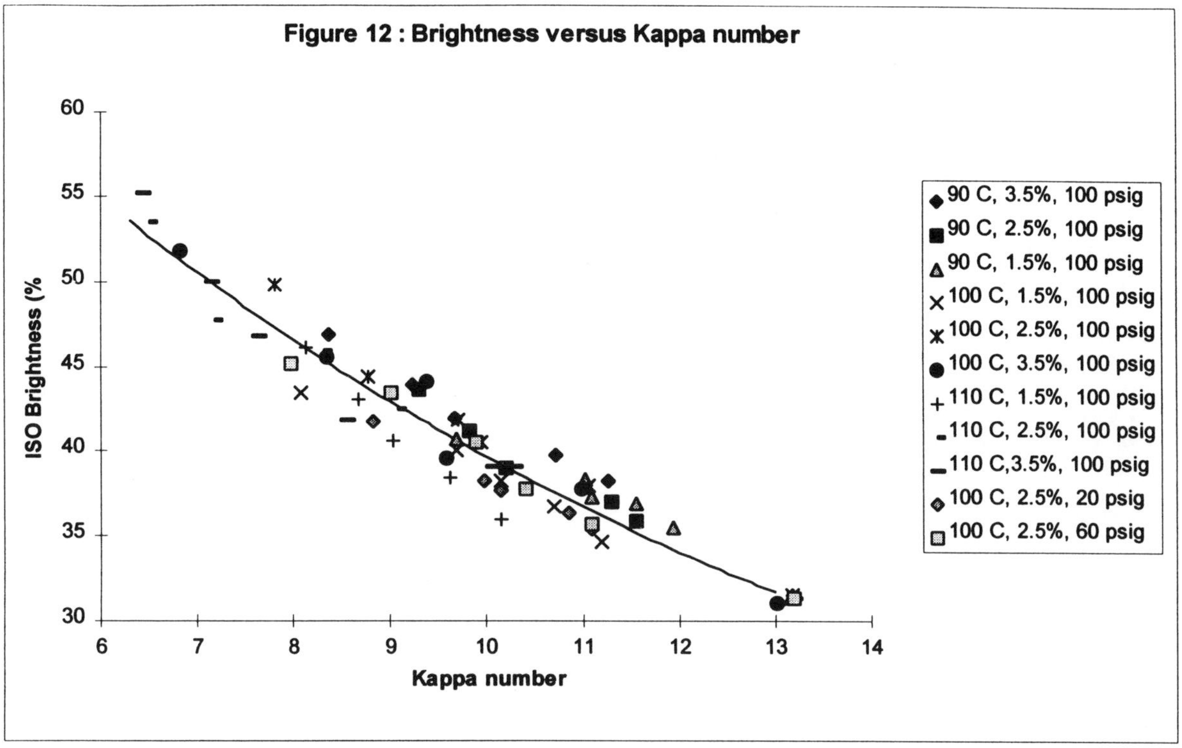
Figure 12 : Brightness versus Kappa number
ISO Brightness (%
Kappa number
90 C, 3.5%, 100 psig
90 C, 2.5%, 100 psig
90 C, 1.5%, 100 psig
100 C, 1.5%, 100 psig
100 C, 2.5%, 100 psig
100 C, 3.5%, 100 psig
110 C, 1.5%, 100 psig
110 C, 2.5%, 100 psig
110 C,3.5%, 100 psig
100 C, 2.5%, 20 psig
100 C, 2.5%, 60 psig

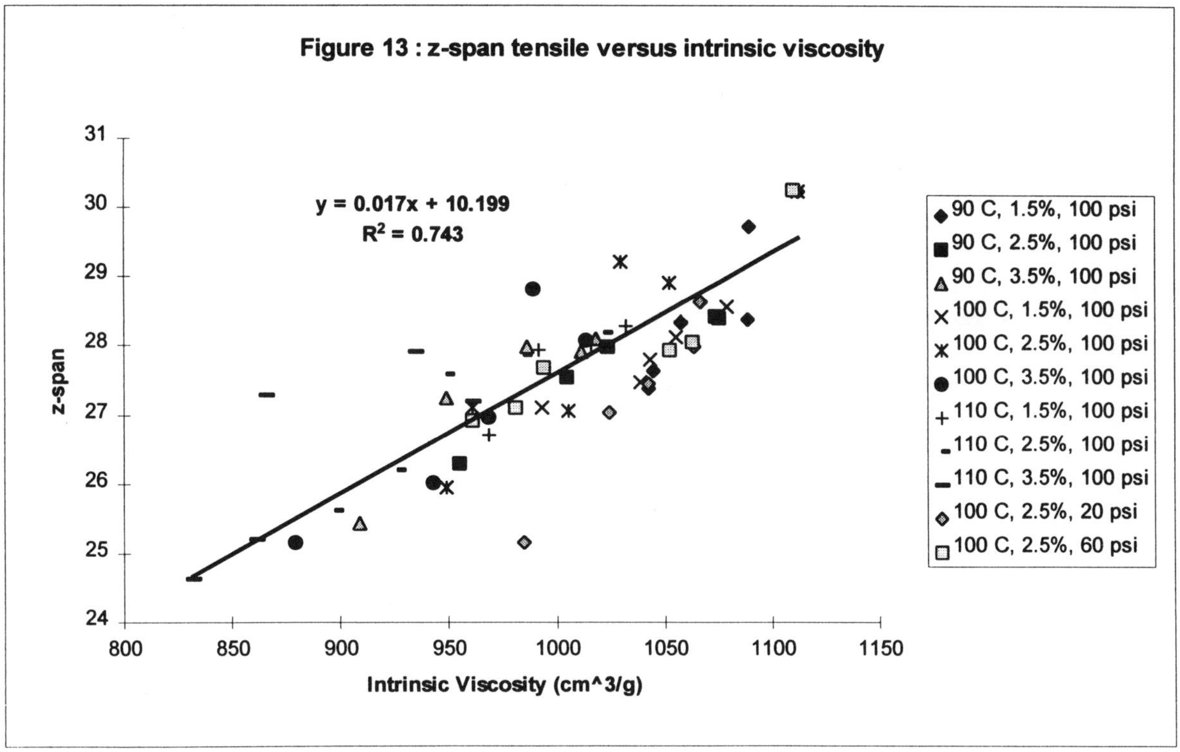
Figure 13 : z-span tensile versus intrinsic viscosity
y = 0.017x + 10.199
R² = 0.743
z-span
Intrinsic Viscosity (cm^3/g)
90 C, 1.5%, 100 psi
90 C, 2.5%, 100 psi
90 C, 3.5%, 100 psi
100 C, 1.5%, 100 psi
100 C, 2.5%, 100 psi
100 C, 3.5%, 100 psi
110 C, 1.5%, 100 psi
110 C, 2.5%, 100 psi
110 C, 3.5%, 100 psi
100 C, 2.5%, 20 psi
100 C, 2.5%, 60 psi

A Fundamental Study of a Novel Extraction Stage for an Organosolv-Based Bleaching Sequence

Brian N. Brogdon, Donald R. Dimmel and Thomas J. McDonough
Institute of Paper Science and Technology, 500 10th Street, N.W., Atlanta, GA 30318

The extent of lignin removal in a caustic extraction stage (E) of a DE bleaching sequence is severely hindered if most of the water is replaced with ethanol. However, an acidic ethanol extraction stage (A_E) may provide an alternative method to alkaline extraction for an ethanol-based organosolv bleaching sequence. We have examined the influence of A_E process variables such as ethanol concentration, acid concentration, temperature, and time on delignification of a D-stage treated pulp. An A_E-stage can remove 20 to 50% of the lignin entering the stage; the specific amount depends on the process conditions. A corresponding aqueous acid stage (A) was ineffective at lignin removal. Delignification during the A_E-stage appears to be controlled by lignin diffusion. High acid concentrations in the A_E-stage caused severe carbohydrate degradation at temperatures ≥100°C; however, the selectivity of the A_E-stage can be significantly improved through acid concentration and temperature adjustments.

INTRODUCTION

A major impetus for using aqueous organic solvents for chemical pulping is to produce easily bleached pulps that are comparable in strength to kraft. Processes such as Alcell®, ASAM, and organocell utilize 15 to 70% alcohol in the pulping media [1-3]. It is believed that these organic solvents promote the dissolution of lignin and/or inhibit non-productive reactions (e.g., condensation) [1,4,5]. Recent research efforts have also examined the washing of organosolv pulps with organic solvent-water solutions. Substantial amounts of lignin are removed from Alcell® pulps if aqueous ethanol solutions are used instead of water [6]. The additional delignification was attributed to the higher lignin solubility in ethanol-water solutions than in water alone [7].

The application of organic solvent-water mixtures in chemical pulp bleaching has been recently investigated. Dramatic improvements in ozone bleaching selectivity of kraft and Alcell® pulps have been observed when ethanol-water or other aqueous organic solutions are employed [8-10]. However, most of these investigations have not examined the use of organic solvent-water in other bleaching stages. The use of organic solvent-water solutions in multiple stages may offer a selective method to easily remove residual lignin similar to organosolv pulping [11].

We have examined "organosolv bleaching" of softwood kraft pulps with the DE delignification sequence [12,13]. Poorer delignification was observed when the caustic extraction (E) was performed with high concentrations of ethanol (≥50%). These results differ from what might be expected on the basis of the results reported for organosolv pulp washing with solvent-water mixtures [6,7]. However, modifications to the residual lignin structure during chlorine dioxide treatment rendered the lignin fragments less soluble in ethanol-water solutions under the alkaline conditions of extraction [13].

Chlorine dioxide reactions generate carboxylic acid functionalities on the residual lignin. In general, most carboxylic acids are more soluble in certain organic solvents, like ethanol, than in water. However, the opposite solubility behavior is observed with sodium salts of carboxylic acids [14]. Thus, it may be possible to extract the oxidized lignin fragments with ethanol under acidic conditions. The objective of this research is to examine acid ethanol-water extraction (A_E) as an alternative method to alkaline extraction for an ethanol-based bleaching delignification sequence.

RESULTS AND DISCUSSION

Preliminary Experiments on Delignification

Several preliminary experiments were performed to establish how process conditions affected A_E delignification. Figure 1 shows the effect of ethanol concentration of the extraction liquor on delignification. Increasing the ethanol concentration facilitated lignin removal from the chlorine dioxide treated pulp. The greatest delignification occurred when the extraction media contained 90% ethanol. An almost negligible amount of lignin was dislodged from the pulp with an aqueous acid extraction (0% EtOH).

The second set of experiments examined the influence of acid concentration on A_E delignification with a 90% ethanol medium. Lignin removal was enhanced as the acid concentration (i.e., HCl) was increased from 10^{-4} to 10^{-1} M (Figure 2). Low concentrations of acid (≤10^{-3}) appeared to have little influence on kappa number reduction.

Additional runs were performed to evaluate how temperature affected A_E lignin removal. Figure 3 indicates that lignin extraction was low at 25°C; however, increasing the temperature from 25 to 100°C strongly increased the extent of delignification. A 40% kappa number reduction occurred when the A_E-stage temperature was increased to 100°C. The kappa number decreased linearly over the temperature range examined.

The dissolution of lignin over time for the A_E-stage was examined at 70, 85, and 100°C (Fig. 4). Increasing the reaction temperature appeared to accelerate the initial kappa number drop observed during the first hour. Afterwards, delignification occurred gradually over the 1- to 5-hour reaction period for all three temperatures. A 14-hour reaction run at 100°C only showed a very slight decrease in lignin content compared to the 5-hour run (11.1 vs. 11.6 kappa, respectively).

Kinetic Analysis of Delignification

A kinetic analysis was performed on the preliminary data to assess possible controlling factors for A_E delignification. Reaction rates were calculated based on the data shown in Figures 3 and 4 using an integrated first-order rate expression with respect to the disappearance of "leachable lignin":

$$\ln\left(\frac{C_l(t)-C_l(\infty)}{C_L(0)-C_L(\infty)}\right)=kt \tag{1}$$

where k is the rate constant; t is the reaction time; $C_L(t)$ is the lign content of the pulp at time t; $C_L(0)$ is the intial lignin concentration in the fiber wall; and $C_L(\infty)$ is the lignin concentration in the pulp at infinite reaction time (i.e., "floor level lignin"). The denominator in Equation 1 represents the initial concentration of removable lignin prior to reaction, whereas the numerator represents the remaining removable lignin at time t.

Kappa numbers were used as representations for the average lignin concentration within the fiber walls. The $C_L(\infty)$ value was estimated to be 11.1 based on the 14-hour extraction at 100°C. The pulp's entering kappa, 19.3, was equated with $C_L(0)$. Estimated rate constants based on Equation 1 are given in Table 1 over the 25 to 100°C temperature range. Equation 1 reasonably fits the delignification kinetics observed based on a first-order relationship with respect to leachable lignin over the 70 to 100°C range.

Table 1. Calculated values for the rate constant k as a function of temperature for an A_E-stage conducted with a 90% ethanol medium with 10^{-2} M [H+].

Temperature (°C)	Rate Constant k from Eq. 1 (s^{-1})	R^2	Data Points
25	0.646×10^{-5}	1	2
40	1.12×10^{-5}	1	2
55	2.75×10^{-5}	1	2
70	3.78×10^{-5}	0.98	5
85	6.20×10^{-5}	0.99	4
100	15.0×10^{-5}	0.98	4

The rates reported in Table 1 for the 25 to 55°C range are tentative estimates based on limited data; they are included here to give an overall general conception of how the rate constant is affected by temperature. An activation energy of 8.9±1.7 Kcal/mole (± values are at the 95% confidence interval) was calculated from this data using an Arrhenius expression). This activation energy is much too low to be attributed to the cleavage of covalent bonds; activation energies for ether linkages, the weakest bonds to cleave in residual lignin, are on the order of 17 to 40 Kcal/mole [15-17].

A second delignification kinetic analysis was performed to determine the contribution of the acid for an A_E-stage. The results in Figure 2, as noted earlier, showed that increasing the acid concentration in the ethanol solution facilitated lignin removal; however, delignification still occurred at very low acid concentrations ($\leq 10^{-3}$ M). A second set of experimental A_E runs was performed varying the acid concentration at 100°C at 2.5 hours (Figure 5). There was a similar delignification profile observed for the 100°C runs as for the 70°C, namely that the delignification levels off at acid concentrations $\leq 10^{-3}$ M. In both cases, the acid concentration at the start and the end of the reactions remained relatively unchanged based on pH measurements.

The acid probably acts as a catalyst during delignification, similar to organosolv pulping processes [1,5]. A more generalized integrated rate expression was written, which incorporates non-catalyzed and acid-catalyzed parallel reactions:where k' and k''

$$\ln\left(\frac{C_L(t)-C_L(\infty)}{C_L(0)-C_L(\infty)}\right)=\left(k'+k''\left[H^+\right]^m\right)t \tag{2}$$

are the rate constants for non-catalyzed and acid-catalyzed reactions; [H+] is the acid concentration; and m is an exponent. The rate constant in Equation 1, k, is an amalgamation of both k' and k'' $[H^+]^m$ at an [H+] of 10^{-2} M. The temperature dependency of this overall k may not be representative of the consecutive parallel reactions occurring for delignification [18]. Thus, the following analysis was performed to assess the activation energies of the non-catalyzed and acid-catalyzed delignification reactions.

Data contained in Figures 2 and 5 were used to estimate the values of k', k'', and m in Equation 2. Plots of the lefthand expression of Equation 2 versus $[H^+]^m$ were made at various m values until a best regression fit was obtained for both temperatures. Once this was accomplished, the values of k' and k'' could be obtained from the ordinate intercept and slope, respectively. The best regression fit for

Table 2. Calculated values for k' and k'' as a function of temperature for an A_E-stage conducted with a 90% ethanol medium.

Temperature (°C)	Rate Constant k' from Eq. 2 (s^{-1})	Rate Constant k'' from Eq. 2 (s^{-1} $M^{-0.25}$)	R^2	Data Points
70	1.47×10^{-5}	7.40×10^{-5}	0.98	4
100	6.98×10^{-5}	15.7×10^{-5}	0.99	5

m based on the 70 and 100°C data was determined to be 0.25. Values of k' and k'' are given in Table 2.

The rate constants in Table 2 were fitted to an Arrhenius expression to yield:

$$k' = 3790e^{-6640/T} \tag{3}$$

$$k'' = 0.861e^{-3210/T} \tag{4}$$

The activation energies for k' and k'' from Equations 3 and 4 were determined to be 13.2 and 6.4 Kcal/mole, respectively. These activation energies for non-catalyzed and acid-catalyzed kinetic expressions are low for intrinsic chemical reactions to be the controlling factor for A_E delignification.

Evaluation of Delignification with Diffusion Theory

The low activation energies determined from the above kinetic analysis suggested that delignification is limited by physico-chemical factors such as diffusion. Diffusion theory has been used to describe the delignification observed with lignin leaching of unbleached kraft pulps under neutral [19,20] and alkaline conditions [21]. A similar approach was undertaken to determine whether or not A_E delignification is a lignin diffusion-controlled process.

The diffusion of lignin fragments from the fiber wall can be modeled as a porous flat plate if the following conditions are assumed: 1) the initial concentration of leachable fragments is constant across the fiber wall, 2) the diffusion of fragments occurs from both the outer and lumen wall, and 3) the concentration of fragments in the bulk solution is negligible in comparison to the concentration within the fiber [19,20].

Fick's Law of transient diffusion relates the concentration of lignin (C_L) as function of distance (x) and time (t):

$$D\frac{\partial^2 C_L(x,t)}{\partial x^2} = \frac{\partial C_L(x,t)}{\partial t} \tag{5}$$

where $\boldsymbol{D}$ is the effective diffusion coefficient [19,20]. Equation 5 can be solved if the following initial and boundary conditions are stipulated:

$$C_L(x,t) = C_L(0) \ for \ -l < x < l \ at \ t = 0 \tag{6}$$

$$C_L(x,t) \approx 0 \ for \ -l \leq x \leq l \ for \ all \ t \tag{7}$$

where $C_L(0)$ is the initial oxidized lignin concentration, and ℓ is one-half of the fiber wall thickness. The solution to Equation 5 can be expressed in terms of an average "leachable lignin" concentration across the fiber wall to yield [19,20]:

$$\gamma(t) = \frac{C_L(t) - C_L(\infty)}{C_L(0) - C_L(\infty)} = 2\sum_{n=o}^{\infty}\left(\frac{2(-1)^n}{(2n+1)\pi}\right)^2 e^{\frac{-(2n+1)^2\pi^2\tau}{4}} \tag{8}$$

where τ is the dimensionless Fourier number ($\boldsymbol{D}t/l^2$), and γ is the ratio of the remaining leachable lignin at time t to the total amount of leachable lignin. Tabulated values of γ versus τ for Equation 8 have been given by Favis [19].

Values of γ were computed based on the kappa numbers in Figure 4, C_L (0) = 19.3 and C_L ( ) = 11.1 for the various temperatures, and the corresponding τ was determined. The value of $\boldsymbol{D}$ was determined from the corresponding τ based on a t of 5 hours and average softwood fiber wall thickness of 5 μm (l = 2.5 x10^{-4} cm) [19-21].

Figure 6 shows how the effective diffusion coefficient is affected by temperature. Diffusion increased somewhat linearly over the 25 to 70°C range; however, higher temperatures (≥70°C) greatly enhanced diffusion. Diffusion-controlled processes generally exhibit a linear dependency between the diffusion coefficient and temperature as is given by the Stokes-Einstein relationship [19-21]:

$$D \propto \frac{T}{\mu} \tag{9}$$

where μ is the solvent's viscosity. The behavior exhibited in Figure 6, however, is not uncommon for diffusion-controlled processes. Li and MacLeod [21], and Favis [19,20] noted dramatic increases in the diffusion coefficient of lignin during leaching of kraft brownstocks after ~70°C under alkaline and neutral conditions, respectively (Figure 7). Likewise, Sprague [22] showed that the diffusion coefficients of dispersed dyes for cellulose acetate fibers increased linearly over the 40 to 80°C range, and suddenly increased above 80°C.

Several explanations have been given for the non-linear temperature dependency of diffusion, such as the one observed in Figure 7. Favis [19,20], and Li and MacLeod [20] attributed the sudden jump in lignin diffusion at 70°C to the thermal softening of lignin and hemicelluloses. However, this explanation cannot explain Sprague's observations [22] of dye diffusion into pure cellulose acetate fibers. Johansson [23] proposed that the dramatic increase in diffusion at ~70 to 100°C was due to the disruption of the hydrogen bonding network of water as the temperature approaches the boiling point. A similar disruption of the hydrogen bonding network with the ethanol-water system [24,25] could be occurring with the A_E-stage.

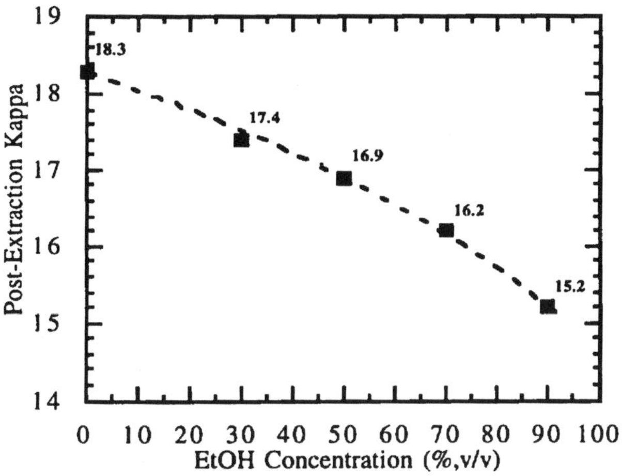

Figure 1. The influence of ethanol concentration on A_E delignification: 0.01 M [H+]; 70°C; and t=5 hrs.

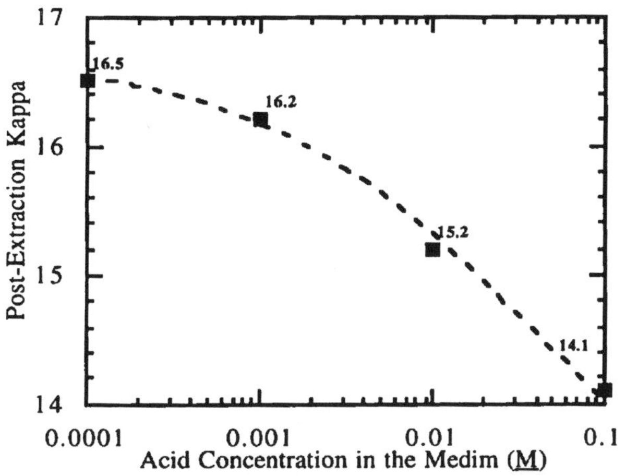

Figure 2. The influence of acid concentration ([H+]) on A_E delignification: 90% EtOH; 70°C; and t=5 hrs.

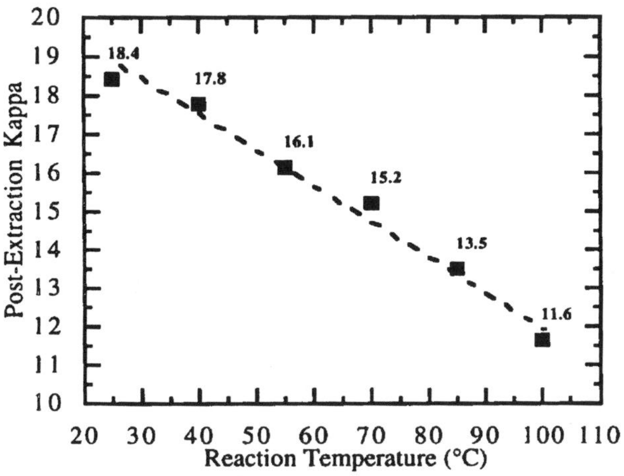

Figure 3. The influence of reaction temperature on A_E delignification: 90% EtOH; 0.01 M [H+]; and t=5 hrs.

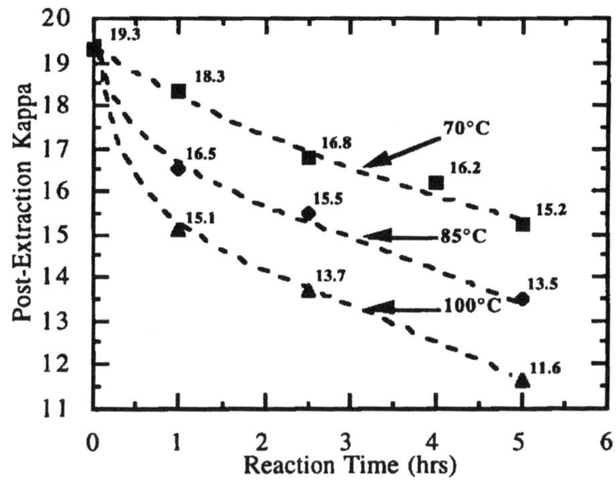

Figure 4. The influence of reaction temperature and time on A_E delignification: 90% EtOH and 0.01 M [H+].

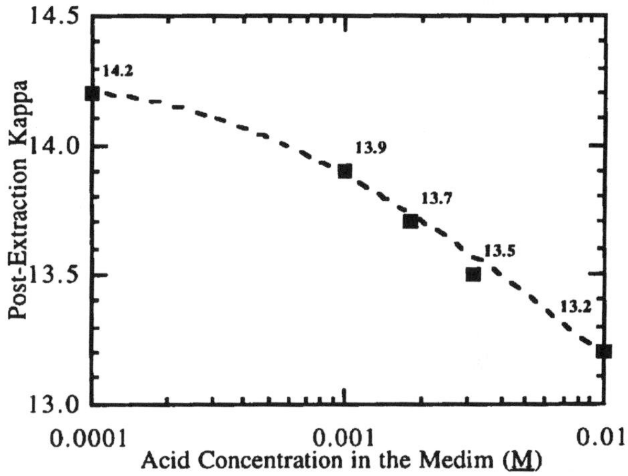

Figure 5. The influence of acid concentration ([H+]) on A_E delignification: 90% EtOH; 100°C; and t=2.5 hrs.

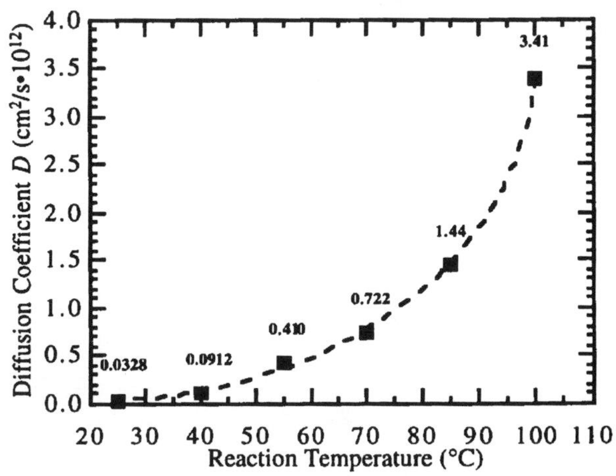

Figure 6. Calculated effective diffusion coefficients for lignin (D) for the A_E-stage at t=5 hrs. Conditions: 90% EtOH and 0.01 M [H+].

The diffusion coefficients in Figure 6 were correlated with temperature using an Arrhenius-type expression [21,26]:

$$D = AT^{0.5}e^{-E_d/RT} \tag{10}$$

Where E_d is the activation energy of diffusion, and A is a pre-exponential factor. The value of E_d was calculated to be 13.2±2.0 Kcal/mole. This E_d is similar to diffusional activation energies calculated for diffusion-controlled processes such as lignin leaching of unbleached pulps (5.5 to 18 Kcal/mole [19,21], and dyeing of cellulosic materials (11 to 35 Kcal/mole [22].

One final analysis was performed to affirm that A_E delignification was diffusion-controlled. The Weisz-Prater criterion, F, represents the ratio of intrinsic chemical reaction and intraparticle diffusion:

$$F = \frac{r_l(t)l^2}{D\gamma(t)} = \frac{k\gamma(t)l^2}{D\gamma(t)} = \frac{kl^2}{D} \tag{11}$$

This simple criterion has been used to define if a process is chemical reaction or diffusion controlled [27]. Values of F that are less than 1 indicate that chemical reactions are controlling, while values of F greater than 1 indicate that particle diffusion is controlling. Calculated F values based on the A_E data obtained are all greater than 1 (Table 3); these F values appear to further corroborate that A_E delignification is a diffusion-controlled process.

Carbohydrate Degradation Kinetics and Selectivity

An important aspect of any bleaching stage is to effectively remove and/or modify the residual lignin without causing significant carbohydrate damage. Thus, the second phase of this study focused on how the A_E-stage affected pulp viscosity and strength. The aggressive conditions utilized during the A_E-stage to promote lignin dissolution may also accentuate the cleavage of acid sensitive glycosidic bonds.

Table 3. Weisz-Prater criterion, F, as a function of temperature for an A_E-stage conducted with a 90% ethanol medium with a 0.01 M [H+].

Temperature (°C)	Rate Constant k from Eq. 1 (s^{-1})	Diffusion Coefficient D from Eq. 8 (cm^2/s)	F from Eq. 11
25	0.646×10^{-5}	0.0328×10^{-12}	12.3
40	1.12×10^{-5}	0.0912×10^{-12}	7.69
55	2.75×10^{-5}	0.410×10^{-12}	4.19
70	3.78×10^{-5}	0.722×10^{-12}	3.27
85	6.20×10^{-5}	1.44×10^{-12}	2.69
100	15.0×10^{-5}	3.41×10^{-12}	2.75

Carbohydrate damage was tracked over time with an A_E-stage performed with an acid concentration, [H^+], of 10^{-2} M (Figure 8). Negligible viscosity reduction was observed over time at 70°C. However, increasing the temperature to 85 or 100°C augmented carbohydrate degradation, especially after 1 hour. Zero-span breaking length measurements were also performed on the above pulps to evaluate fiber strength (Figure 9); these measurements closely mirrored the trends observed with viscosity. The viscosity data were coupled with the delignification data in Figure 4 to yield a selectivity plot (Figure 10). This plot shows that A_E selectivity steadily decreases as the temperature of the stage is increased from 70 to 100°C.

Several kinetic analyses were performed to evaluate how reaction temperature and acid concentration affected pulp viscosity. The data in Figure 8 were initially fitted to an integrated first-order rate equation with respect to pulp viscosity (η):

$$\ln\left(\frac{\eta(t)}{\eta(0)}\right) = k_c t \tag{12}$$

The activation energy for the rate constant in Equation 12, k_c, was determined to be 24.2 Kcal/mole. This activation energy is similar to 20 to 30 Kcal/mole values reported for cellulose degradation during acid sulfite pulping [15-17].

Additional experiments were performed with different acid concentrations at 85 and 100°C (Table 4). Decreasing the acid concentration reduced the viscosity loss associated with the extraction stage. The overall rate constant, k_c, appeared to increase linearly as the acid concentration increased; this linear dependence between k_c and [H^+] has been noted in the literature for acid degradation of cellulosic materialmaterials [16,28].

Table 4. Calculated k_c values from 5-hour viscosity values for an A_E-stage at various temperatures and acid concentrations. [Stage conducted with 90% EtOH; initial D-stage treated pulp viscosity of 23.8 cP.]

Temperature (°C)	[H^+] (M)	Viscosity (cP)	Rate Constant k_c from Eq. 12 (s^{-1})
85	1×10^{-2}	16.0	2.11×10^{-5}
85	1×10^{-4}	21.2	0.643×10^{-5}
100	1×10^{-2}	8.7	5.59×10^{-5}
100	4×10^{-3}	14.1	2.89×10^{-5}
100	1×10^{-4}	21.1	0.661×10^{-5}

Equation (12) was modified to incorporate the catalytic [H^+]

$$\ln\left(\frac{\eta(t)}{\eta(0)}\right) = \left(k'_c + k''_c\left[H^+\right]\right)t \tag{13}$$

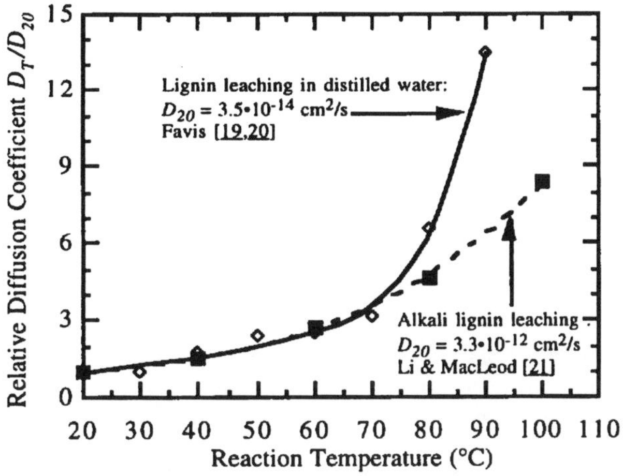

Figure 7. Relative effective diffusion coefficients for lignin (D) during leaching under neutral and alkaline conditions with softwood kraft pulps [19-21].

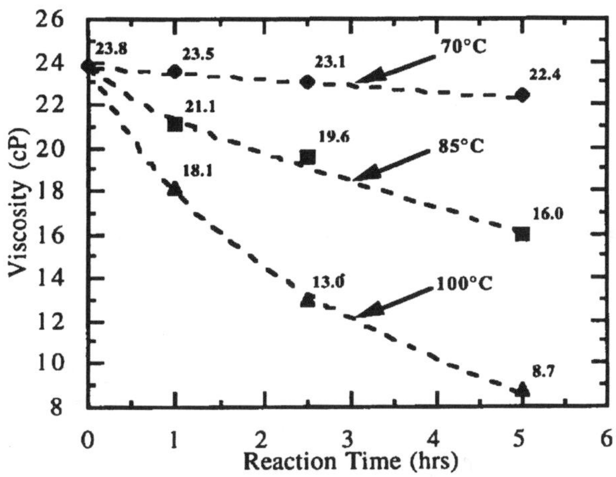

Figure 8. Carbohydrate degradation at various T for an A_E-stage. Conditions: 90% EtOH and 0.01 M [H+].

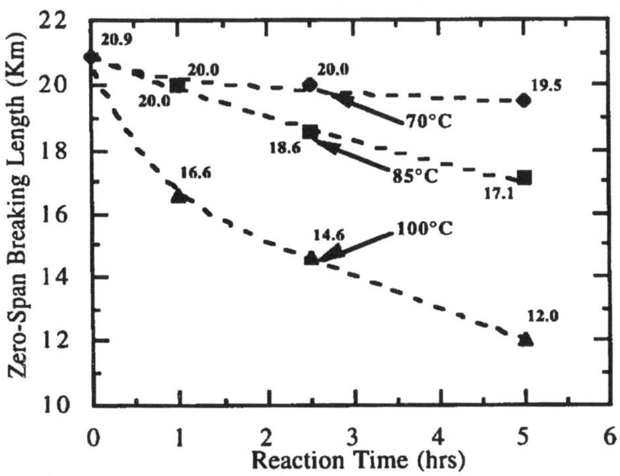

Figure 9. Fiber strength degradation at various T for an A_E-stage. Conditions: 90% EtOH and 0.01 M [H+].

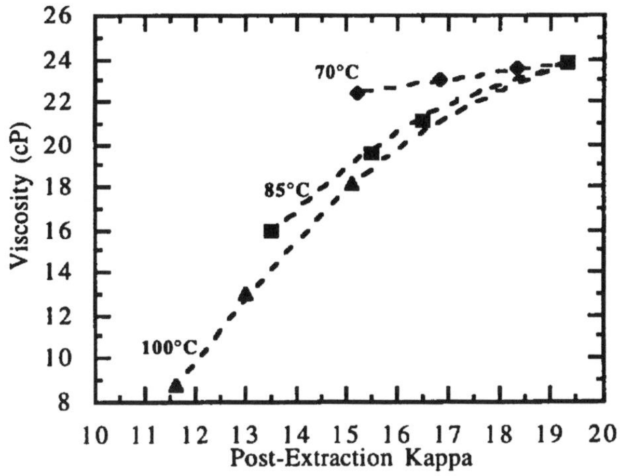

Figure 10. Selectivity plot at various T for an A_E-stage. Conditions: 90% EtOH and 0.01 M [H+].

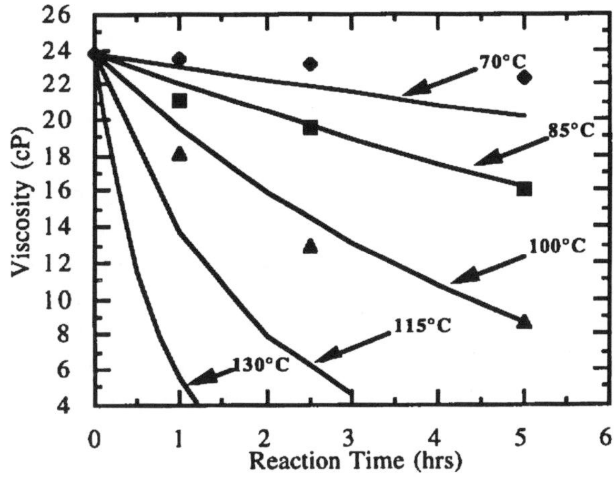

Figure 11. Predicted viscosity loss (solid lines) from Eqs. 13 - 15 compared to the data in Fig. 7 (symbols). Conditions: 90% EtOH and 0.01 M [H+].

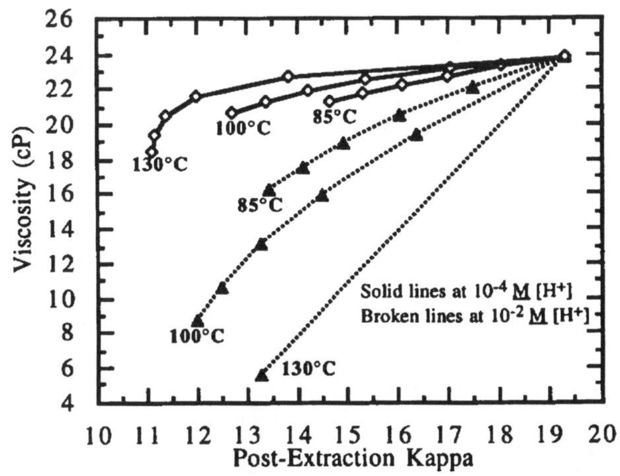

Figure 12. Predicted selectivites at various T and [H+] with kinetic model Eqs. 2 & 13. Symbols are at one-hour reaction increments.

term in the overall rate constant:where k'_c and k''_c are the non-catalyzed and acid-catalyzed carbohydrate reactions, respectively. Equation 13 indicates that carbohydrate damage can occur through non-catalyzed reactions. This is not a surprising result. In their lignin leaching study, Li and MacLeod [21] reported a decrease in pulp viscosity when a kraft brownstock was subjected to high temperature extractions (≥100°C) in deionized water at near neutral pH conditions. The data contained in Table 4 were used to determine how temperature affected the rate constants in Equation 13:

$$k'_c = 4.29x10^{-4} e^{-1520/T} \tag{14}$$

$$k''_c = 8.87x10^{9} e^{-10.500/T} \tag{15}$$

The activation energies of k'_c and k''_c were determined to be 3.1 and 20.9 Kcal/mole, respectively. Figure 10 compares the above model (Eq. 13 with Eqs. 14 and 15) to the data in Figure 8. The model slightly over predicts the viscosity drop with the 70°C data (Figure 11); however, the model reasonably fits the data, considering the limited number of observations used to develop the model.

Improving A_E Bleaching Selectivity

Both carbohydrate (Eq. 13) and delignification (Eq. 2) models were developed to help predict A_E selectivity and to optimize process conditions. Figure 12 illustrates how the models were used to examine the effect of $[H^+]$ and reaction temperature on bleaching selectivity. Increasing the extraction temperature severely decreased the selectivity of the process at high $[H^+]$ (10^{-2} M). Yet, increasing the reaction temperature greatly improved A_E selectivity when the $[H^+]$ is low (10^{-4} M). This interesting selectivity relationship between acid concentration and temperature can be readily seen by comparing acid-catalyzed and non-catalyzed rate constants from the delignification and carbohydrate degradation models.

The results from the above analysis were utilized to better optimize the A_E-stage. Several A_E extractions were performed at various times at 130°C and 10^{-4} M $[H^+]$ (Table 5). Approximately 40% of the incoming lignin was removed within the first hour of reaction using the above conditions. Essentially, all of the soluble lignin was removed within 5 hours. There was a nominal reduction in pulp viscosity and the zero-span strength. Overall, the higher temperatures and lower acid concentrations vastly improved A_E selectivity.

Table 5. Changes in kappa, viscosity, and zero-span for an A_E-stage conducted at 130°C with 90% EtOH and 10^{-4} M $[H^+]$ with a D-stage treated pulp.

Reaction time (hrs)	Kappa Number	Viscosity (cP)	Zero-Span (Km)
0	18.8	23.8	20.9
1	11.1	22.7	19.3
2.5	10.6	22.8	19.2
5	9.3	22.1	19.1
14	9.3	-	-

A_E Comparisons to Aqueous Acid (A) and Caustic (E) Extractions

An aqueous A-stage, as was noted earlier in the preliminary investigation, was ineffective at delignification of a chlorine dioxide treated pulp. A-stages were conducted with an 18.8 kappa D-stage pulp at 70 and 100°C for 5 hours. Increasing the temperature an A-stage with 0.01 M $[H^+]$ from 70 to 100°C increased the kappa number reduction (17.1 vs. 15.1 kappa, respectively). The majority of this delignification occurred within the first hour; extending the extraction time did not increase lignin removal. Overall, an A-stage is not as proficient as an A_E-stage for lignin extraction.

Unlike the A-stage, caustic extraction is very effective at removing the oxidized lignin from the D-stage treated pulp. An E-stage conducted at 60°C for one minute can reduce a 19.5 kappa pulp to 12.1 [29]. The A_E-stage under selective conditions, on the other hand, requires higher temperatures (≥100°C) and longer periods of time to approach this level of lignin removal.

Implications of A_E-Stage in an Organosolv-Based Bleaching Sequence

The results of this study indicate than an A_E-stage, based on process conditions, can be substituted for an E-stage [12] to extract lignin in an ethanol-based bleaching sequence. An appealing aspect of the A_E-stage is the elimination of alternating acid and alkaline bleaching stages. A partial organ-osolv bleaching sequence could be conducted on the acidic side using other oxidants besides chlorine dioxide, such as ozone and peracids. Effluents from oxidative and A_E stages can be combined and evaporated to recover the ethanol [30]. Recovered organic residues could be burned for their heating value or sold as potential high value byproducts similar to other organosolv lignins [1,30,31]. Another potential benefit of the ethanol-based sequence is the reduction of bleach plant water usage and its subsequent treatment.

CONCLUSIONS

This study investigated acid ethanol extraction (A_E) as an alternative to alkaline extraction for a chlorine dioxide treated brownstock for an "organosolv bleaching" sequence. Preliminary experiments showed that A_E lignin removal increased significantly as the temperature, ethanol content, and acid concentration were increased. A_E delignification appeared to be diffusion-controlled based on kinetic and diffusion theory analyses. Cellulose degradation and fiber strength loss for an A_E treatment were accentuated when the $[H^+]$ in the media was high (~10^{-3} to 10^{-2} M). Preliminary kinetic models for lignin and carbohydrate removal were developed to predict bleaching selectivity. The models showed that selectivity could be improved by decreasing the $[H^+]$ to ~10^{-4} M and increasing the reaction temperature to >100°C; experiments performed at these conditions confirmed the predictions. However, further experimental work is needed to refine the kinetic models to increase their accuracy. The results of this study suggested that an A_E-stage can be used as an alternative extraction method to caustic extraction in an ethanol-based organosolv bleaching sequence.

EXPERIMENTAL PROCEDURES

A thoroughly washed conventional softwood kraft pulp (southern pine) was obtained from a mill and had a kappa number of 29.3. Aqueous chlorine dioxide bleaches were performed on the pulp at the given conditions: 1.79% ClO_2 on o.d. pulp (0.16 kappa factor), 3.0% consistency, initial pH of 4.2 prior to ClO_2 addition, 45°C, and 30 minutes reaction time. [A 1 M HCl solution was used to acidify the pulp.] After the ClO_2 treatment, the pulp was washed with copious amounts of deionized water until the filtrate was clear. The resulting treatment yielded a pulp with kappa of 18.8 to 19.5, and a bleach filtrate with a 2.4 pH and no residual ClO_2.

Acid ethanol extractions were performed on 2.5 o.d. gram samples of the aqueous D-stage treated pulp. Experiments were conducted at low consistencies (0.50 to 1.25%) in 90% ethanol medium unless otherwise noted. Low temperature reactions (< 70°C) were conducted in sealed containers, which were heated using a water bath. High temperature reactions (≥70°C) were performed with a 300-mL Parr bomb reactor. Generally, the high temperature reactions (≥100°C) required 10 minutes to heat up to the specified reaction temperature. The pulps were extracted for the required period, cooled to 65 to 70°C (if needed), and drained of extraction effluent to ~25% consistency [29].

Micro-kappa number analyses (TAPPI Useful Method UM-246) were performed on the dried samples of the extracted pulps. Pulps subjected to dry zero-span measurements (TAPPI Standard T-231 cm-85) were washed with an aqueous sodium bicarbonate solution (~8.5 pH) to neutralize the residual acid from extraction. Unused paper strips from the zero-span analyses were utilized for pulp viscosity measurements (TAPPI Standard T-230 om-89).

Alkaline extraction runs with the D-stage treated pulp were conducted with the appropriate medium at the following conditions: 2.58% NaOH on o.d. pulp, 10% consistency, and 60°C. Additional details of E-stage experiments have been reported in our previous study [12].

NOMENCLATURE

Bleaching stages

A	aqueous acid extraction stage
A_E	acid ethanol extraction stage
D	chlorine dioxide stage
E	caustic extraction stage

General

$[H^+]$	acid concentration, (M, moles/L)
R	ideal gas constant, (Kcal/mole•K)
T	temperature, (°C or K)
t	time, (s)

Delignification

A	pre-exponential factor for Arrhenius-type expression for diffusion, (Kcal/mole•$K^{0.5}$)
$C_L(t)$	average lignin concentration in the fiber at time t, (denoted by kappa number)
$\boldsymbol{D}$	effective diffusion coefficient, (cm^2/s)
$\boldsymbol{D}_T/\boldsymbol{D}_{20}$	relative diffusion coefficient, ratio of effective diffusion coefficient at some T to effective diffusion coefficient at 20°C
E_d	diffusional activation energy, (Kcal/mole)
F	Weisz-Prater criterion (or Weisz modulus) [Eq. (11), see ref. [32]]
$\gamma(t)$	ratio of removable lignin at time t to total amount of removable lignin [see Eq. (8)]
k	overall delignification rate constant, (s^{-1})
k''	non-catalyzed delignification rate constant, (s^{-1})
k''	acid-catalyzed delignification rate constant, (s^{-1} $M^{-0.25}$)
l	one-half average fiber wall thickness, (cm)
m	exponent
$r_l(t)$	rate of lignin removal at t, $-d\gamma/dt$, (s^{-1})
τ	Fourier Number, ($\boldsymbol{D}t/\ell^2$)
x	distance from the fiber wall center, (cm)

Carbohydrate Degradation

$\eta(t)$	pulp viscosity at time t, (cP)

k_c	overall carbohydrate degradation rate constant, (s^{-1})
k'_c	non-catalyzed carbohydrate degradation rate constant, (s^{-1})
k''_c	acid-catalyzed carbohydrate degradation rate constant, (s^{-1} $\underline{M}^{-1}$)

ACKNOWLEDGMENTS

The authors thank the Institute of Paper Science and Technology and its Member Companies for supporting this research. The generous assistance of C.M. Norton with the pulp viscosity measurements is gratefully acknowledged. Portions of this work were used by B.N. Brogdon as partial fulfillment for the requirements of the Ph.D. degree at the Institute of Paper Science and Technology.

LITERATURE CITED

1. **Hergert, H.L., and Pye, E.K.,** *1992 TAPPI Solvent Pulping Symp.,* Boston, MA, TAPPI Proc., p. 9 (1992).

2. **Lora, J.H., and Aziz, S.,** *Tappi J.,* **68** (8):94 (1985).

3. **Borgards, A., Pratt, R., Kordsachia, O., Odermatt, J., and Hunter, W.D.,** *1993 TAPPI Pulping Conf.,* Atlanta, GA, TAPPI Proc., p. 629 (1993).

4. **Brogdon, B.N., and Dimmel, D.R.,** *J. Wood Chem. Tech.,* **16** (3):297 (1996).

5. **McDonough, T.J.,** *Tappi J.,* **76** (8):186 (1993).

6. **Ni, Y., and van Heiningen, A.H.P.,** *Tappi J.,* **79** (3):239 (1996).

7. **Ni, Y., and Hu, Q.,** *J. Appl. Polym. Sci.,* **57**:1441 (1995).

8. **van Heiningen, A.R.P., and Ni, Y.,** U.S. Pat. #5,354,423 (Oct. 11, 1994).

9. **Kang, G.J., Ni, Y., and van Heiningen, A.R.P.,** *82nd Annual Meeting, Tech. Section CPPA*, Montréal, Québec, CPPA Proc., p. B303 (1996).

10. **Solinas, M., and Murphy, T.,** *Pulp & Paper,* **70** (3):133 (1996).

11. **Sundquist, J., and Poppius-Levlin, K.,** *1992 TAPPI Solvent Pulping Symp.,* Boston, MA, TAPPI Proc., p. 45 (1992).

12. **Brogdon, B.N., Dimmel, D.R., McDonough, T.J., and Bryant, P.S.,** *1996 Intl. Pulp Bleaching Conf.*, Washington, DC, TAPPI Proc., p. 541 (1996).

13. **Brogdon, B.N., Dimmel, D.R., and McDonough, T.J.,** *1996 TAPPI Pulping Conf.,* Nashville, TN, TAPPI Proc., p. 35 (1996).

14. **Morrison, R.T., and Boyd, R.N.,** *Organic Chemistry,* Fourth Ed., John Wiley & Sons, Inc., New York, Chapt. 19 (1991).

15. **Rydholm, S.A.,** *Pulping Processes*, Interscience Publishers, New York, Chapt. 9 (1965).

16. **Ingruber, O.,** "Process Measurement and Control," In *Pulp and Paper Manufacture, Vol. 4: Sulfite Science & Technology*, Ed. by O. Ingruber, M. Kocurek, and A. Wong, Joint Textbook Committee of the Paper Industry, Atlanta (1985).

17. **Vuorinen, T., Teleman, A., Fagerström, P., Buchert, J., and Tenkanen, M.,** *1996 Intl. Pulp Bleaching Conf.*, Washington, DC, TAPPI Proc., p. 43 (1996).

18. **Ni, Y., Kubes, G.J., and van Heiningen, A.R.P.,** *J. Pulp Paper Sci.,* **16** (1):J13 (1990).

19. **Favis, B.D.,** *The Leaching of Lignin Macromolecules from Pulp Fibers During Washing*, Ph.D. Dissertation, McGill University, Montréal (1981).

20. **Favis, B.D., Willis, J.M., and Goring, D.A.I.,** *J. Wood Chem. Tech.,* **3** (1):1 (1983).

21. **Li, J., and MacLeod, J.M.,** *J. Pulp Paper Sci.,* **19** (2):J85 (1993).

22. **Sprague, B.S.,** *J. Polymer Sci., Part C,* **20**:159 (1967).

23. **Johansson, A.,** *J. Pulp Paper Sci.,* **11** (3):J78 (1985).

24. **Coccia, A., Indovina, P., Podo, F., and Viti, K.,** *Chem. Phys.,* 7:30 (1975).

25. **Laiken, N., and Hemethy, G.,** *J. Chem. Phys.,* **74**:3501 (1970).

26. **Wong, C.F., and Hayduk, W.,** *Can. J. Chem. Eng.,* **68**:849 (1990).

27. **Pugliese, S.C., and McDonough, T.J.,** *Tappi J.,* **72** (3):159 (1989).

28. **Hansen, S.M.,** *The Role of Solvent in Aqueous-Organic Solvent Delignifiction of Southern Pine*, Ph.D. Dissertation,

University of Alabama, University, p. 24 (1981).

29. **Brogdon, B.N.,** *Effects of Ethanol Media on Chlorine Dioxide and Extraction Stages for Kraft Pulp Bleaching*, Ph.D. Dissertation, IPST, Atlanta, (1997).

30. **Lora, J., Winner, S.R., and Pye, E.K.,** *AIChE 1989 and 1990 Forest Products Symp. Proc.,* Ed. by J.D. Lisius, TAPPI Press, Atlanta, p. 35 (1992).

31. **Lora, J., and Pye, E.K.,** *1992 TAPPI Solvent Pulping Symp.,* Boston, MA, TAPPI Proc., p. 27 (1992).

32. **Levenspiel, O.,** *The Chemical Reactor Omnibook*, Oregon State University Book Stores, Inc., Corvallis, Chapt. 22 (1979).

Physico-Chemical Processes Limiting Lignin Removal During Bleaching Extraction

Brian N. Brogdon, Donald R. Dimmel and Thomas J. McDonough
Institute of Paper Science and Technology, 500 10th Street N. W., Atlanta, GA 30318

*Delignification of chlorine dioxide treated pulps with an acid ethanol extraction stage (A_E) was less effective than an aqueous caustic extraction stage (E). Calculated effective lignin diffusion coefficients (**D**) from A_E-stages were generally much lower than those for an aqueous E-stage. The slow lignin removal process of the A_E-stage was analogous to aqueous lignin leaching of unbleached pulps. These two extraction processes seemed to have similar diffusion limiting mechanisms, namely lignin association within the fiber wall. The difference in delignification efficiency of an A_E-stage versus an aqueous E-stage is presumably related to how each process disrupts the associated lignin complex. An aqueous E-stage, unlike the A_E-stage, can breakup associated complexes by the electrostatic repulsion of ionized acid groups.*

INTRODUCTION

Our ongoing investigations of "organosolv bleaching" have yielded some interesting data on how delignification is affected when the lignin extraction stage medium is changed from aqueous to ethanol-water solutions [1]. An earlier report has examined the effect of ethanol-water media on the E-stage of a DE partial bleaching sequence [2]. Extraction delignification was strongly inhibited when an aqueous medium was switched to an ethanol medium (Fig. 1). A subsequent study determined that the solubility of oxidized lignin in alkali was limiting delignification with ethanol-water media [3].

These results prompted us to examine an alternative extraction stage for chlorine dioxide bleached pulp, namely an acidic ethanol extraction stage (A_E-stage) [4]. Under the acidic conditions ($[H^+] \geq 10^{-5}$ M), the oxidized lignin, containing carboxylic groups, is expected to be more soluble in ethanol-water solutions than in water. This expectation is based on the solubility behavior of carboxylic acids and on the dissolution of lignin with acid organosolv pulping processes [1,3-6]. More delignification could be achieved with a 90% ethanol medium than with water under acidic conditions ($10^{-4} \leq [H^+] \leq 10^{-2}$ M). However, the A_E-stage requires high temperatures and longer reaction times to achieve the same post-extracted kappa number as an aqueous E-stage [1,4].

Recently, we have examined lignin removal in an ethanol-based DA_E partial bleaching sequence [1,7]. The results indicate that

Table 1. Residual lignin removal with aqueous- and ethanol-based partial bleaching sequences [1,7].

Treatment	Bleaching Medium	Residual Lignin (%)
Unbleached	-	4.63
D	Water	2.99
DE	Water	1.37
D	90% EtOH	3.02
DA_E	90% EtOH	2.18

25 to 30% of the residual lignin entering the A_E-stage can be removed (Table 1). An aqueous E-stage in a DE partial bleaching sequence, on the other hand, can remove 55% of the entering residual lignin. Again, an aqueous E-stage is more efficient at delignification than an A_E-stage.

These studies have indicated some significant differences in the ability of various media to extract oxidized lignin [2-4,7]. What makes an aqueous E-stage more efficient at delignification than an A_E-stage? The answer could be related to physico-chemical factors that have been proposed for an analogous extraction process, aqueous lignin leaching from unbleached pulps [8-13]. In this paper, we examine the similarities between the A_E-stage and lignin leaching, and determine what possible factors are contributing to the lower delignification efficiency of the A_E-stage.

RESULTS AND DISCUSSION

Effective Lignin Diffusion Coefficients in Various Media

Comparisons of Effective Diffusion Coefficients

Separate studies have estimated and analyzed the effective diffusion coefficients (***D***) of oxidized lignin during extraction with acidic [4] and alkaline ethanol-water [2] media. These ***D*** values are compiled in Table 2 along with the estimated ***D*** values of aqueous E-stage [2] and from other alcohol-water media, and with the reported ***D*** values for aqueous lignin leaching of unbleached pulps [9,12]. Most of the lignin extractions performed under alkaline conditions had equivalent ***D*** values, namely on the order of 10^{-11} cm^2/s. The ***D*** values for acidic or neutral conditions were similar ($\sim 10^{-13}$ cm^2/s) and were much lower than the ***D*** values for alkaline conditions. These comparisons suggested that the removal of dissolvable lignins from unbleached and D-stage treated pulps responded to alkalinity in a similar way, irrespective of the medium.

Table 2. Comparisons of *D* from different lignin extraction processes with various media.

Pulp	Ref.	Extraction Medium	Temp. (°C)	*D* ($cm^2/s \cdot 10^{12}$)
D-stage	[2]	Water/0.07 M NaOH	60	9.9
D-stage	[2]	75% EtOH/0.07 M NaOH	60	4.7
D-stage	[4]	90% EtOH/0.01 M HCl	55	0.41
D-stage	[1]	90% *n*-PrOH/0.01 M HCl	55	0.41
D-stage	[1]	90% *t*-BuOH/0.01 M HCl	55	0.33
Kraft	[12]	Water/0.2 M NaOH	60	8.8
Kraft	[9]	Water/near neutral pH	60	0.097

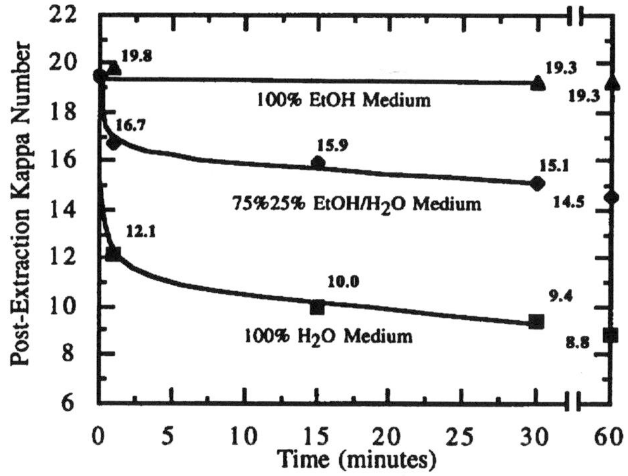

Figure 1. Low consistency (0.5%) alkaline extraction of a 19.5 kappa D-stage treated pulp with aqueous and ethanol-water solutions.

Enhanced Diffusion at Elevated Temperatures

Increasing the temperature of an A_E-stage to 100°C increased the *D* to approximately the same order of magnitude as alkaline extraction processes at 60°C (Table 3). Likewise, acid extractions performed with other alcohols exhibited an order of magnitude increase in *D* when the temperature was increased to 100°C. This effect was analogous to the increase in *D* with increased temperature to 90°C for lignin leaching of unbleached pulps (Table 3).

Table 3. Effect of elevated *T* on *D* for acid alcohol extractions and brownstock lignin leaching.

Pulp	Ref.	Extraction Medium	Temp. (°C)	*D* ($cm^2/s \cdot 10^{12}$)
D-stage	[4]	90% EtOH/0.01 M HCl	100	3.4
D-stage	[1]	90% *n*-PrOH/0.01 M HCl	100	3.7
D-stage	[1]	90% *t*-BuOH/0.01 M HCl	100	1.5
Kraft	[9]	Water/near neutral pH	90	0.47

The striking resemblance between acid alcohol extraction and aqueous leaching of unbleached pulp is better seen in the *D* behavior over a temperature range (Figures 2 and 3). Both processes showed the same augmentation of diffusion occurring at temperatures ≥70°C, indicating that these processes presumably have similar limiting mechanisms. Acid ethanol extraction and lignin leaching have been characterized as being diffusion-controlled, and both processes have similar activation energies (13 to 20 kcal/mole) [4,9]. Diffusion activation energies for the other acid alcohol extractions were also low (~7 to 10 kcal/mole) [1], confirming that delignification is limited to other factors besides the alcohol medium.

Review of Limiting Processes for Aqueous Lignin Leaching

Several researchers have hypothesized various factors that could be limiting the diffusion of lignin from the fiber wall during aqueous lignin leaching of brownstocks [8-13]. These particular mechanisms are discussed and are related to our A_E-stage studies in the following sections.

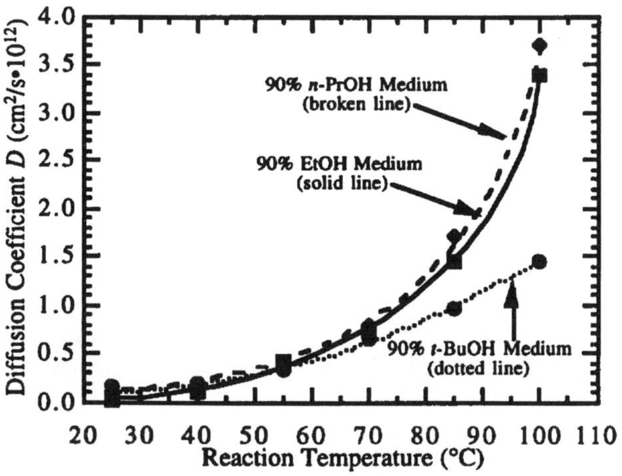

Figure 2. Calculated effective diffusion coefficients for lignin *(D)* for various acid alcohol extractions at $t = 5$ hrs.

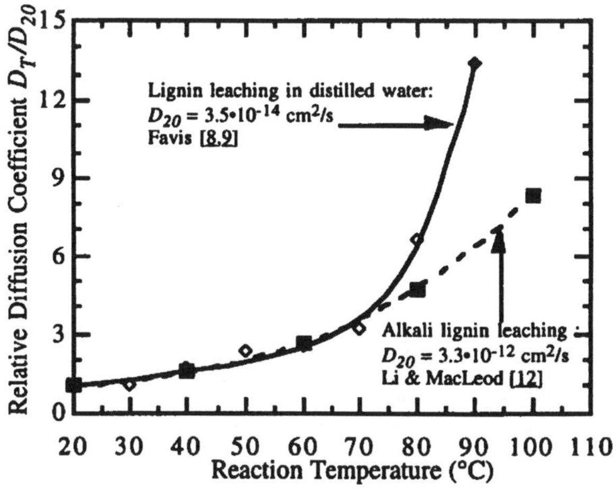

Figure 3. Relative effective diffusion coefficients for lignin during leaching under neutral and alkaline conditions with softwood kraft pulps [8,9,12].

Fiber Pore Size Restrictions

Favis et al. [8-11] proposed that lignin diffusion, under neutral conditions, is limited by size restrictions of the fiber pores. Lignin fragments are entrapped within the fiber wall as a consequence of their molecular diameters being approximately the same size or larger than the diameters of the fiber pores. This particular model was based on the observed correlations between the sizes of lignin fragments (estimated from molecular weight distribution data) and the sizes of fiber pores during various wood pulping processes [14]. Favis noted in one of his reports, however, that the spherical diameters of the leached lignin fragments (~10 nm), at various wash times, were generally larger than the median pore size (4 nm) [15]. This apparent inconsistency was resolved if the fragment shape was assumed to be an oblate ellipsoid [8,11,15].

Another key piece of evidence that Favis used to substantiate his pore restriction model [8,11] was the effect that electrolytes had on the lignin leaching rate [10]. The leaching rate was strongly inhibited when the ionic strength of the aqueous medium was increased with inorganic salts; this effect was intensified for divalent cations (e.g., Mg^{+2}) versus monovalent cations, (e.g., Na^{+}). In general, cations have been shown to decrease in fiber swelling by as much as 20% [16]. Favis claimed that the inhibited lignin diffusion with electrolytes was due to the reduction in fiber pore size (~4%) brought about by a 10% reduction in fiber swelling [8,11]. His previous data interpretation suggested that the average fragment size was close to the average pore size. Favis' diffusion model predicted that lignin leaching would be eliminated if the pore size decreased by 4%. This particular model, however, neglects the effect that cations have on the solubility of lignins [3,17].

Self-Association of Lignin

Another possible limiting mechanism for lignin diffusion is lignin association [12,13]. Association can affect lignin solubility or colloidal stability in a medium [17,18]. Reports of lignin association have been discussed extensively in the literature, especially in the determination of molecular weight distributions of dissolved lignins [19-24].

The major forces that can cause lignin fragments to associate with one another include hydrogen bonding and stereoregular association [19]. Hydrogen bonding forces are typically very strong (~5 kcal/mol) due to the short distance between the dipoles. This type of interaction can occur between phenolic, aliphatic hydroxyl and ether groups, or between hydroxyl groups and the π-electron systems of aromatic moieties [25]. Stereoregular association is caused by van der Waals forces arising from the HOMO-LUMO interactions of the π-orbitals between fragments. Furthermore, Sarkanen et al. [20,22] have indicated that there can be different levels, or modes, of lignin association. On the smallest level, discrete lignin fragments can associate with one another to form an ensemble. Afterwards, the ensembles can also associate with one another and form larger complexes. Evidence of such modes of association comes from the kinetics of kraft lignin association/dissociation in various media [20,25].

The association/dissociation of lignin fragments is also strongly influenced by several factors, such as alkalinity and inorganic constituent concentrations in the medium. Kraft lignins in aqueous media typically exhibit low levels of association at high pH (>13). According to Rudatin et al. [19], the majority of kraft lignin association occurs when the pH of the solution is lowered from 13 to 11; below pH 10 there was little additional association. Rudatin et al. have ascribed the pH dependency of association to the protonation of the phenoxide ions and the subsequent hydrogen bonding of phenolic groups with other electronegative elements. Studies performed by Lindström [17] have shown that multivalent cations can cause aqueous kraft lignin sols to associate with one another and precipitate out of solution. The multivalent cations can act as ionic bridges between anionic groups that would otherwise repel one another.

In alkaline leaching studies of unbleached pulps, Li et al. [13] have observed that ***D*** values increase from 10^{-15} to 10^{-9} cm^2/s, and the rates of lignin removal increase as the pH value is increased from neutral to 14. They proposed that lignin leaching is limited by the associated state of the lignin within the fiber [12,13]. Increasing the alkali concentration increases the mobility of the associated lignin by causing the associated complex to breakup into smaller and smaller ensembles. The smaller ensembles can more easily diffuse out of the fiber wall than the larger associated complex. Li et al. [12,13] have examined the effect of fiber swelling and various electrolytes on alkaline leaching. However, their results indicated no correlations between the reduction of lignin diffusion and the reduction in pore size that was caused by electrolytes.

Lignin Association with Cellulose

The association of lignin fragments with cellulose (i.e., lignin adsorption onto cellulose) could also limit lignin diffusion during extraction. This particular limiting mechanism has been considered as a minor factor by Li and MacLeod [12]. Lignin fragments could associate with cellulose through hydrogen bonding of the cellulose hydroxyl groups with the electronegative elements or the π-electron systems of lignin [25]. Evidence of this type of lignin-cellulose association in the literature is sketchy at best. An analogous system is the association (adsorption) of aromatic dyes by cellulose substrates; however, most of the textile literature suggests weak interactions between aromatic dyes and cellulose hydroxyl groups in aqueous solutions [26].

Severtson [27] has examined the sorption of 2,4-dichloro-phenol and 2,4,5-trichlorophenol onto softwood kraft fibers in aqueous solutions. His results indicated that pH and lignin concentration of the fiber substrate strongly affected chlorophenolate adsorption; it is low at >7 pH and high at <6 pH. This effect was attributed to the ionization of the chlorophenols (pK_a ~7-8). The chlorophenols adsorbed very strongly onto high lignin content fibers (kappa 105) as opposed to negligible adsorption onto cotton linters (kappa ~0). Similar behavior has also been reported with the adsorption of aromatic compounds onto low and high lignin pulps [28,29].

Limiting Processes for the Acid Ethanol Extraction Stage

A number of limiting factors have been hypothesized for lignin diffusion from unbleached pulp fibers in water. Each of the factors was analyzed to see if any applied to our observations with A_E-stage treatment of D-stage treated pulps.

Favis' pore restriction model was eliminated as a factor based on some of our earlier caustic extraction results in ethanol-water solutions [2]. The sizes of the extracted lignin fragments (<1.4 nm) were generally smaller than the average size of the fiber pores (4-6 nm). The observed decreases in fiber swelling in ethanol-water solutions, ~12%, would roughly decrease the average pore size by ~4%. This small reduction in pore size with ethanol-water solutions was determined to contribute negligibly to lignin diffusion resistance. In addition, there was no correlation between decreased fiber swelling and decreased $\boldsymbol{D}$, even in other aqueous de-swelling media [2]. The same basic argument can be made for the A_E-stage. The sizes of the lignin fragments will presumably be similar to those observed in aqueous D- or E-stage effluents (1-3 nm) [1,2], and the slight reduction in pore size from fiber de-swelling should not contribute significantly to diffusion resistance. Actual measurement of the molecular weights and sizes of the A_E-stage dissolved lignins would be needed to further substantiate this assumption.

Oxidized lignin association with cellulose (i.e., lignin absorption directly onto cellulose) was also eliminated as a major limiting mechanism in diffusion resistance with the A_E-stage. Evidence from the literature suggests that the association interactions among cellulose hydroxyl groups and aromatic moieties are very weak, and that the absorption of aromatic molecules onto fibers is strongly dependent upon the residual lignin contained within the fiber [27]. It is likely that the hydroxyl groups on the cellulose are strongly hydrogen bonded with water or ethanol molecules in the media [26]. This tightly held layer of solvent molecules will be difficult for a lignin fragment to displace and associate directly with the cellulose substrate. It is more probable that lignin association with cellulose is indirect and occurs through association with lignin that is chemically bonded with the carbohydrates.

The association of oxidized lignin appears to be the major limitation in lignin diffusion with the A_E-stage. Current lignin leaching studies with unbleached pulps strongly indicate that lignin association is limiting its removal from the fiber wall in aqueous media [12,13]. Various degrees of lignin association in non-aqueous and organic solvent-water solutions have been documented in the literature [21-24]. Furthermore, the oxidized lignin of D-stage pulp contains a higher concentration of carboxylic acid and other oxygen containing groups than the original kraft lignin. This will increase the number of hydrogen bonding sites on the lignin fragments and possibly increase the probability of lignin association. Berry [30] and Jokela and Salkinoja-Salonen [31] have cited possible evidence of oxidized lignin association with isolated extraction lignins.

In addition, the observed $\boldsymbol{D}$ enhancement at higher temperatures ($\geq$ 70°C) could be another indication of lignin association for the A_E-stage (Figure 2). A previous study [4] suggested that this sudden increase in $\boldsymbol{D}$ was related to the decreased hydrogen bonding of medium as the temperature approached the boiling point. This particular mechanism was proposed by Johansson [32] to explain the enhanced $\boldsymbol{D}$ for brownstock leaching in water (Figure 3). Although such a mechanism is plausible for acid alcohol extraction, one would expect the onset of this enhancement to change with each extraction medium because the boiling point of each medium is different. Instead, all the acid alcohol extractions have basically the same onset of enhanced $\boldsymbol{D}$ as the aqueous lignin leaching system (~70°C). An alternative explanation is that the increase in the extraction temperature ($\geq$ 70°C) is diminishing the hydrogen bonding that causes the lignin to associate into complexes. The breakup of these complexes into smaller, more mobile ensembles would presumably decrease the diffusion resistance [1].

Explanations for Oxidized Lignin Extraction Efficiencies

Lignin association may also explain why an A_E-stage is less efficient than an aqueous E-stage. The alkali in an aqueous E-stage ionizes the carboxylic groups in the oxidized lignin; these ionized groups repel one another and disrupt the associated lignin complexes into smaller fragments. Li et al. [12,13] have cited this electrostatic repulsion mechanism for the effectiveness of brownstock lignin leaching under alkaline conditions. Such a lignin dispersing mechanism will not occur under the acidic conditions of an A_E-stage; the acid groups on the lignin will not be ionized to a significant extent.

Our previous studies have indicated the ineffectiveness of caustic extraction performed with ethanol-water media due to solubility problems [2,3]. This factor can also be rationalized in terms of lignin association and solvation effects. Ethanol is not as good as water at stabilizing separated ion pairs; the bulky alkyl group of ethanol interferes with the ion-dipole interactions that stabilize charge-separated ions in solution. It is likely that the ionized carboxylic anions (ROO^-) and Na^+ exist as an intimate, non-solvated ion pair (ROO^-Na^+) in ethanol-water solutions [1]. Thus, the lignin remains as an associated complex with the electrostatic repulsive ROO^- groups effectively shielded by Na^+.

CONCLUSIONS

Lignin removal from a D-stage treated pulp is less efficient in an A_E-stage than an aqueous E-stage. The strong similarities between acid ethanol extraction and aqueous brownstock lignin leaching suggest that both lignin diffusion processes are limited by a common mechanism. Several researchers have hypothesized that lignin removal during brownstock leaching is restricted by the association of lignin inside the fiber wall. Improvements in aqueous lignin diffusion with higher alkali concentrations appear to be related to the disruption of lignin complexes by the electrostatic repulsion of ionized groups. A similar type of mechanism probably occurs in an aqueous E-stage. The acid conditions of the A_E-stage, however, do not promote the dissolution of associated lignin by an electrostatic repulsion mechanism. In addition, ethanol-water solutions are less effective at stabilizing ion pairs than water. Consequently, E-stages performed in ethanol-water solutions will be less effective due to the shielding of ionized acid groups by the corresponding counterions.

EXPERIMENTAL PROCEDURES

A thoroughly washed conventional softwood kraft pulp (southern pine) was obtained from a mill and had a kappa number ~30. The specific conditions of the D-stage, A_E-stage, and E-stage have been previously reported [1-4,7].

Acid alcohol extractions (*n*-PrOH and *t*-BuOH) were performed on 2.5 o.d. gram samples of the aqueous D-stage treated pulp that had a kappa of 18.8. Experiments were conducted at 1.25% consistency in 90% ethanol medium with 0.01 M HCl. Low temperature reactions (<70°C) were conducted in sealed containers, which were heated using a water bath. High temperature reactions (≥70°C) were performed in a Parr bomb reactor. Generally, the high temperature reactions (100°C) required 10 minutes to heat up to the specified reaction temperature. The pulps were extracted for the required period, cooled to 65 to 70°C (if needed), and drained of extraction effluent to ~25% consistency. The estimation of ***D*** from the experimental data has been previously explained [1,2,4].

NOMENCLATURE

A_E	acid ethanol extraction stage
D	chlorine dioxide stage
$\boldsymbol{D}$	effective diffusion coefficient, (cm^2/s)
$\boldsymbol{D_T/D_{20}}$	relative diffusion coefficient, ratio of $\boldsymbol{D}$ at some T to $\boldsymbol{D}$ at 20°C
E	caustic extraction stage
T	reaction temperature, (°C)

ACKNOWLEDGMENTS

The authors thank the Institute of Paper Science and Technology and its Member Companies for supporting this research. Portions of this work were used by B.N. Brogdon as partial fulfillment for the requirements of the Ph.D. degree at the Institute of Paper Science and Technology.

LITERATURE CITED

1. **Brogdon, B.N.,** *Effects of Ethanol Media on Chlorine Dioxide and Extraction Stages for Kraft Pulp Bleaching,* Ph.D. Dissertation, IPST, Atlanta, (1997).

2. **Brogdon, B.N., Dimmel, D.R., McDonough, T.J., and Bryant, P.S.,** *1996 International Pulp Bleaching Conference,* Washington, DC, TAPPI Proc., p. 541 (1996).

3. **Brogdon, B.N., Dimmel, D.R., and McDonough, T.J.,** *1996 TAPPI Pulping Conf.,* Nashville, TN, TAPPI Proc., p. 35 (1996).

4. **Brogdon, B.N., Dimmel, D.R., and McDonough, T.J.,** In *Further Advances in the Forest Products Industries: The 1996 Forest Products Symp.,* Ed. by P.W. Hart, AIChE, New York, (1997).

5. **Ni, Y., and van Heiningen, A.H.P.,** *Tappi J.,* **79** (3):239 (1996).

6. **Ni, Y., and Hu, Q.,** *J. Appl. Polym. Sci.,* **57**:1441 (1995).

7. **Brogdon, B.N., Dimmel, D.R., and McDonough, T.J.,** In *9th Intl. Symp. on Wood and Pulping Chem.,* Montréal, Québec, June 9-12, CPPA/TAPPI Oral Proc., p. A2-1(1997).

8. **Favis, B.D.,** *The Leaching of Lignin Macromolecules from Pulp Fibers During Washing,* Ph.D. Dissertation, McGill University, Montréal (1981).

9. **Favis, B.D., Willis, J.M., and Goring, D.A.I.,** *J. Wood Chem. Tech.,* **3**:1 (1983).

10. **Favis, B.D., Willis, J.M., and Goring, D.A.I.,** *J. Wood Chem. Tech.,* **3**:9 (1983).

11. **Favis, B.D., and Goring, D.A.I.,** *J. Pulp Paper Sci.,* **10:**J139 (1984).

12. **Li, J., and MacLeod, J.M.,** *J. Pulp Paper Sci.,* **19**:J85 (1993).

13. **Li, J., Phoenix, A., and MacLeod, J.M.,** *Pulp Washing '96,* Vancouver, BC, CPPA/TAPPI Proc., p. 139 (1996).

14. **Ahlgren, P.A., Yean, W.Q., and Goring, D.A.I.,** *Tappi J.,* **54** (5):737 (1971).

15. **Favis, B.D., Yean, W.Q., and Goring, D.A.I.,** *J. Wood Chem. Tech.,* **4**:317 (1984).

16. **Lindström, T., and Carlsson, G.,** *Svensk Papperstidn.,* **85** (3):R14 (1982).

17. **Lindström, T.,** *Colloid Polym. Sci.,* **258**:168 (1980).

18. **Lindström, T.,** *Colloid Polym. Sci.,* **257**:277 (1979).

19. **Rudatin, S., Sen, Y.L., and Woerner, D.L.,** In *Lignin, Properties and Materials,* Eds. W.G. Glasser, and S. Sarkanen, ACS Symp. Series 379, ACS, Washington, DC, p. 144 (1989).

20. **Dutta, S., Garver, T.M., and Sarkanen, S.,** In *Lignin, Properties and Materials,* Eds. W.G. Glasser, and S. Sarkanen, ACS Symp. Series 379, ACS, Washington, DC, p. 154 (1989).

21. **Benko, J.,** *Tappi J.,* **47** (8):508 (1964).

22. **Sarkanen, S., Teller, D.C., Hall, J., and McCarthy, J.L.,** *Macromolecules,* **14**:426 (1981).

23. **Connors, W.J., Sarkanen, S., and McCarthy, J.L.,** *Holzforschung,* **34**:80 (1980).

24. **Chum, H.L., Johnson, D.K., Tucker, M.P., and Himmel, M.E.,** *Holzforschung,* **41**:97 (1987).

25. **Yoshida, Z., Osawa, E., and Oda, R.,** *J. Phys. Chem.,* **68**:2896 (1964).

26. **Giles, C.H.,** "Adsorption of Dyes," In *Adsorption from Solution at the Solid/Liquid Interface,* Eds. G.D. Parfitt, and C.H. Rochester, Academic Press, New York, (1983).

27. **Severtson, S.J.,** *Sorption of 2,4-Dichlorophenol and 2,4,5-Trichlorophenol by Softwood Kraft Fibers*, Ph.D. Dissertation, IPST, Atlanta, (1995).

28. **Garbaromo, D.R., and Lion, L.W.,** *Environ. Sci. Tech.,* **20**:1263 (1986).

29. **Daube, A.K., Mohammed, R.K., Dimmel, D.R., and McDonough, T.J.,** *Environ. Sci. Tech.,* **26**:1324 (1986).

30. **Berry, R.** "(Oxidative) Alkaline Extraction," In *Pulp Bleaching: Principles and Practice,* Eds. C.W. Dence, and D.W. Reeve, TAPPI Press, Atlanta, pp. 296-297 (1996).

31. **Jokela, J.K., and Salkinoja-Salonen, M.,** *Environ. Sci. Tech.,* **26**:1190 (1992).

32. **Johansson, A.,** *J. Pulp Paper Sci.,* **11**:J78 (1985).

Distribution of Mass Flows in Black Liquor Sprays

H. Jeff Empie, Steven Lien and Douglas B. Samuels
Institute of Paper Science and Technology, Atlanta, GA 30318

Black liquor has been sprayed through standard splashplate and V-jet nozzles at typical nozzle conditions using an enclosed spray chamber. The sheets of liquor that issue from the nozzles have been characterized with respect to distribution of mass flows within the total spray angle. A maximum mass flux occurs at the spray centerline which coincides with the nozzle axis. The mass flux exhibits a parabolic dependence on angular position, decreasing in magnitude out to the edge of the sheet. At the very outer edge was a thick, slow-moving rim. Model predictions of the centerline flux and the full spray angle agree to within 10% of the experimental values for the splashplate nozzle. Results will be useful in modeling combustion processes in kraft recovery boilers where a detailed knowledge of the black liquor spray characteristics is required.

INTRODUCTION

An essential and valuable by-product of the kraft pulping process is black liquor, an aqueous solution of spent inorganic pulping chemicals and organic species extracted from the wood. Included in the organics are dissolved lignin and degraded carbohydrates. For economic and environmental reasons, kraft black liquor is put through a recovery process where the energy content provided by the organics is reclaimed in the form of process steam, and the inorganics are regenerated and recycled to the pulping operation. Typically, 95-97% of the inorganic chemicals are recovered, and the steam generated satisfies over 40% of a modern kraft mill's energy needs [1].

The most important unit operation employed in the recovery cycle of the kraft pulping process is combustion of black liquor in the recovery boiler. This step is initiated by spraying the concentrated liquor through one of several types of commercial nozzles, the most common being the splashplate, the V-jet, and the swirl cone. The liquor issues from the nozzle as a thin sheet, which subsequently breaks up into droplets. These droplets then go through the sequential processes of drying, pyrolysis and gasification, combustion, chemical reduction, and coalescence of the molten inorganic salts (2). The rates at which these physical and chemical processes occur are highly dependent upon the size and size distribution of the droplets formed from the spray. The smaller the droplet, the greater the surface area per unit mass of liquor and, hence, the greater the rates of heat and mass transfer between the liquor particles and the furnace gases. While this is desirable for increasing recovery boiler capacity, it is offset by higher entrainment and carry-over, which are characteristic of small particles in an upward flowing turbulent gas stream. Inevitably, the result is accelerated fouling of the relatively cool boiler tubes and more rapid plugging of the heat transfer section of the boiler.

To increase throughput while minimizing tube fouling and plugging rates, improved understanding and control of the liquor spraying step is needed. Not only are drop size and size distribution important in this regard, but also of concern is the spatial mass distribution of the liquor across the furnace cross-section. An uneven distribution could lead to localized regions of intense combustion and result in an unwanted hot spot. Such an occurrence can accelerate corrosion rates of the furnace walls and result in metal failure and premature shutdown.

The distribution of liquor within the furnace firing zone is not known and will not be in the near future because of the extreme difficulty of taking measurements in this turbulent, particulate-laden, corrosive, inaccessible, high temperature region. Characterization is best achieved through mathematical modeling using computational fluid dynamics and fundamental kinetic and transport rate data.

The key to successfully modeling the recovery boiler furnace zone is by starting with an accurate understanding of black liquor nozzle performance. Drop size and size distribution data are of course paramount. With regard to spatial distribution, it is also important to know not only the total angle of the spray sheet produced initially by a given nozzle and liquor flow conditions, but also how the mass flow of black liquor varies with the angular position within the sheet. Data are reported below on the mass flow distribution of liquor sprays from two different commercial nozzles, the Babcock & Wilcox (B&W) 12/45 splashplate and the Spraying Systems (SS) V-jet 11/65.

EXPERIMENTAL

Black liquor was sprayed continuously through the B&W splashplate and SS V-jet nozzles using a heated, recirculating pump-around loop system (Fig. 1). The central component of the system was a spray chamber which served the dual purpose of providing a visible spray pattern while containing and storing the liquor inventory. The stainless steel chamber was 2.1

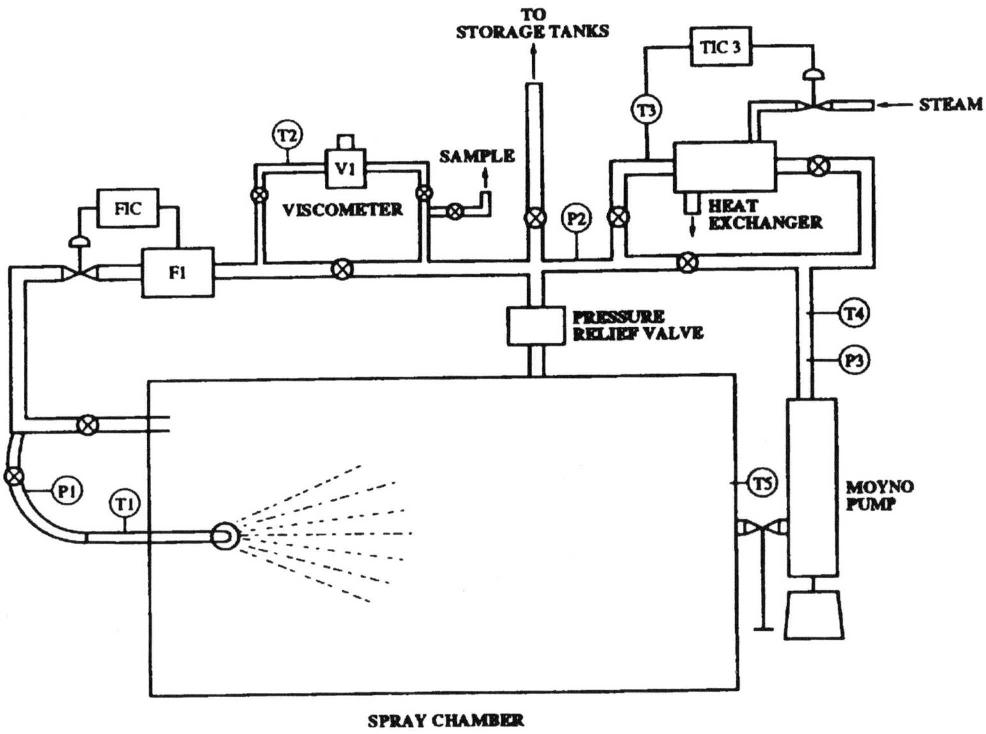

Fig. 1 Schematic of Black Liquor Spray System.

m long, 1.8 m deep, and 3.0 m high with a V-shaped, heated bottom section that could store up to 1100 liters of liquor at temperatures up to 100°C. The front and rear walls contained 1.5m x 1.2m tempered glass windows to allow backlighting and videotaping of the liquor spray sheet and droplets. The side of the chamber had an opening for the spray nozzle which was oriented to deliver the liquor sheet in a horizontal trajectory. (When videotaping to get dropsize measurements, the liquor sheet is oriented parallel to the windows.)

A two-stage Moyno pump circulated the liquor from the spray chamber through a spiral heat exchanger, which could raise the liquor temperature by 20°C at 190 l/min, then through a Brookfield viscometer and back to the chamber. When the liquor was at the desired conditions for testing, the bypass loop was closed and the flow directed through the nozzle. The system had the capacity to deliver up to 150 l/min (as measured by a Foxboro magnetic flowmeter) of liquor to the nozzle at 550 kPa (80 psi). A continuous water addition line was included in the pump-around loop to replace any water vapor which escaped from the hot liquor spray, thereby maintaining a constant liquor solids content. Because of odor control considerations, the spray chamber was operated under a slight negative draft. An ID fan drew air in through eight adjustable dampers, after which this gas flow plus any volatiles coming from the liquor spray were pulled through a carbon absorber (provided by Westvaco Corp.) before being exhausted to the atmosphere.

The two nozzles used in this work were a B&W 12/45 splashplate (Fig. 2) and a SS 11/65 V-jet (Fig. 3). Splashplate type nozzles are widely used for black liquor spraying. They feature a straight tapered nozzle section which causes the black liquor flow to impinge on a flat plate oriented at angles to the nozzle axis ranging from 35° to 55° (in this case 45°). The first two digits in the nozzle designation (i.e.12) refer to the nozzle exit diameter in thirty-seconds of an inch. In the SS V-jet designation, the first two digits are again the exit diameter in 32nds of an inch; the second two digits represent the total spray angle of the sheet of liquor issuing from the nozzle.

Samples of sprayed liquor were collected by using two different types of patternators. The first arrangement used a single box with a large 7.6 x 7.6-cm. opening. A trap door on the front of the box could be opened for a fixed length of time (generally 5 to 10 seconds) and then quickly closed. This was used with the SS V-jet nozzle, where samples were collected at a distance of about 84 cm. from the nozzle.

A second patternator was constructed for the B&W splashplate nozzle. Nine 2.5 x 2.5-cm. sampling boxes were arranged on a semicircular frame at 20° intervals. The radial distance from the nozzle to the front edge of each box was fixed at 11.5 cm. The sampling boxes cut the liquor sheet into sections which could be collected simultaneously for a fixed time and accurately weighted to determine the mass flow rate at the nine different angles. Because the angle and position

Table 1 - Spray Angle for the Test Nozzles

Test No.	Nozzle	Liquor Viscosity (cP)	Nozzle Pressure (kPa)	Reynolds Number	Spray Angle (°)
30A	B&W 12/45	219	123	492	140
30B	"	229	239	684	170
31A	"	50	126	2697	150
31B	"	66	241	2862	166
28A	SS 11/65	55	75	2120	57
28B	"	68	137	2333	60
27A	"	346	76	321	40

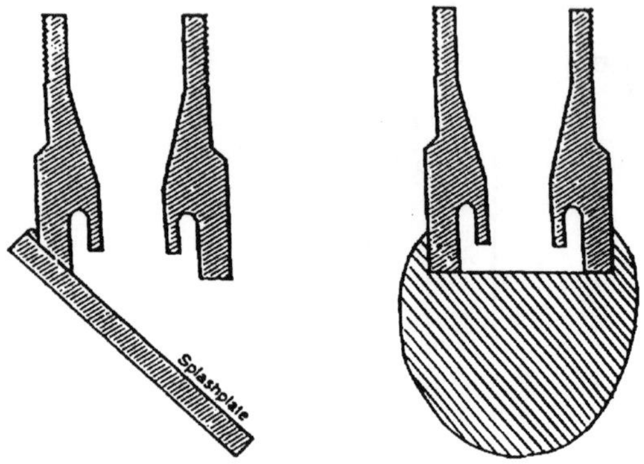

Figure 2. Splashplate Nozzle.

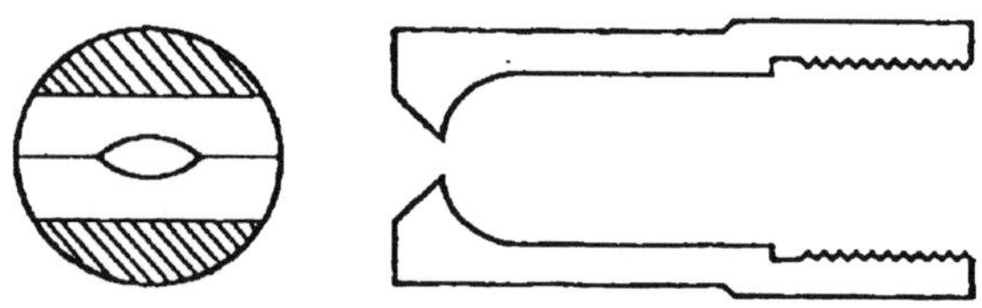

Figure 3 V-jet Nozzle.

the sampler were fixed with respect to the nozzle, these samples were slightly more reproducible than those collected with the larger single-box sampling device, which had to be repositioned for each sample.

For each data point, two to four samples at each angular position were collected and averaged. Measurements over the total spray angle were taken at five points for the SS V-jet and nine for the B&W 12/45 splashplate. The test conditions are listed in Table 1. In all cases at a given spray angle the entire liquor sheet inthe direction perpendicular to the sheet surface was collected.

This work was done at relatively low solids concentrations (55% to 65%), which allowed the experiments to be performed at lower temperatures (50°C to 85°C). This was an important safety consideration, since the experimental procedures for these tests required close physical access to the spray nozzle. Changes in solids concentration are not expected to have a large influence on the mass flux distribution.

Four tests were performed with the B&W 12/45 splashplate (Figs. 4-7), while three were done with the SS V-jet nozzle (Figs. 8-10). The spray angle (α) represents the angle bounded by the left and right outer edges of the spray sheet. The overall angle determinations were relatively inexact and have an estimated confidence limit of $\pm$ 10°. At the extreme outer edge of the sheet there existed a rim of liquor that was thicker than the sheet just inside the rim. Hence, the mass flow just inside the spray angle was larger than expected, while just beyond it the flow was essentially zero.

The spray angle for the splashplate is much wider than for the V-jet and approaches 180° at high flow rates and low viscosities. The SS V-jet nozzle is classified as a 65° nozzle by the manufacturer, and the measured values do approach this value at high flows and low viscosities. The variation of the spray angle seen in these tests is typical behavior for pressure nozzles. The spray angle will increase with operating pressure until it reaches a maximum value; beyond this point the spray angle will decrease slightly with increasing pressure.

In order to quantify how the mass flow varies with angular position in the spray sheet, it is necessary to define a mass flow factor (MFF) as the ratio of the measured flow at a given angular position to the average flow over the total spray angle [3]. Because each sampling box has a fixed width, the measured mass flow represents an average over the angle, $\Delta\theta$, where the position angle, θ , is defined as the angular position in the spray sheet measured with respect to the centerline of the splashplate. Hence, the MFF becomes the ratio of the mass flux into the box at angle ($\theta \pm \Delta\theta/2$) to the mass flux averaged over the spray angle, α.

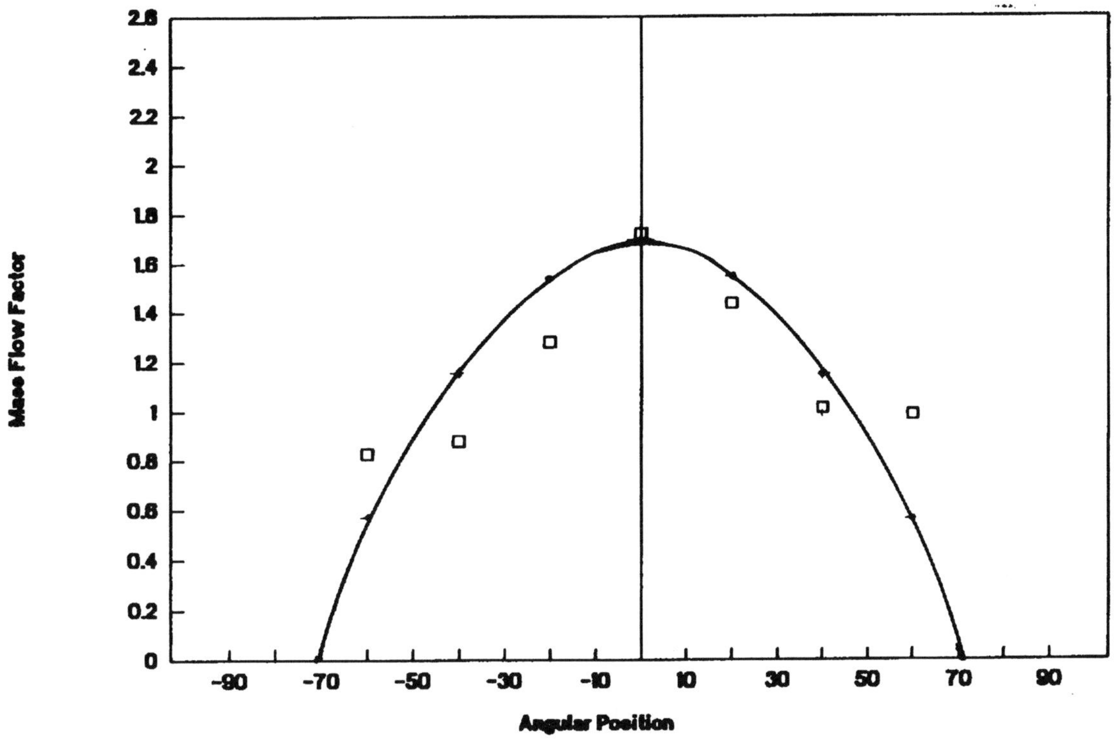

Fig. 4 - Mass Flow Distribution for Test 30A

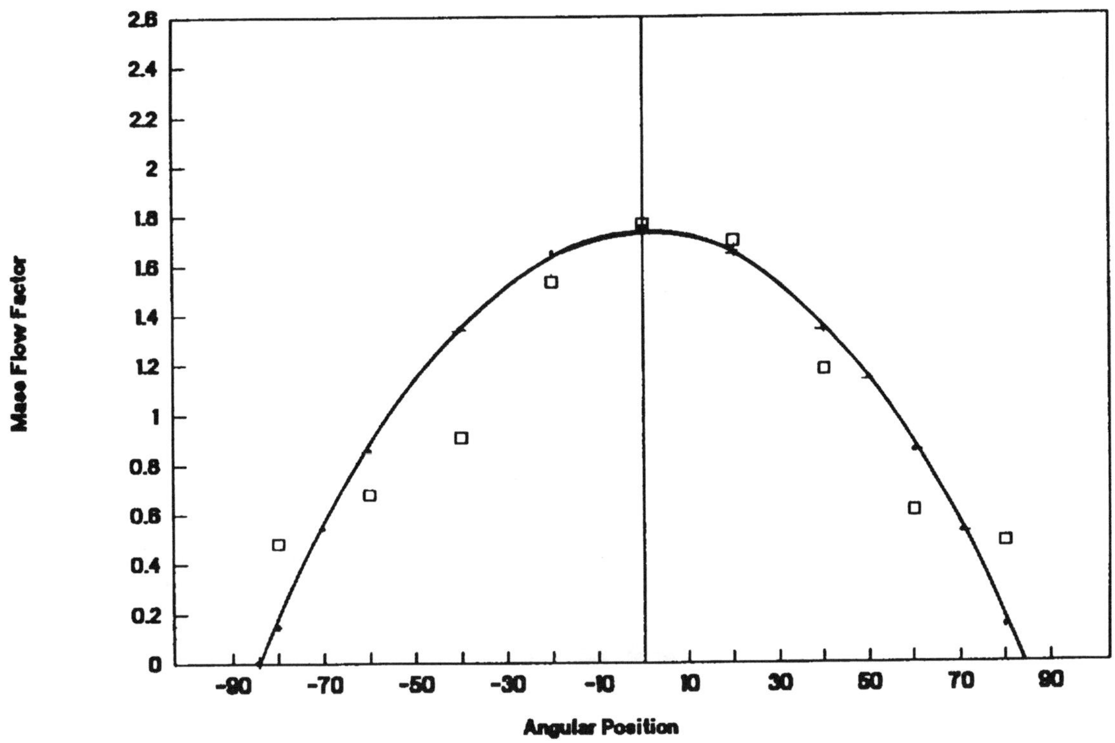

Fig. 5 - Mass Flow Distribution for Test 30B

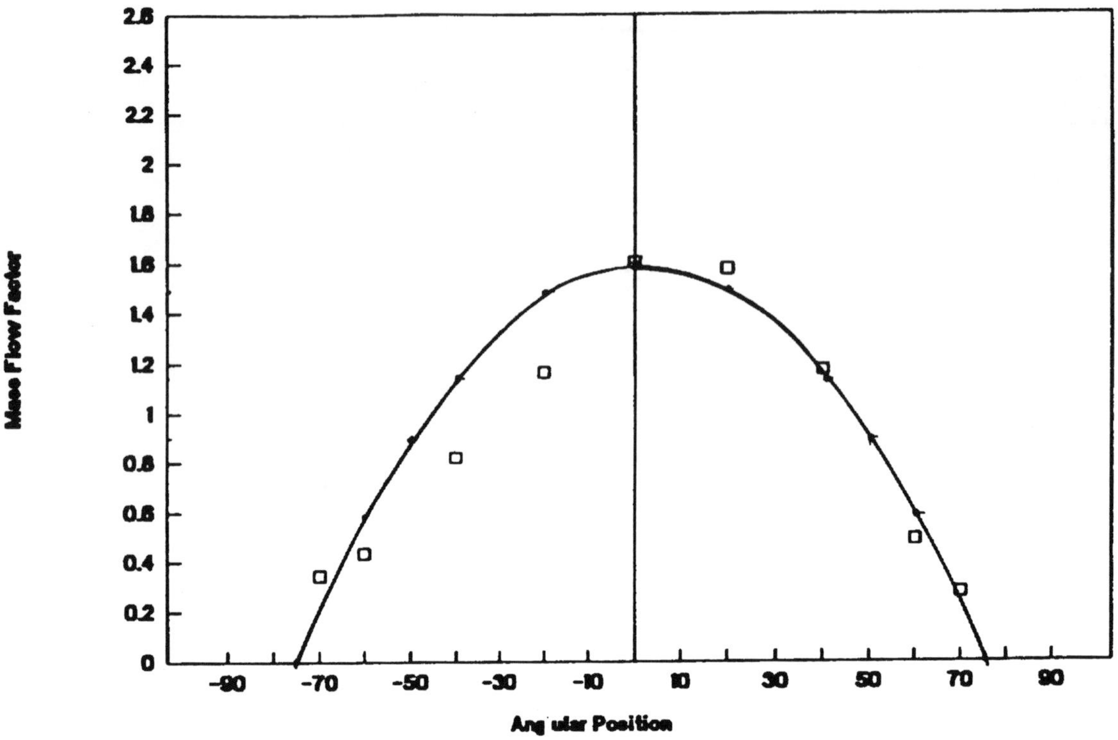

Fig. 6 - Mass Flow Distribution for Test 31A

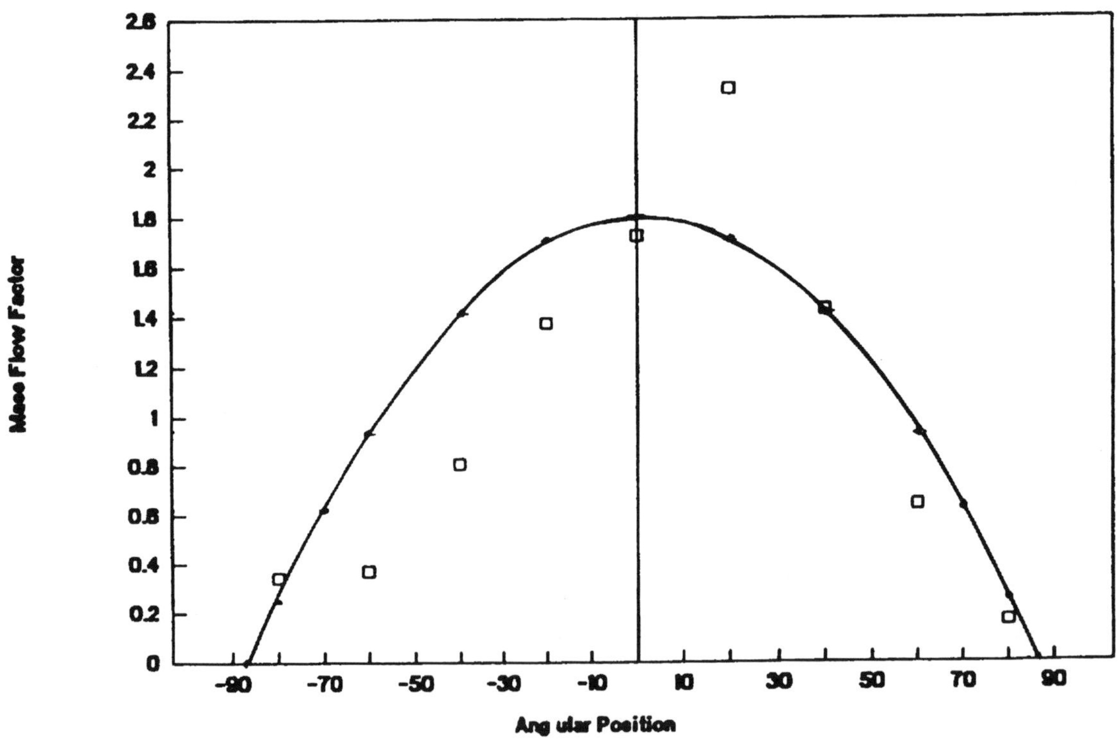

Fig. 7 - Mass Flow Distribution for Test 31B

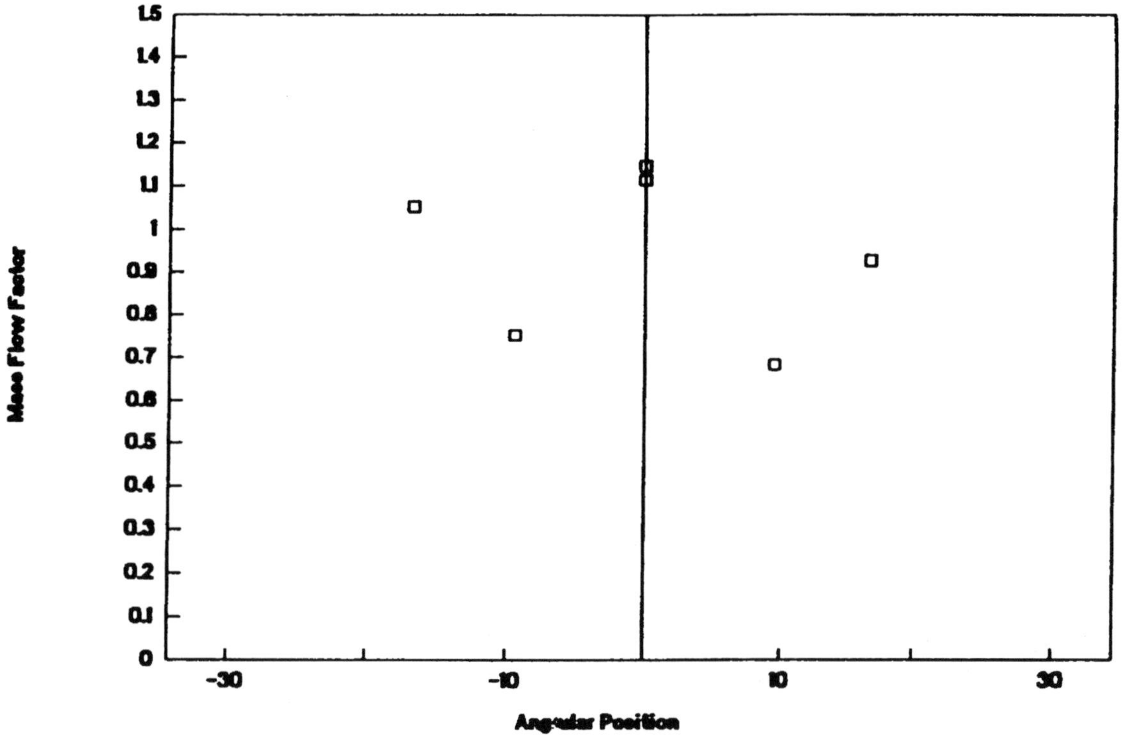

Fig. 8 - Mass Flow Distribution for Test 27A.

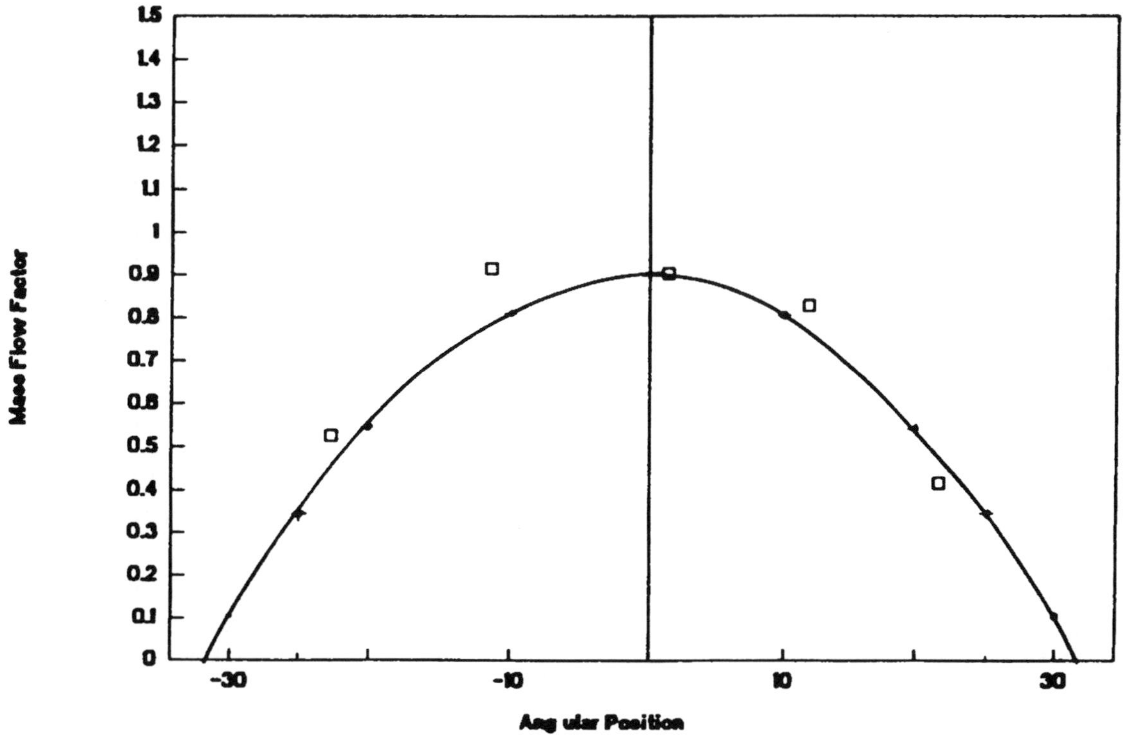

Fig. 9 - Mass Flow Distribution for Test 28A.

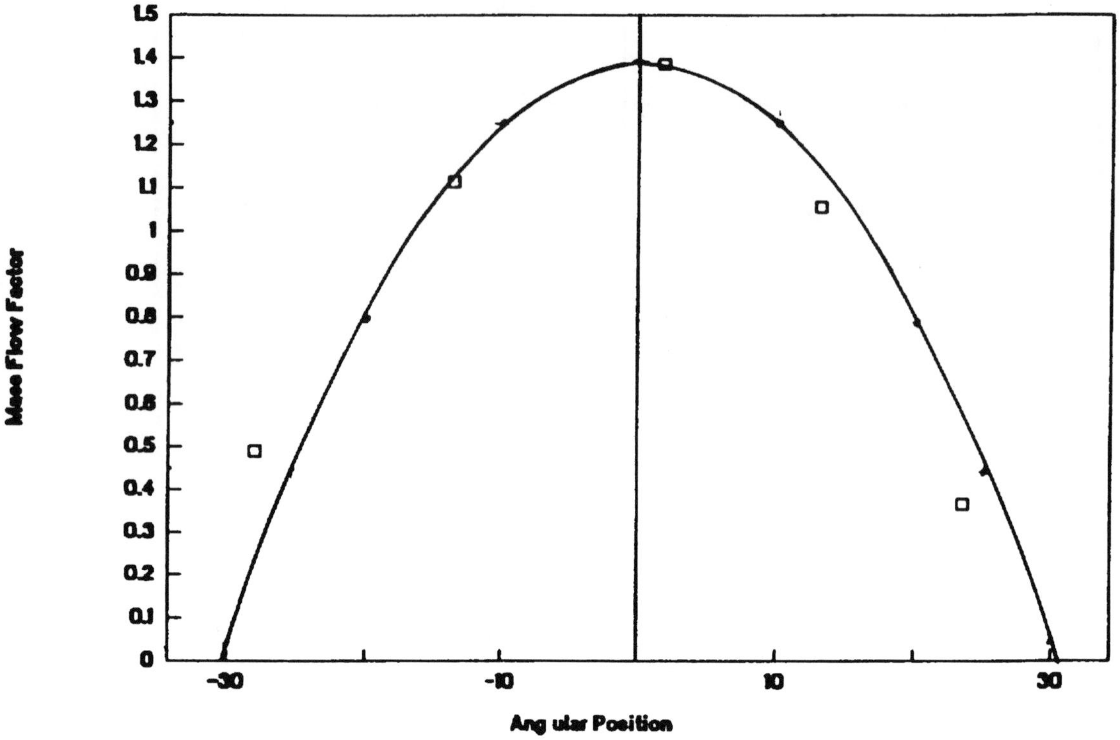

Fig. 10 - Mass Flow Distribution for Test 28B.

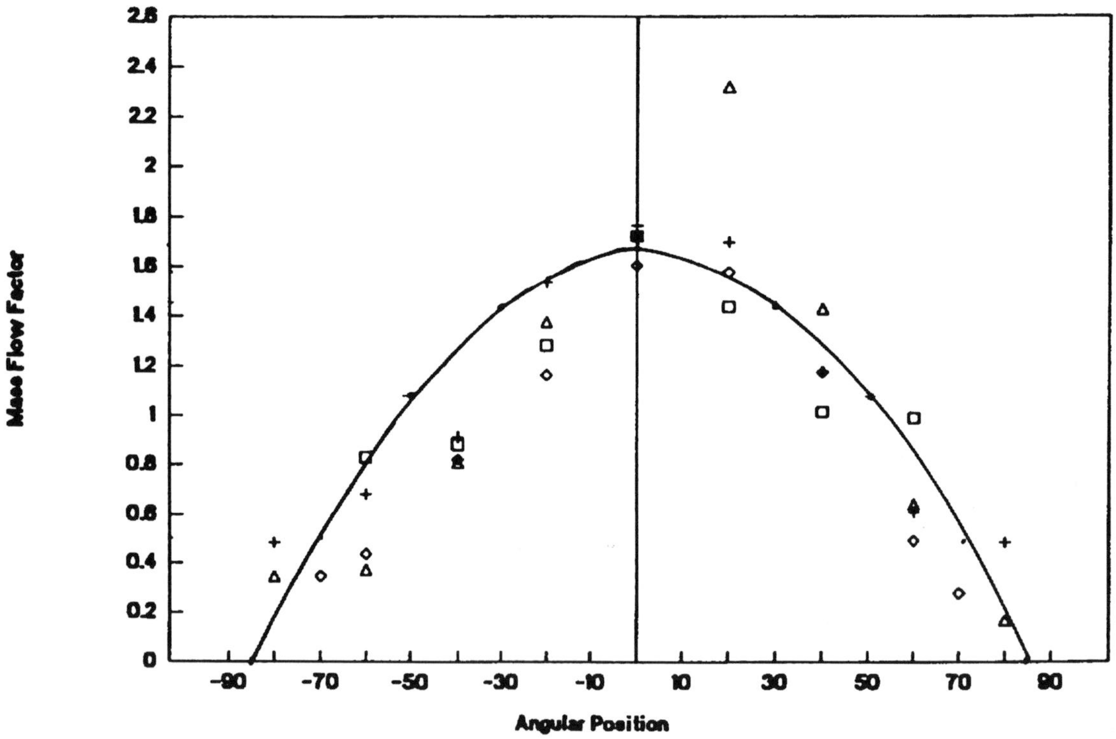

Fig. 11 - Mass Flow Distribution for Combined Splashplate

$$MFF = \frac{\Delta M/\Delta\theta}{M_o/\alpha} \quad (1)$$

where: ΔM = mass flow rate into sample box (g/s)
θ = position angle of sample box (^{o})
$\Delta\theta$ = angle subtended by sample box (^{o})
M_o = total liquor flow rate (g/s)
α = spray angle (^{o})

The MFF is plotted as a function of the position angle in Figs. 4-7 for the B&W splashplate nozzle. All four curves have a similar overall shape which does not appear to be strongly influenced by variations in viscosity or flow rate over the ranges tested. From visual observations, one could see that there was a central core in the angular range from $-15^o < \theta < 15^o$ having a relatively uniform flow rate. Outside this region, the flow rate dropped off steadily almost to the outer edge. At the very edge of the sheet, there was a thick rim. Hence, just inside the spray (half) angle, the MFF did not tail off, but beyond the spray (half) angle the mass flux abruptly dropped to zero.

Tests 31A and 31B showed unusually high flows at about $+20^o$ from the centerline. These high flows were relatively stable phenomena which remained in these sprays for several minutes and showed up consistently in the mass measurements at this location. The cause of this may be due to a secondary flow of black liquor which circulates around the tip of the nozzle and falls back on to the spray sheet. This secondary flow is always present with splashplate nozzles. It is created by backwards flow on the splashplate starting at the point of impingement, followed by flow around the recess that encircles the nozzle tip and by formation of a separate flow off the splashplate. The point where it contacts the sheet is a function of nozzle orientation. Unlike the video work done to determine drop size where the spray sheet is in a vertical plane, the mass flow measurements were made with the spray pattern in a horizontal orientation. The mass flow of this secondary stream was relatively small, but it appeared to have the effect of shifting the splashplate flow off to one side of the splashplate centerline, most likely because the splashplate was not exactly horizontal.

The spray patterns for the V-jet as shown in Figs. 8-10 were more sensitive to variations in the test conditions. At low viscosity and pressure (55 cP, 75 kPa), Fig. 9 shows the mass distribution to be close to plug flow over much of the spray angle. At a higher flow rate (68 cP, 137 kPa), not only did the spray angle increase (Fig. 10), but there was also a higher mass flow at the centerline and a steeper profile. At high viscosity (Test 27A: 346 cP, 76 kPa), the spray mass distribution exhibited an unusual notched pattern (c.f. Fig.8). Visual observation showed this spray to have a much narrower angle than at low viscosity (Test 28A: 68 cP, 14 kPa) with a thick outer rim and a thin region inside this rim. The reason for this is not known explicitly. Droplet formation is known to be a strong function of the balance between viscous, momentum, and surface tension forces (3). For Tests 27A and 28A, the viscosities differed by a factor of seven, but the flows had nearly the same momentum (c.f.,Table 1). Hence, the spray angle at high viscosity may be dominated by viscous forces, whereas at low viscosity, momentum forces are stonger and may cause a more energetic breakup.

An empirical correlation of MFF as a function of angular position (θ) would be useful for predicting mass flow distribution for a nozzle under specified operating conditions. A regression analysis indicated that either a linear or a parabolic dependence would produce a reasonable estimate of the MFF profile. Although the linear correlation has the advantage of simplicity, the slope of the "curve" at $\theta = 0$ has a discontinuity, whereas the experimental curve is rather flat in this region. The parabolic correlation satisfies this criterion at the centerline, and hence, it has been used to represent the data.

If the MFF is expressed in the form:

$$MFF = A - B\,\theta^2 \quad (2)$$

the constants A and B can be approximated, knowing the MFF at $\theta = 0$ and at some other angular position, θ (1). Hence, at $\theta = 0$, MFF = A; at $\theta = \theta$ (1), MFF = MFF(1), and $B = [A - MFF(1)]/\theta(1)^2$. Results are shown in Table 2 and are represented by the curves in Figs. 4-7,9-11.

Table 2 - Constants for MFF Correlation

Run	MFF @ θ = 0	MFF @ θ(1)	θ(1)	A	B	α
30A	1.7	0.85	50	1.7	0.00034	157
30B	1.75	0.60	68	1.75	0.00025	190
31A	1.6	0.40	65	1.6	0.00028	160
31B	1.8	0.25	80	1.8	0.00024	200
Splashplate Composite	1.68	0.5	70	1.68	0.00024	184
28A	0.9	0.55	20	0.9	0.00088	-
28B	1.4	0.45	25	1.4	0.00152	56

In modeling the MFF as a parabolic function, it will be shown that the experimental constants, A and B, and the spray angle, α, are not all independent. In fact, experimentally determining A and B allows calculation of the expected spray angle, α. This is very desirable, since experimental values of α could not be determined exactly.

From eq.(1),

$$MFF = \frac{\Delta M/\Delta\theta}{M_o/\alpha} = \frac{\rho\, v\, z\, R\, \Delta\theta/\Delta\theta}{\rho\, v_a\, z_a\, R\, \alpha/\alpha} = \frac{v\, z}{v_a\, z_a} \quad (3)$$

where: ρ = liquor density
v = liquor sheet velocity at angle θ and distance R from

the nozzle

v_a = average liquor sheet velocity

z = sheet thickness at angle θ and distance R from nozzle

z_a = average sheet thickness at distance R from nozzle

R = radial distance from nozzle

Therefore:

$$\frac{v\,z}{v_a\,z_a} = A - B\,\theta^2 \qquad (4)$$

But

$$\rho\, v_n\, \pi\, D_n^2/4 = \rho\, v_a\, z_a\, R\, \alpha$$

$$v_a\, z_a = \frac{v_n\, \pi\, D_2^n}{4\, R\, \alpha} \qquad (5)$$

where: v_n = liquor velocity at nozzle

D_n = nozzle diameter

Therefore:

$$v\, z\, R\, d\theta = \frac{v_n\, \pi\, D_n^2}{4\alpha}\,(A - B\,\theta^2)\, d\theta \qquad (6)$$

Integrating both sides of eq.(6) between the limits of $\pm\,\alpha/2$, the result is

$$A = 1 + B\,\alpha^2/12$$

or

$$\alpha = \pm\,(12(A-1)/B)^{1/2} \qquad (7)$$

Values of α calculated from eq.(7) are listed in Table 2 and can be compared with the experimentally observed values for α listed in Table 1. The predicted values and are generally within 20% of each other.

Since the model defines the spray (half)angle as the angular position ($\pm\,\alpha/2$) where the product, v z, goes to zero, eq.(6) with $\theta = \alpha/2$ reduces to:

$$0 = A - B(\alpha^2)/4$$

or

$$B = 4A/\alpha^2$$

and from eq. (7):

$$A = 1 + \frac{4\,A}{12} = 1.5$$

Hence, the model fundamentally predicts the MFF at $\theta = 0$ to be 1.5. Experimentally determined values generally fall within 20% of the theoretical value.

CONCLUSIONS

1. Measurements of the local distribution of black liquor flow in the spray pattern from sheet-forming nozzles were made as a function of the angle from the sheet centerline. Qualitative visual observations of the liquor sheets from both the splashplate and V-jet nozzles showed that there was a central core with a fairly uniform flow rate in the area within about 15° of the sheet centerline. Outside this region, the flow dropped off steadily almost to the outer edge of the sheet. At the very edge was a thick rim.

2. Modeling the mass flux profile as a parabolic function of angular position, with the flux exhibiting a maximum at the centerline and decreasing with increasing angular position, predicted the mass flux at the sheet centerline to be 1.5 times the average flux for the entire sheet cross section at the same radial distance from the tip of the nozzle. Model predictions of both the centerline flux and the full spray angle agreed to within 20% of the experimental values for the splashplate nozzle.

3. These results should provide key input to mathematical models of the combustion processes occurring in kraft recovery boilers.

LITERATURE CITED

1. **Green, R.P. and Hough, G.** (eds.), "Chemical Recovery in the Alkaline Pulping Processes," 3rd ed., p. 2, TAPPI Press, Atlanta, GA (1992).

2. **Hupa, M., Solin, P., and Hyoty, P.**, "Combustion Behavior of Black Liquor Droplets," Proc. Intl. Chem. Recov. Conf. (New Orleans), TAPPI/CPPA, 335-344 (1985).

3. **Empie, H.J., Lien, S.J., Yang, W., Samuels, D.B., and Adams, T.N.**, "Kraft Black Liquor Delivery Systems Report No. 3," U.S. DOE Contract No. DE-FC02-88CE40839, IPST (December 1991).

Characterization of Black Liquor Char Gasification Using Thermogravimetry and Molecular Beam Mass Spectrometry

K.J. Wåg, W.J. Frederick*, D.C. Dayton and S.S. Kelley****

Oregon State University, Chemical Engineering Department, Corvallis, OR

*The Institute of Paper Science and Technology, Atlanta, GA

**National Renewable Energy laboratory, Golden, CO

Black liquor char gasification was investigated in mixtures of CO_2, CO, and He using a thermobalance to characterize the gasi-'cation behavior at atmospheric pressure. A molecular beam mass spectrometer was coupled to the thermobalance in order to moni-ɔr the evolved gases. The heating rate was 10°C/min to a maximum temperature of 1000°C. The CO_2/CO ratio was varied between .1 and 10. In an effort to study the decomposition of alkali carbonate, the gasification gases were turned off at 700°C compensating ʼith Helium. At low CO_2/CO ratios, the char decomposed to 10% of its original mass (at 1000°C). However, at high CO_2/CO ratios ʼie thermograms showed a decomposition rate that was clearly slower above 800°C. This was likely a result of that more carboxylic rgano-alkali sites were preserved in the char during heat-up to 700°C. The measured content of inorganic carbonate (at 1000°C) in ʼie residue supports this conclusion. It could also be deduced from the overall reaction mechanism that CO_2 reacting with the char ıcreased the amount of oxidized catalytic sites. Hence, there are more carboxylate groups at high CO_2/CO ratios preserving inorganic arbonate. At low CO_2/CO ratios, there are less carboxylate groups allowing inorganic carbonate to decompose. The molecular beam ıass spectrometer revealed that CO was the main product gas during decomposition of organo-alkali structures and that CO_2 was trongly adsorbed during the same process.

NTRODUCTION

ʼhe study of black liquor char gasification is of interest to many ıdustrial and academic parties. Since black liquor is an important i-product and fuel in pulp and paper making, the understanding of ıe basic combustion process is a constant focus of research, due to ıe possibilities for substantial savings through design and opera-onal improvements. There are many similarities between alkali-atalyzed char combustion and black liquor char combustion, such s the mechanism for carbon removal. However, some notable dif-ɜrences exist, such as the sulfate reduction processes. These phe-omena have been studied (for black liquor char) using thermo-ravimetry in CO_2 and CO under pressurized conditions [1] but not et at atmospheric pressure. This paper is an attempt to fill this gap s well as to improve the understanding of the overall gasification ıechanism by applying thermogravimetry-mass spectrometry for ıe first time to the study of black liquor char gasification in CO_2 nd CO.

BJECTIVES

ʼhe specific objective is to obtain new data for black liquor char asification in CO_2 and CO atmospheres to improve the under-tanding of the overall char gasification process. The following uestion was particularly addressed: what is the mechanism by ʼhich alkali carbonate reduction is suppressed? This includes ıe detection of gases evolved from an atmospheric thermobalance by using a molecular beam mass spectrometer to preserve highly reactive species. It also includes obtaining the concentration of carbonate in the char residue at different gasification temperatures and gas conditions.

EXPERIMENTAL PROCEDURES

Isothermal thermogravimetric experiments have been conducted previously for gasification of black liquor below the melting point of the inorganics [2]. This paper is an attempt to characterize the gasification of black liquor char with CO_2 and CO using non-isothermal thermogravimetry both below and above the melting point of the inorganic matter. When chemical reactions are involved, useful information is obtained by coupling a thermogravimetric apparatus (TGA) to an on-line gas analyzer. This can provide accurate data on the nature and rate of evolution of decomposition products by detecting CO and CO_2 in the gases evolved. A quadrupole molecular beam mass spectrometer (MBMS) is very useful for gas analysis due to its capability to preserve and detect highly reactive and condensable gas species.

The instruments used was a Thermogravimetric Analyzer TGA 2950, commonly known as a thermobalance, manufactured by TA Instruments Inc., New Castle, Delaware. The instrument was controlled by a special operating system TA version 8.6. All thermobalances contain the following components: a specimen holder, weight measuring system, furnace, temperature program-

Table 1. TGA 2950 operating parameters [3].

Temperature range	25-1000°C
Thermocouple	Platinel II
Heating rate	0.1 to 100°C/min
Sample pan	Platinum 100 μL
Weighing capacity	1.0 g
Balance resolution	0.1 μg
Accuracy	$< \pm 0.1\%$
Purge gas	N_2
Reacting gases	CO_2 and CO
Total flow rate	100 sccm

To measure the weight loss of black liquor char samples during gasification, the sample pan was first tared and a sample of black liquor char (20 mg) was poured into it. The pan is placed in the pan hole on the sample platform. The wire in the bottom of the pan was aligned with the groove in the panhole so that the pan could be picked up by a wire hook attached to the automatic recording balance. The sample hook and pan are shown in Figure 1. The sample is weighed continuously during the experiment either isothermally as a function of time, or non-isothermally as the sample is heated to higher temperatures at a pre-set constant heating rate. In this study the experiments were done non-isothermally at a heating rate of 10°C per minute.

Calibration

Although the basic principle of thermogravimetry is simple, various tests had to be performed to ensure that precise data were obtained. Pieces of indium and aluminum were run to calibrate the operation of the thermocouple. Even though the temperature sensor was not in direct contact with the sample pan, it was assumed that the measured temperature was the same as the sample temperature.

Buoyancy Effects in TGA

A blank run with the pan only was performed to check the effects of the density change of the reacting gases during heating. The weight increase was 0.06 mg which is approximately 0.5% of the measured maximum weight change of 11.6 mg. The buoyancy effect from the change in sample size was tested. Equation 1 gives the mass of the displaced gas resulting in a decrease in weight according to Archimedes' principle that a body in a fluid is buoyed up by a force equal to the weight of the

$$m = \frac{pM\Delta V}{RT} \tag{1}$$

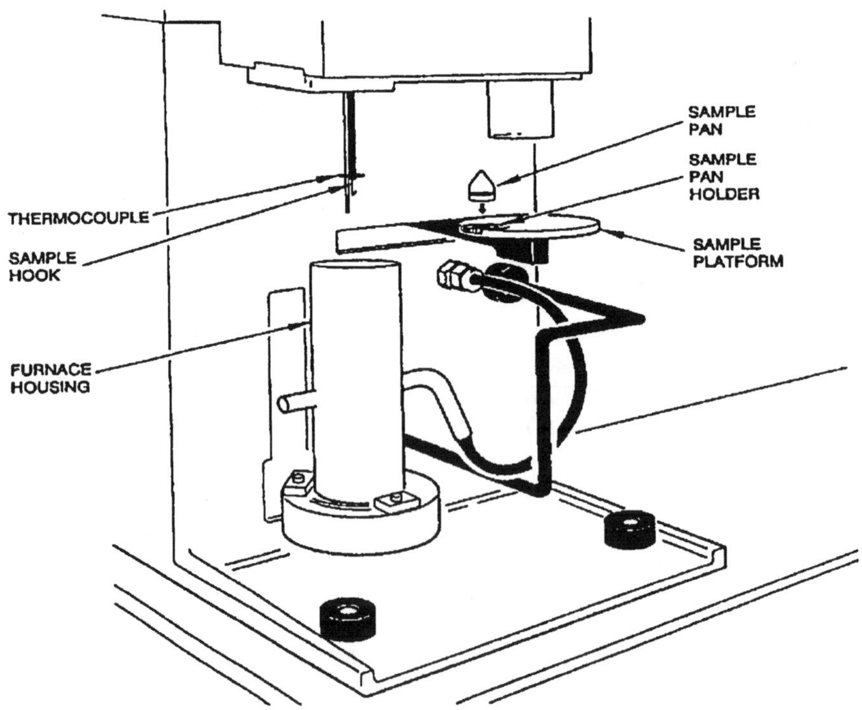

Figure 1. Sample pan, thermocouple and furnace of TGA 2950 [3]

Table 2. The elemental composition of the studied chars. Carbon data was measured by [4]. Sulfur data was taken from [5]. The valence ratio for the black liquor solids (BLS) was ~2.4 and the char ~1.4. Data are in weight-%.

	C	Na^+	K^+	$SO_4^=$	$S_2O_3^=$	$SO_3^=$	$S^=$	Cl^-	$CO_3^=$
BLS	35.0±0.1	22.7±0.1	0.62±0.1	2.4±0.1	4.45±0.1	0.0	0.0	0.67±0.1	8.1±0.1
char52	28.6±0.1	17.0±0.1	0.25±0.1	1.94±0.1	0.13±0.1	0.75±0.1	1.10±0.1	0.66±0.1	-
char70	31.1±0.1	20.0±0.1	0.43±0.1	1.72±0.1	-	-	-	0.90±0.1	16.5±0.1

quation 1 shows that the mass of the displaced nitrogen was .11 mg at 300 K and 0.27 mg at 1273 K (max. volume dis- lacement = 100 (μL). Considering the offsetting effect of the ghter gas during heat up, the maximum weight change is there- ore = 0.27 mg (decreased weight due to Archimedes' law) - .06mg (increased weight due to less dense gas) = 0.21 mg. This 1.8 % of the measured maximum weight change of 11.6 mg hich was considered negligible.

IATERIAL STUDIED

he black liquor char samples used in this study were obtained y pyrolyzing dry black liquor 90-125 (μm diameter particles in itrogen at 900°C in a laminar entrained-flow reactor (LEFR) ith a residence time of 0.8-0.87 s. A detailed description of the evice is given in reference [3]. The char samples were crushed into small particles ~10-50 microns before being placed into the respective sample pans for the TGA analyses. This procedure was necessary to have a sufficient amount of sample in the thermobalance (~20 mg). The elemental composition for the studied chars are given in Table 2. The equivalence ratios for the chars were around 1.4 and 2.4 for black liquor solids.

The BET surface area for this char was about 10 m^2/g, fairly low but typical for black liquor chars. The average pore diameter was 210-230 Å. The average pore volume was between 0.042 and 0.054 cm^3/g. An illustration of a typical molecular structure of active sites in black liquor char is shown in Figure 2.

Figure 2 illustrates the active sites that consist of free carbon sites and three types of catalytic organo-sodium moieties: carboxylate, phenolate, and a sodium atom attached to a six-atom

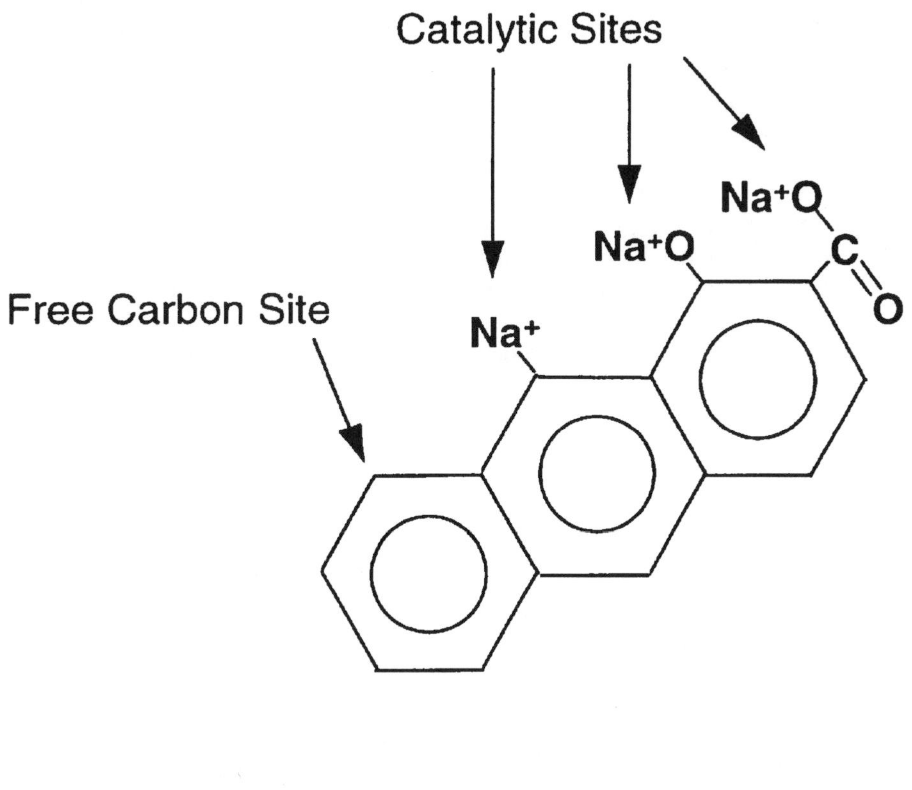

igure 2. Black liquor char contains phenolic ring-systems with free carbon sites and catalytic sites, the latter sonsisting of catalytic ·gano-sodium moities.

phenolic ring. The alkali atom could potentially be potassium, but since its concentration in black liquor char is very low compared to sodium it will merely noted that organo-potassium groups exist in small quantities in comparison to organo-sodium groups.

ASSESSMENT OF RESULTS

The thermogravimetric experiments were made at a heating rate of 10°C/min with different gas compositions, 5-50% CO_2 and CO in helium or nitrogen, for different runs. The experiments were performed at a total pressure of 1 atm with a temperature increasing from about 50°C to 1000°C. All runs were made in triplicate but some of them had to be discarded due to experimental errors such as the sample pan hitting the reactor walls. Figure 3 shows a typical experimental thermogram where the weight decreases as the temperature increases at a constant rate of 10°C/min.

In region I of Figure 3, the sample weight decreases due to drying. In region II, at about 140°C, there appears to be a shoulder that will be explained when the DSC is discussed. In region III, between 150 and 600°C, there is a weight loss of about 2 mg corresponding to about 20% of the total weight change. It is due to carbonization.

The main species found in the char residues were sodium, carbonate, sulfateIn region I of Figure 3, the sample weight decreases due to drying. In region II, at about 140°C, there appears to be a shoulder that will be explained when the DSC is discussed. In region III, between 150 and 600°C, there is a weight loss of about 2 mg corresponding to about 20% of the total weight change. It is due to carbonization, a slow release of volatile matter as well as transformation and formation of catalytic moieties before reaching temperatures where significant carbon gasification with CO_2 occurs. Beyond 800°C, the weight loss has usually leveled off. Step V is discussed in more detail in the Suppression Effects section.

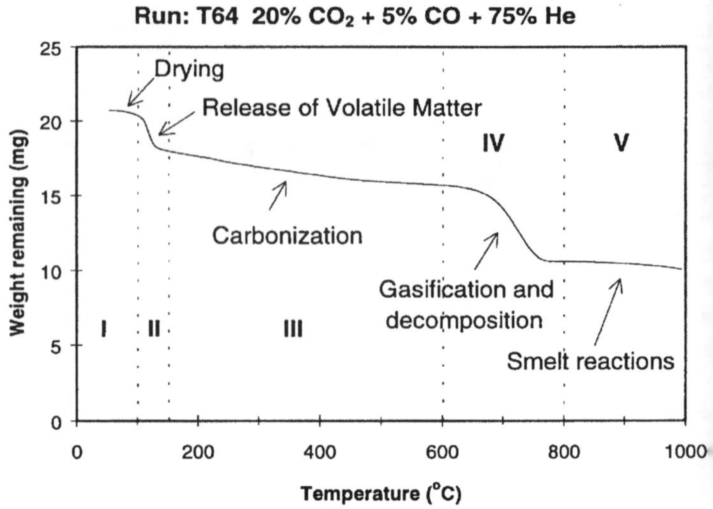

Figure 3. An example of a thermogram shows five distinguishable mass loss regions. The heating rate = 10°C per min.

The average equivalence ratio was 2.0 with a standard deviation of 0.85. An equivalence ratio of 2 means that half of the Na and K are present as acidic functional groups in the char (carboxylates and phenolates as shown in Figure 2). The balance closure was reasonable since it was not possible to account for the organic ions. The corresponding cation to anion equivalence ratio was 1.3 for the ungasified char and 2.4 for the black liquor solids. Reference [6] shows the data and the conditions at which these results were obtained. The high standard deviation was caused by a small volume of the residue solution. The reason why the equivalence ratio was ~2 is because it was not possible to measure the amount of organ-alkali species (Figure 2). This is because these groups would not form $CO_3^{=}$ ions in the dissolved solution. In effect, the carbonate measured was exclusively from inorganic salts present in the char. No sulfide was detected, because the sample was hot when exposed to the ambient atmosphere which would oxidize the sulfur to sulfate. This is the same phenomenon as found earlier [7].

Char Residue Composition and Valence Balance

The main species found in the char residues were sodium, carbonate, sulfate, potassium, chloride, and thiosulfate in the order of decreasing abundance. Reference [6] gives the concentrations of Na, K, Cl, $SO_4^{=}$, $S_2O_3^{=}$, $CO_3^{=}$, and the cation/anion equivalence ratio based on these species. The equivalence ratio was calculated using equation 2.

Carbonate Content in Char Residue

It was of interest to measure the variation of carbonate content of the char heated in 100% nitrogen. The change in carbonate content was obtained by running experiments at a constant heating rate and terminating the run at 450, 700, 800, 900, and 1000°C. When the respective temperature was reached, the experiment was terminated and the sample residue was dis-

$$\text{Equivalence ratio} = \frac{\frac{\text{ppm Na}}{23} + \frac{\text{ppm K}}{39.1}}{\frac{\text{ppm Cl}}{36.5} + 2\times\frac{\text{ppm SO}_4}{96} + 2\times\frac{\text{ppm S}_2\text{O}_3}{112} + 2\times\frac{\text{ppm CO}_3}{60}} \quad (2)$$

solved in a test tube into deionized water using an ultrasonic bath. The oxidation of the residue could not be prevented upon termination of the experiment because the TGA would always lower the reactor (at time of termination), exposing the residue to ambient air. However, the effect on the content of inorganic carbonate was small (if not negligible) because most of the organic carbon was depleted at time of termination of experiment. Assuming that the amount of carbonate did not change during cooling to room temperature and that the ion concentration of carbonate in the solution is representative of that in the solid char residue, the results obtained indicate that the carbonate concentration goes through a maximum at around 700^oC as shown in Figure 4.

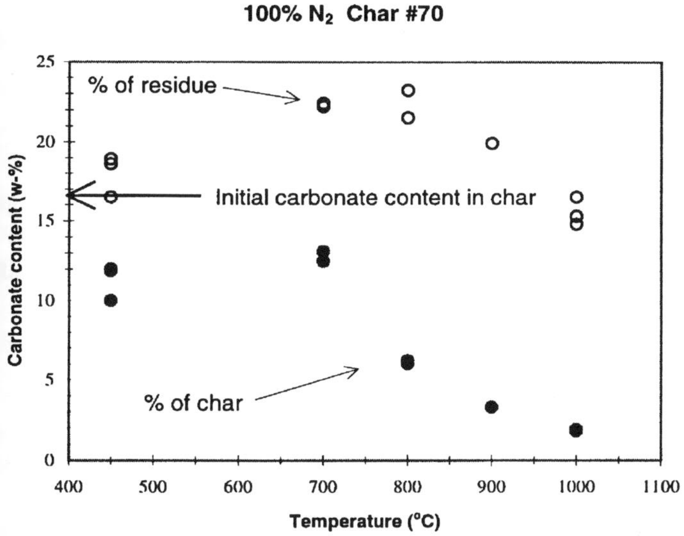

Figure 4. The concentration of carbonate goes through a maximum in the char residue. Each data point corresponds to one experiment. Each condition was repeated at least once, usually twice.

The initial carbonate content was around 16.5 w-% in the black liquor char (Table 2). Thus, the carbonate content decreased between 25 and 450^oC with respect to the initialchar mass. Intuitively one wouldnot expect that the carbonate concentration would decrease to ~11 w-% by 450^oC. This phenomenon is under investigation. The subsequent increase in carbonate content can be explained in terms of the overall gasification mechanism. Polysaccharides and lignin are the organic matter in wood that are the precursors of black liquor char. Constituents from these species decompose to form inorganic carbonate during carbonization between 450 and 700^oC. This is consistent with the results of Stewart et al [8] who showed that alkali benzoate decomposes between 400 and 600^oC to form condensed ring organic molecules, alkali carbonate, and carbon dioxide.

Mechanism for Carbonate Formation and Depletion

The reaction scheme in Figure 5 shows the proposed overall mechanism of carbonate formation and the subsequent reduction of sulfate and decomposition of carbonate. The carbonate

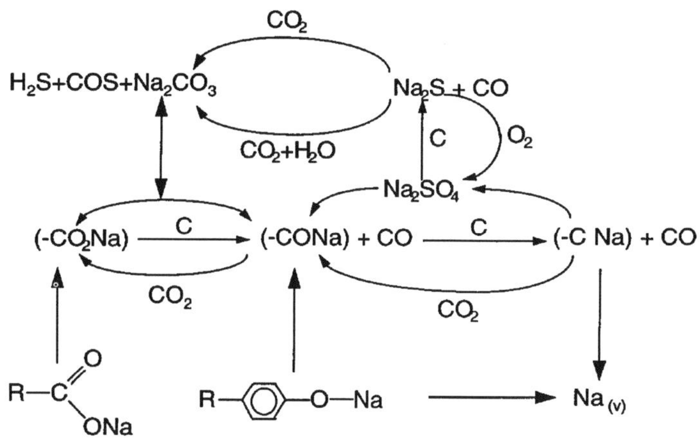

Figure 5. In the overall gasification mechanism, CO is formed from carboxylic and phenolic groups. These groups are formed from inorganic carbonate, lignin and polysac charinic precursors.

decomposition process is based on the widely accepted mechanism for catalyzed CO_2 gasification by Sams et al [9], where the active sites consist of alkali phenolates and carboxylates:)<C-Na,) >C-O-Na, and)<C=O-O-Na. These groups act as important gasification intermediates in decomposition of inorganic carbonate and in disintegration of lignin and polysaccharinic constituents. The overall scheme also includes the sulfate reduction mechanism which is another important process in black liquor char gasification processes. The sulfate-sulfide cycle as presented by Grace and coworkers [10] provides reduced sulfur species that can regenerate carbonate in the presence of CO_2 and water vapor. The major sodium and sulfur release pathways are also indicated.

The reaction mechanism in Figure 5 helps explain the data shown in Figure 4. When a temperature of around 700^oC is reached, the decomposition rate of sodium carbonate offsets the formation rate of carbonate from alkali carboxylates. At higher temperatures the decomposition rate increases faster than the formation rate. Hence, the carbonate concentration decreases which can be clearly seen in Figure 4. It is also evident that CO_2 and CO are important intermediate species in the reaction scheme in Figure 5. In fact, CO is the same product species from catalyzed CO_2 gasification. Therefore, an attempt was made to assess their role as suppression agents during decomposition of alkali carbonate.

Suppression Effects of Carbonate Decomposition in CO_2 and CO

One of the objectives was to investigate to what extent the suppression effects of $CO_2$and CO were important during decomposition of alkali carbonate. A first estimation was made by equilibrium calculations. Two cases were considered. One that had an equivalent amount of moles of compounds corresponding to the stoichiometric constants in reaction3, and the other case according to reaction 4. The Gibbs' free energy was minimized using the software HSC Chemistry [11].

$$Na_2CO_3 + C \rightarrow 2Na + CO + CO_2 \quad (3)$$

$$Na_2CO_3 + 2C \rightarrow 2Na + 3CO \quad (4)$$

Details on calculating the results shown in Figure 6 are given in reference [6].

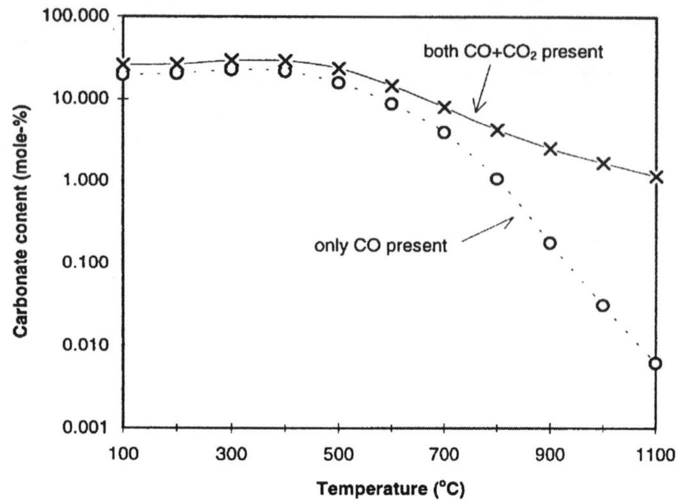

Figure 6. When both CO_2 and CO were present, the carbonate content was much higher at chemical equilibrium compared to when only CO was present. Details of the calculations can be found in reference [6]. Figure 6 can't be directly compared with Figure 4, because the former is given in mole-% and the latter in weight-%.

The equilibrium calculations showed that carbonate is much more stable when both CO_2 and CO were present according to reaction 3. If no CO_2 is present, then the carbonate is much less stable. However, equilibrium calculations do not always represent real systems especially when kinetic limitations are important. Therefore, thermogravimetric experiments were carried out in an attempt to study whether a model of a real system would behave similarly.

TGa experiments were performed at different CO_2/CO ratios at a constant heating rate of 10°C per min till 450, 700, 1000°C, when each run was terminated at the corresponding temperature. The gas concentrations were varied between 5 and 50% CO_2 with Helium as the balance. The resulting char residues were extracted with deionized water and analyzed for the carbonate ion. Further details about the analyses can be found in reference [6]. At temperatures below 600°C, the experimental carbonate contents as shown in Table 3 were ina greement with the equilibrium calculations that predicted no remarkable difference in carbonate content.

However, the experiments that were terminated at 700 and 1000°C did not have a lower carbonate content at low CO_2/CO ratio and not very dependent on the final temperature. Another way of studying the suppression effects is to compare the thermograms at different gas compositions: 100% He, 20%CO_2+5%CO, and 5% CO_2+20%CO. This is shown in Figure 7.

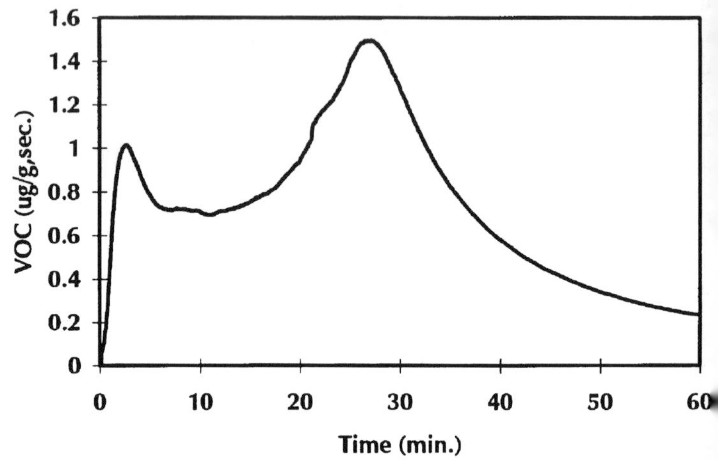

Figure 7. The thermograms in different gas compositions show that the total weight loss is much higher in 100% He compared to when CO_2 and CO were present. The reproducibility of these runs is reported in reference [6].

Figure 7 shows that the thermograms are virtually the same until 750°C after which the sample in an inert (only Helium containing) atmosphere continues to lose weight to below 10% of its original weight, leaving a light gray ash residue. The carbonate remaining in the residue was 11.9 w-%. The high CO_2 run leveled off at 50% weight remaining of the initial mass leaving a gray salt-like residue with some pink color that probably indicates sulfide. The low CO_2 run started to level off at around 45% weight remaining but starting t lose weight rapidly at 925°C. The residue was grey (residual carbon) and salt like but with no pink color. The carbonate remaining in the residue was between 22.9 and 27.7 w-% for the high and low CO_2 experiments as shown in Table 3.

Even though Figure 7 indicates that CO_2would inhibit the decomposition of organo-sodium groups more than CO, the sig-

Table 3. The carbonate content in the char residue is independent of the CO_2/CO ratio at 450, 700 and 1000°C. A complete listing a 100°C is given in reference [6].

CO_2/CO ratio	0.1	0.25	1	4	10
$CO_3^{=}$ in residence at 450°C (w-%)	21.0±0.2	-	-	-	22.2±0.2
$CO_3^{=}$ in residence at 700°C (w-%)	-	20.3±0.2	19.3±0.2	24.6±0.2	-
$CO_3^{=}$ in residence at 100°C (w-%)	-	27.0±0.2	22.9±0.2	27.7±0.2	-

nificance of the observed difference was not entirely clear. This is because it is likely that more carbon is gasified the higher the CO_2 content. The higher the CO_2 content, the faster the carbon would be depleted. In aninert atmosphere, however, there is no driving force to remove the carbon besides the carbonate decomposition process. Thus, there would be more volatilization of sodium and less inorganic matter left in the residue. Therefore, the weight remaining would be the lowest in 100% Helium.

To address this issue, several experiments were performed where the CO_2 and CO flows were turned off at 700°C and then immediately replaced with Helium to maintain a constant total flow rate. Since the total carbonate is approximately the same at 700oC in 100% Helium as when CO_2 and CO were present (as shown in Figure 6 and Table 3), then the carbonate in the residue should be the same after turning off the gasification gases at 700°C and allowing the experiment proceed to 1000°C. The result is shown in Figure 8.

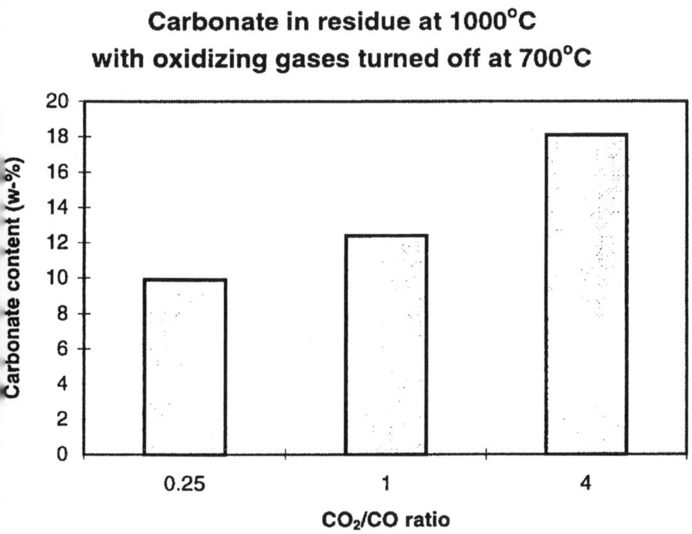

Figure 8. The carbonate content in the residue is strongly dependent on CO_2/CO ratio at 1000oC when CO_2 and CO are turned off at 700°C and switched to 100% helium.

Figure 8 shows that the carbonate in the residue is vastly different at 100°C when CO_2 and CO were turned off at 700°C. Even though the carbonate content was the same when terminating the runs at 700°C for different CO_2/CO ratios (Table 3),this does not give any information specifically about the organic carbonate (phenolates and carboxylates in char) since these species were undetectable as discussed in the ion equivalence balance section. The initial carbonate molecular structures have to be different at the respective CO_2/CO ratio at 700°C.This means that there has to be a different population distribution of the carboxylic and phenolic groups. A plausable explanation for this could be that at high CO_2/CO ratios (-CNa) shifts to (-CONa), and (-CONa) shifts to (-CO_2Na), which inhibits and inhibits the devolatilization of elemental sodium in Figure 5. The carboxylic groups have to become phenolic groups before they can decompose to CO, (-CNa), and eventually to elemental sodium. Therefore, there has to be more carboxylate groups at high CO_2/CO ratios than at low. Figure 9 shows the thermograms for the same experiments as in Figure 8.

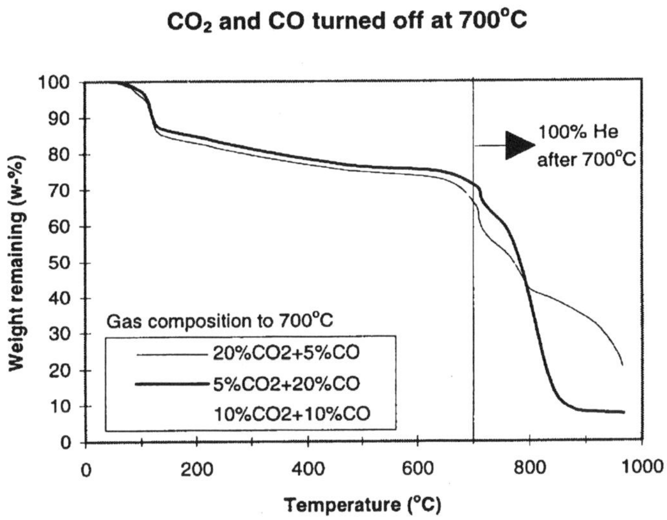

Figure 9. When CO_2 and CO were turned off at 700°C, the thermograms were different depending on the CO_2/CO ratio.

Figure 9 shows that the thermograms were different depending on the CO_2/CO ratio. At a high CO_2/CO ratio the rate of weight loss was much slower than at a low ratio. The thermograms should all be the same if the carboxylate concentration was independent of CO_2 and CO. In other words, the overall reaction mechanism (Figure 6) must be unaffected by CO_2 and CO, if the thermograms in Figure 9 were all the same. Since they are not, we conclude that there is a shift in active site distribution of carboxylate and phenolic compounds as the gas composition changes. Reference [6] includes data on the effect of carbon depletion when the gases are turned off at different temperatures.

Boudouard Equilibrium

In order to understand the overall gasification mechanism one has to consider the Boudouard equilibrium. The Boudouard equilibrium is the equilibrium of reaction 5.

$$CO_2 + C \leftrightarrow 2CO \qquad (5)$$

with the equilibrium constant defined by equation 6

$$K_{eq} = \frac{P_{CO}^{\ 2}}{P_{CO_2}} \qquad (6)$$

The procedure for calculating the chemical equilibrium is given in reference [6] and calculation details in reference [11]. The result is given in Figure 10.

Figure 10 shows that the higher the CO partial pressure, the higher the temperature for the onset of carbon decomposition. The carbon decomposition temperatures of 560 and 670°C for 5

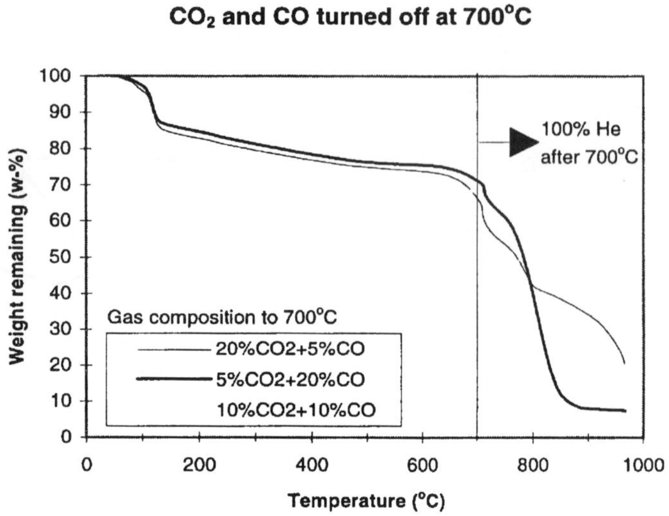

Figure 10. The Boudouard equilibrium shows that the cross over point to carbon decomposition occurs between 560 and 750°C depending on CO partial pressure.

and 20 % CO, respectively, agree well with the measured thermograms. At a fixed CO_2 partial pressure, say 0.1 bar, the higher the CO partial pressure, the higher the temperature is required for the decomposition of carbon to commence.

Summary of Suppression Effect Results

Based on these results, there is least amount of resistance to decomposition of inorganic carbonate when no CO_2 and CO are present. This is due to the fact that the carbon, necessary to reduce the inorganic matter, has been depleted faster in high CO_2 atmospheres. This effect shows up as an apparent suppression effect. In fact, the decomposition of inorganic carbonate is coupled to the overall carbon removal mechanism as shown in Figure 5. In this view, the suppression effect of CO_2 is actually a redistribution of catalytic sites from phenolic to carboxylate groups which has an impact on the rate at which carbon is removed. This hypothesis will have to be verified by experimental methods such as NMR and IR spectroscopy.

MBMS Spectra

A quadrupole molecular beam mass spectrometer (MBMS) was used to monitor the off gases to study the gasification mechanism of black liquor char. A free molecular jet beam is especially suitable when condensible gases such as elemental soldium are monitored. The thermobalance was connected with a 1 meter tygon line which was heated to 100°C. However, the highly reactive gases such as elemental sodium condensed before reaching the MBMS orifice. Therefore, only the more stable (permanent) gases were detected such as CO, CO_2, and water vapor. Figure 11 shows a typical spectrum in 100% helium. The intensity was normalized with respect to background argon (1sccm).

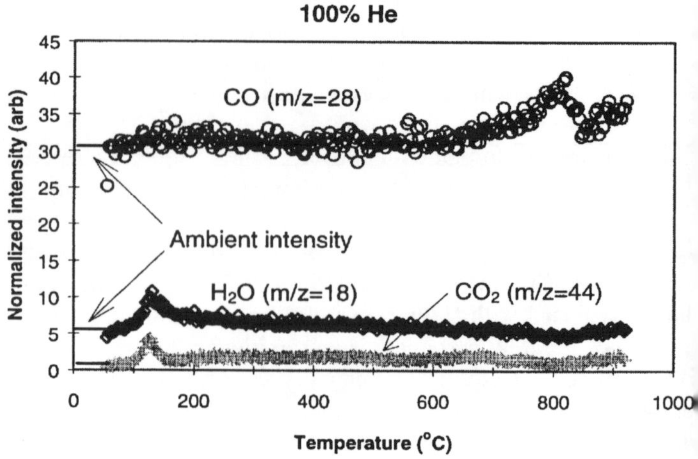

Figure 11. The MBMS spectra in 100% He show that humidit and adsorbed CO_2 are all released by 150°C. The CO peak is du to carbonate decomposition.

The data in Figure 11 shows that water evaporates right after 100°C and that some CO_2 release coincides. This could be adsorbed CO_2 from the atmosphere. The release of CO is fairly constant until a temperature of around 650°C is reached, increasing steadily at 700 and reaching its peak at 800°C after which it rapidly decreases. The data in Figure 11 shows that water evaporates right after 100°C and that some CO_2 release coincides. This could be adsorbed CO_2 from the stmosphere. The release of CO is fairly constant until a temperature of around 650°C is reached, increasing steadily at 700 and reachin its peck at 800°C after which it rapidly decreases. The release o CO at 700-8000°C verifies that CO is the main product gas during carbonate decomposition. This peak for CO releases coincides with a rapid weight loss in the thermogram in Figure 7. The question as to whether the same was true when CO_2 and CO were present was also examined. The results are shown in Figure 12.

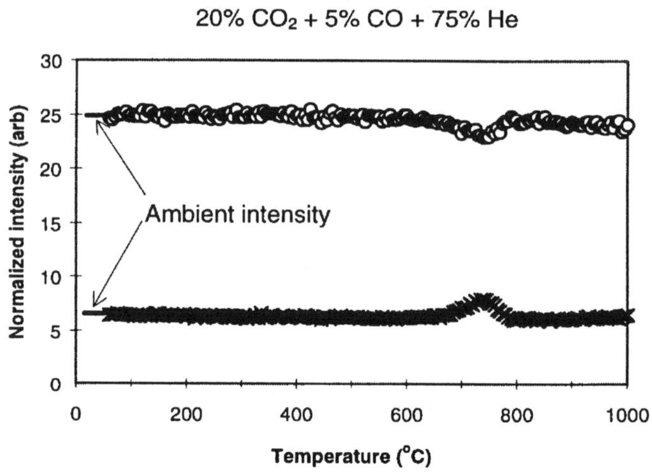

Figure 12. MBMS spectra in 20% CO_2 + 5% CO + 75% He shov that CO is released and CO_2 consumed when carbonate decomposes

Figure 12 shows that CO is generated by the gasification process and that CO_2 is consumed. The blip of CO is the product gas shown in Figure 5, and the blip of CO_2 is the reactant consumed in Figure 5. Similar results were obtained in low CO_2 to CO ratios. The release of water vapor is not shown due to the dominating signal from CO_2 and CO. The signal from water vapor is similar to that in Figure 11. The normalized intensities in Figures 11 and 12 are not directly comparable due to different gas atmospheres, i.e. the intensity of CO2 in Figure 12 is higher than for CO because there is 20% CO_2 in the reacting gases.

CONCLUSIONS

The main conclusion of this paper is that what is generally known as carbonate reduction in black liquor char gasification is really two distinct processes. One is the decomposition of inorganic alkali carbonate and the other is the decomposition of organo-alkali moieties. Since the carbon removal mechanism is part of the decomposition process, as illustrated by the overall reaction mechanism in Figure 5, one can't talk about "carbonate reduction" as an independent process from carbon removal. As a consequence, what is known as the "suppression effect in carbonate reduction" is probably mostly due to a redistribution of organo-alkali sites.

Thermogravimetric experiments with a black liquor char showed that the total weight loss was much higher in inert conditions compared to when CO_2 and CO were present. This may be due to more rapidly consumed organic carbon that is needed to deplete the inorganic matter. The weight loss history was nearly the same until 700°C was reached, at both high and low CO_2/CO ratios. Further experiments revealed that the chemical structures in the char were probably different (Figure 9). The measured inorganic carbonate is a precursor to organic carbonate: the carboxylate and phenolate moieties. At high CO_2/CO ratios the concentrations of inorganic carbonate indicated that there may have been a lot more carboxylic groups present in the char, and at low CO_2/CO ratios more phenolic groups (Figure 8). The catalytic moieties decompose in two steps: first, the carboxylate forms an oxidized phenolate, and then, the oxidized phenolate becomes a reduced phenolate giving off elemental sodium. A shift in site distribution occurs when CO_2 is reacting with the char increasing the amount of oxidized catalytic moieties. The more CO_2 is present, the faster does the increase occur.

MBMS spectra confirmed that CO was the main product gas and CO_2 the main reacting gas during decomposition of organo-alkali groups. However, in order to completely validate the proposed overall reaction mechanism, the experimental setup should be improved to better preserve condensable species, i.e. elemental sodium and potassium. Descriptions of such systems can be found in the literature [8,12].

ACKNOWLEDGMENTS

The National Renewable Energy Laboratory is acknowledged for providing the facilities and equipment necessary to obtain the data presented. The Associated Western Universities Inc. is acknowledged for financial support. Mr. Joe Patrick is acknowledged for technical support.

NOTATION

Symbol	*Description, Units*
K_{eq}	Boudouard equilibrium constant, bar
M	molecular weight, g/mol
m	mass of displaced gas, mg
p, P_i	absolute pressure of gas or partial pressure of CO_2 and CO, Pa or atm
R	ideal gas constant, J mol^{-1} K^{-1}
T	particle temperature, K
V	volume of displaced gas by TG sample pan, μL

LITERATURE CITED

1. **Frederick, W.J., Wåg, K.J., Hupa, M.,** "Rate and Mechanism of Black Liquor Char Gasification With CO_2 at Elevated Pressures," *Ind.Eng.Chem.Res.*, Vol.3, No.8, p.1747-1753 (1993).

2. **Li, J.,** "Pyrolysis and CO_2 Gasification of Black Liquor," M.S. thesis, McGill U. (1986).

3. **TA Instruments Inc.,** "Thermal Analyst 2100 Operators Manual Version 8.6," New Castle, Delaware (Sept. 1992).

4. **Weyerheuser Analytical & Testing Services,** Federal Way, WA (Sept. 20, 1994).

5. **Sricharoenchaikul, V.,** "Sulfur Species Species Transformations and Sulfate Reduction During Pyrolysis of Kraft Black Liquor," M.S. thesis, Oregon State University (1995).

6. **Wåg, K.,** "Characterization and Modeling of Black Liquor Char Combustion Processes," Ph.D. thesis, Chapters 3, 6 and Appendix 13, Oregon State University (1996).

7. **Sricharoenchaikul, V., Frederick, W. J., Kymäläinen, M., Grace, T. M.,** "Sulfur Species Transformations and Sulfate Reduction During Pyrolysis of Kraft Black Liquor," Proceedings of TAPPI-CPPA International Chemical Recovery Conference, Toronto, Canada (1995).

8. **Stewart, G.W., Chakrabarti, A., Moore, W.R.,** "Reactions of Alkali Containing Species in Combustion Streams," USDOE Proceedings: High-Temperature, High-Pressure Particulate and Alkali Control in Coal Combustion

Process Streams. Science Applications Inc., McLean, VA, p.275-300 (1981).

9. **Sams, D.A., Shadman, F.,** "Mechanism of Potassium Catalyzed Carbon/CO_2 Reaction," *AIChE J.*, Vol.37, No.7, p.1132-1137 (1986).

10. **Grace, T.M., Cameron, J.H., Clay, D.T.,** "Char Burning Project 3473-6," Summary Technical Report to American Paper Institute Recovery Boiler Committee, (Feb. 22, 1985).

11. **Outokumpu Research Ltd.,** "HSC Chemistry for Windows Version 2.03," Pori, Finland (1994).

12. **Behrens, R., Jr.,** "New Simultaneous Thermogravimetriy and Modulated Molecular Beam Mass Spectrometry Apparatus for Quantitative Thermal Decomposition Studies," *Rev.Sci.Instrum.*, 58(3), Am. Inst. of Physics (March 1987).

A Fundamental Study of Carryover Deposition in Kraft Recovery Boilers

Ted Mao, David C.S. Kuhn and Honghi Tran
Pulp & Paper Centre, and Department of Chemical Engineering and Applied Chemistry
University of Toronto, 200 College St., Toronto, M5S 1A4, Canada

The tendency of molten carryover droplets to be captured by convective heat transfer surfaces upon impact determines the carryover fouling rate in kraft recovery boilers, but little information is available to predict this tendency. This work is the first to study the dynamic capture of liquid droplets upon impact on solid surfaces as applied to molten carryover deposition.

Deformation, spread, recoil, and/or rebound of a droplet upon impact on solid surfaces were quantitatively studied using a fast shutter-speed Charge Coupled Device camera. The static contact angle and viscosity were found to have the strongest effect on the tendency of a droplet to deposit or to rebound. The tendency to rebound increases with an increase in the contact angle, but decreases with increasing viscosity.

A semi-empirical spread model was developed to predict the maximum spread as a function of the Reynolds number, the Weber number and the static contact angle. A rebound model was then developed to predict the tendency of a liquid droplet to either deposit or rebound as a function of maximum spread and static contact angle. Both model predictions agree well with experimental results.

INTRODUCTION

Carryover deposition can cause severe fouling of convective heat transfer surfaces in kraft recovery boilers. The accumulation of fireside deposits reduces heat transfer efficiency, impedes flue gas flow, corrodes tube surfaces, and in severe cases, leads to complete plugging and an unscheduled shutdown of the boiler.

Carryover deposition is determined by two processes: (i) particle transport from the flue gas to heat transfer surfaces, and (ii) particle capture by these surfaces upon impact. The former, which is related to aerodynamic effects, has been studied extensively [1-4]. Carryover is transported to heat transfer surfaces by inertial impaction. The latter, which is related to particle material properties and impact dynamics, has been investigated only to a limited extent focusing mostly on micron size solid particles [5-8]. The capture of molten carryover upon impact has not been studied before, and its deposition tendency determines the carryover fouling rate for a given carryover load.

In a kraft recovery boiler, because of its hostile environment and highly unstable conditions, it is extremely difficult to study systematically the dynamic capture of carryover in an actual boiler. Since the behavior of a molten carryover droplet as it strikes a surface is governed by the physical principles of liquid droplet impact, liquid droplets analogous to molten carryover were used to investigate the impact dynamics of carryover. The impact of liquid droplets on solid surfaces has been studied in a wide variety of applications such as ink jet printing, spray coating, precision solder dispensing and spray cooling [9-16]. However most prior work only concerns spreading, and few systematic experimental studies were carried out to investigate the capture and rebound phenomena. Therefore, the present work aims at characterizing the dynamic capture behavior of liquid droplets upon impact on solid surface and to develop a model to predict their deposition or rebound which can be then used to calculate the tendency of molten carryover to either stick or rebound in kraft recovery boilers.

EXPERIMENTAL SETUP AND PROCEDURE

The experimental setup for the droplet impact study is shown schematically in Figure 1. A droplet was generated by forcing

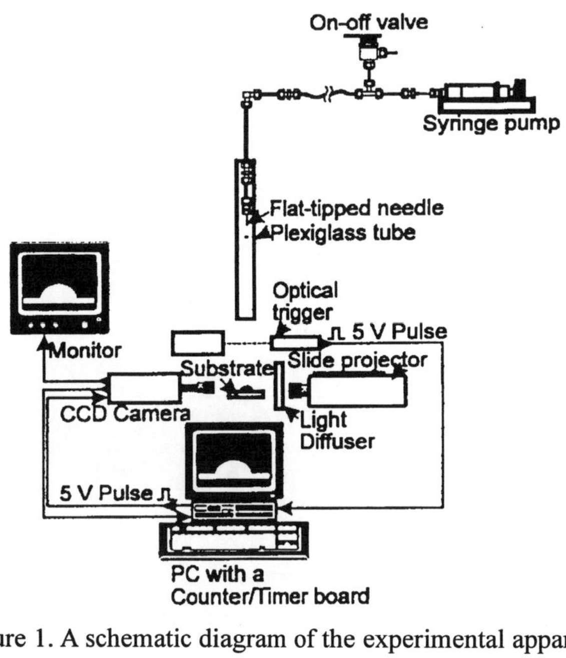

Figure 1. A schematic diagram of the experimental apparatus of droplet impact.

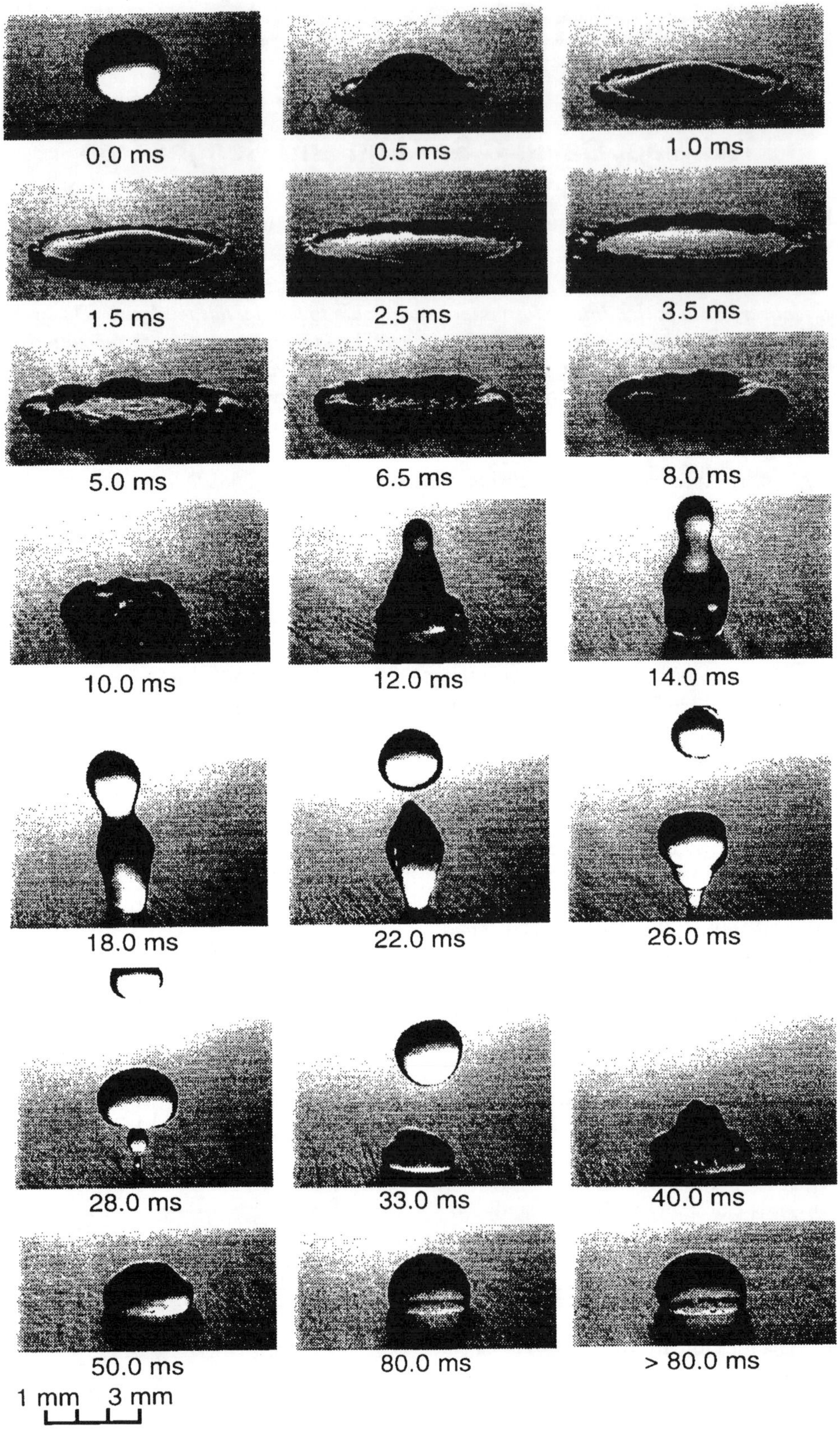

Figure 2. Impact sequence of a 2.7 mm water droplet on a paraffin wax surface at a velocity of 1.58 m/s.

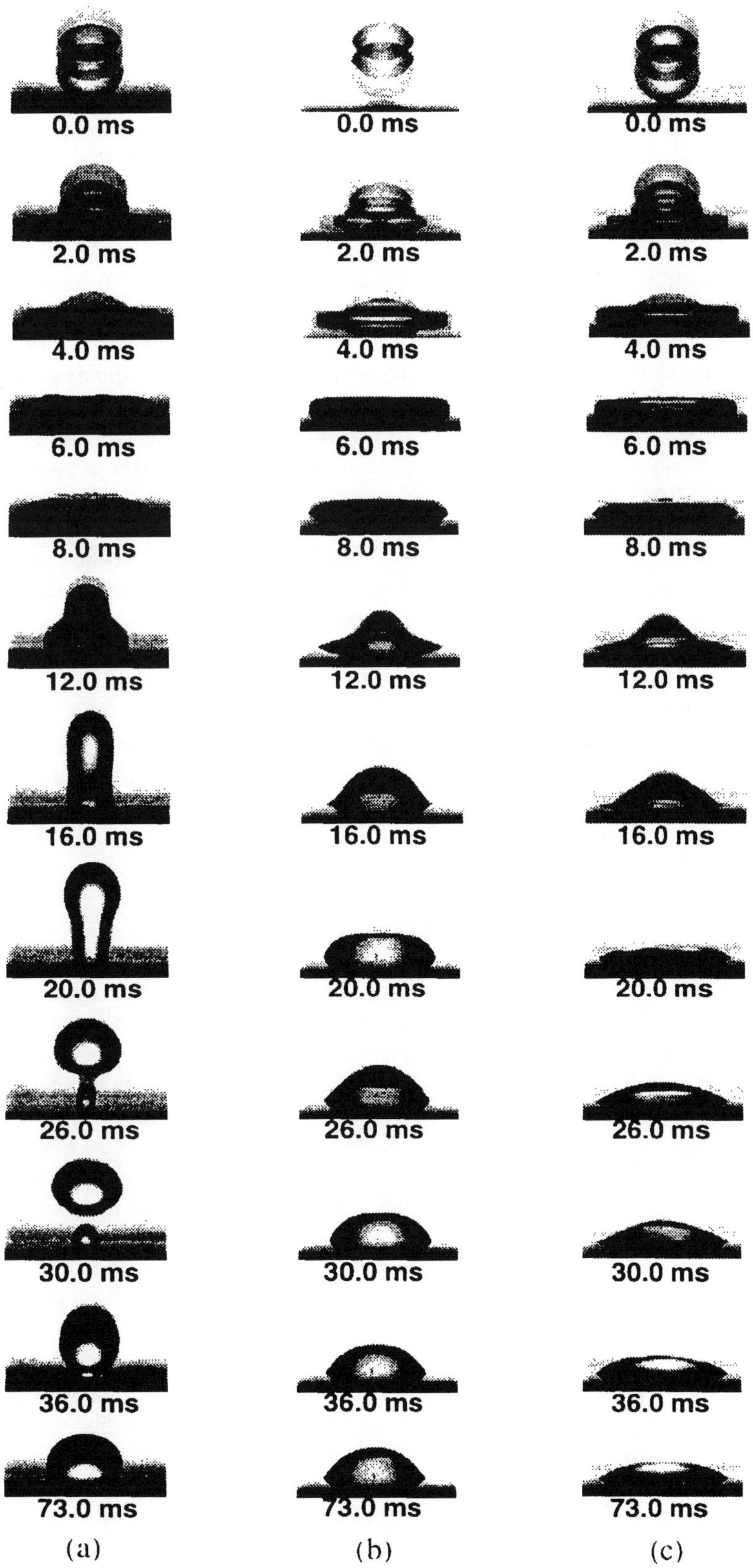

Figure 3. Impact of a 2.7 mm water droplet on (i) the paraffin wax block; (ii) the stainless steel plate; (iii) the glass slide.

the test fluid through a flat-tipped stainless steel needle (25G5/8) using a syringe pump (Model 100, KD Scientific Corp.). The liquid flow rate was adjusted so that individual droplets first formed at the tip of the needle, then detached and fell due to the force of gravity. The droplet was enclosed in a vertical Plexiglass tube with an outer diameter of 50.8 mm (2 in) to protect it from air currents. Before a droplet struck a substrate surface, it moved through an infrared beam between the emitter and the detector of an optical trigger (MAZOF VIS II Trigger System). The 5 V pulse generated by the trigger went to a timer/counter board (CIO-CTRO5, ComputerBoards, Inc.) in an IBM compatible personal computer, which sent a 5 V pulse to activate the shutter of a Charge Coupled Device (CCD) camera after a preset time delay. The time delay was varied so that images were taken at different times after the initial droplet impact. The images were stored in a digital format and downloaded to the computer for analysis. A sequence of pictures describing the impact process was created from individual images of many droplets taken at different time delays.

The fast-shutter CCD camera (FlashCam, Optikon Corp.) has a multiple exposure feature and is capable of taking pictures at shutter speeds as fast as 1µs. Up to 10 images can be superimposed on a single frame with a time delay of 1µs to 1 ms between individual exposures. Using the multiple exposure feature, droplet impact velocity and the rate of deformation upon impact were precisely determined.

Mixtures of water and sucrose were used as test fluids. The liquid viscosity was measured using a HAAKE Rotovisco RV12 viscometer. Its density was determined by the weight of a volumetric flask.

Three types of substrates were used throughout the study: a smooth glass slide, a polished stainless steel plate and a paraffin wax block. The stainless steel plate was first polished using 600 grit emery cloth and then brought to a mirror finish with a 6µm Leco premium diamond suspension (Leco Corp.). The paraffin wax block was polished using 600 grit emery cloth.

RESULTS AND DISCUSSION

Impaction Process

Figure 2 shows the impact sequence of a 2.7 mm water droplet on a paraffin wax surface at room temperature. The camera is aimed 15° below the horizontal in order to obtain clear images of the droplet spread upon impact. The impact velocity was 1.58 m/s with an associated droplet Reynolds number, $Re = \rho uD/\mu$, of 4257, and a Weber number, $We = \rho u^2 D/\gamma_{LV}$, of 92. The static contact angle was 97°.

At t = 0.0 ms, just before impact, the droplet is clearly spherical in shape. It deforms and spreads rapidly upon impact in the radial direction. A raised rim is formed at the periphery at t = 0.5 ms due to surface tension forces which limit the spread and decelerate the radial flow close to the periphery. After t = 2.5 ms, the rate of spread slows down, and the central liquid pool thins out and continues to feed liquid into the rim, which forms a series of liquid lobes. At t = 3.5 ms, the droplet reaches its maximum spread and is momentarily at rest because of the depletion of kinetic energy. The initial impact energy of the droplet is dissipated in overcoming viscous flow and in producing new surface area.

After the droplet reaches maximum spread, pairs of liquid lobes at the periphery join, and a 10-lobed annulus is formed. This moves back toward the centre, i.e., recoils (t = 5.0 ms), as a result of the large surface energy of the droplet-solid system. At the same time, concentric ripples propagate inward on the surface of the central lake of liquid. From t = 5.0 to 10.0 ms, the area of liquid-solid contact shrinks rapidly causing the lobes in the rim to merge and the rim to swell, and to move inward until the rim reaches the center. This marks the beginning of the rebound of the droplet from the solid surface. At t = 12.0 ms, the liquid rises up from the centre and assumes a roughly conical shape as the area of liquid-solid contact further decreases. At t = 22.0 ms, the tip of the cone detaches and forms a separate subdroplet. Finally, the contact area shrinks to zero as the whole droplet rebounds from the wax surface (t = 28.0 ms).

Under the influence of gravity, the droplets fall back to the wax surface, and further droplet spread and recoil sequences occur with decreasing amplitude and frequency for approximately 40 ms until the droplet excess energy is dissipated. Equilibrium is finally achieved at t ~ 80.0 ms.

Effect of Substrate Surface

Figure 3 shows the effect of the substrate surface on the impact process. Separate 2.7 mm water droplets impact on a paraffin wax block, a stainless steel plate and a glass slide at a velocity of 0.85 m/s. The camera is aligned to the horizontal, and is set to take three exposures of a droplet upon impact.

From 0.0 ms to 6.0 ms, the droplet deformation is similar in these three cases. Each reaches the maximum spread at t = 6.0 ms, and each attains almost the same radius. Therefore, the deformation and spreading of the droplet from the point of impact to the maximum spread is dominated by inertial and viscous forces, and the effect of the substrate is minor.

In Figure 3, after t = 6.0 ms, the flattened droplet starts to recoil. The droplet on the wax surface recoils much faster than the droplets on the other two surfaces. After t = 12.0 ms, the water droplet bounces off the wax surface. In the case of the stainless steel substrate, only spread-recoil oscillations are observed, and in the case of the glass substrate, there are smaller and less frequent spread-recoil oscillations.

In the equilibrium state, the shape of the liquid meniscus is different for different substrates: the static contact angles of the water droplet on the wax, stainless steel and glass surfaces are 97°,

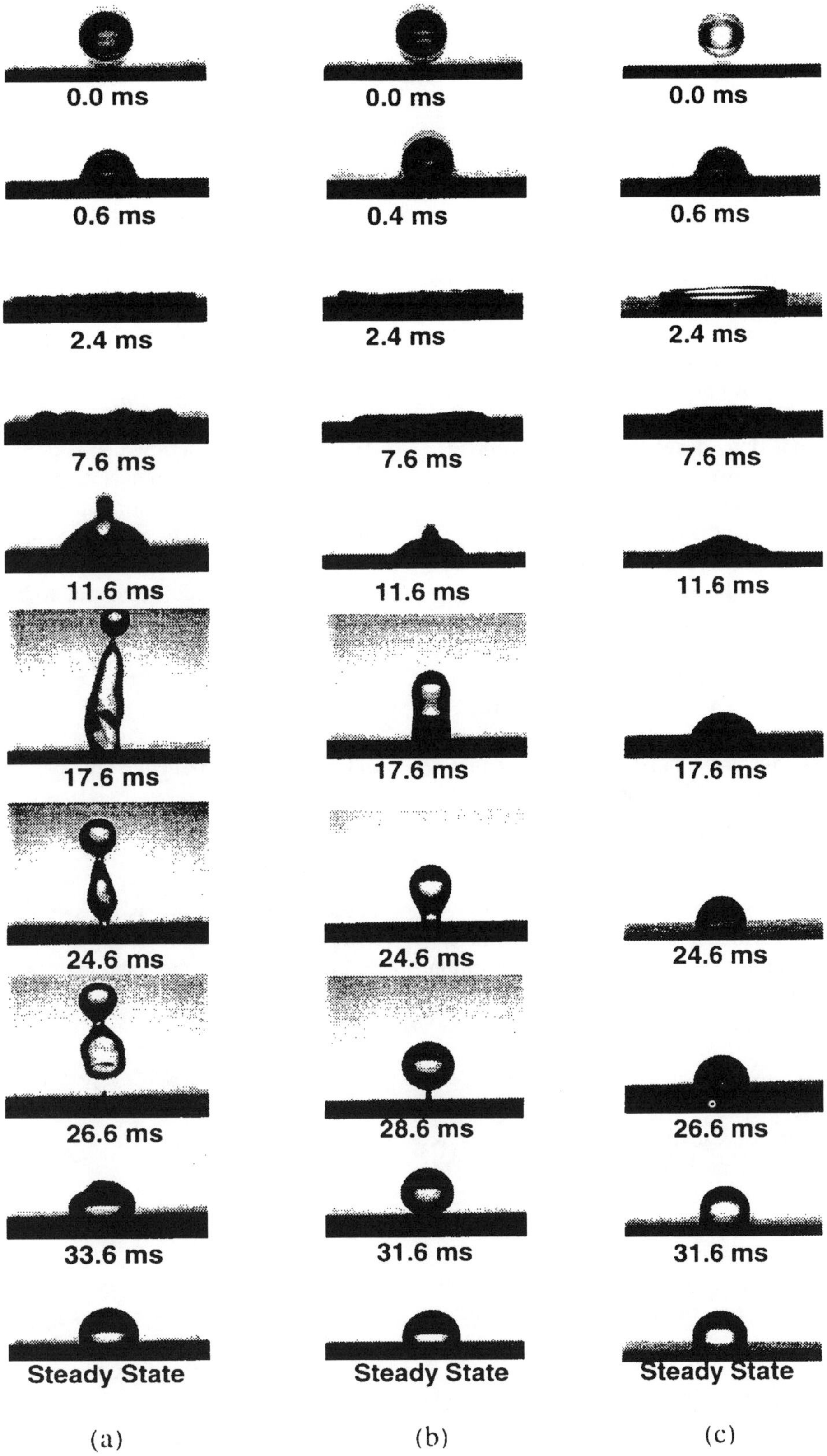

Figure 4. Impact of (i) a water droplet (1.0 mPa ·s), D=2.7 mm, U_{Imp}=2.0 m/s; (ii) a 20% sucrose solution droplet (2.0 mPa ·s), D=2.6 mm, U_{Imp}=2.0 m/s; (iii) a 50% sucrose solution droplet (16.4 mPa ·s), D=2.5mm, U_{Imp}=2.0 m/s on the paraffin wax surface.

67° and 37°. The tendency to rebound increases with increasing contact angle value.

Effect of Viscosity

Droplet viscosity is an important variable affecting the impact process. In Figure 4, droplets with viscosities of 1.0 mPa ·s, 2.0 mPa ·s and 16.4 mPa ·s impact on the paraffin wax surface at a velocity of 2.0 m/s. No significant difference is seen during the initial rapid deformation stage (t < 0.6 ms) because the inertial force of impact is much larger than the viscous force. However, as the droplets approach their maximum spread and the rate of spread decreases, the viscous effect begins to play a dominant role. The maximum spread for a given impact energy decreases with increased viscosity, as shown in Figure 4 at t = 2.4 ms.

A droplet with higher viscosity produces smaller maximum spread; it thus has less energy available for recoil and/or rebound. Moreover, its excess surface energy must overcome higher viscous dissipation during the recoil and/or rebound. The tendency of droplets to recoil and rebound is therefore less for higher viscosity liquids than for lower viscosity liquids. Figure 4 shows that there is significant rebound in the case where the droplet viscosity is 1.0 mPa ·s, whereas the droplet with a viscosity of 2.0 mPa ·s only just bounces off the surface. In contrast, the droplet with a viscosity of 16.4 mPa ·s undergoes limited recoil and no rebound. At the end of the impact process, the droplets in all three cases have approximately the same contact angle on the paraffin wax surface. Therefore, the tendency to rebound decreases with increasing viscosity.

Theoretical Analysis

The impact process can be divided into the four distinct stages, as shown schematically in Figure 5. Based on the conservation of energy between stages (a) and (b), and stages (b) and (c), models of maximum spread and rebound were developed.

Stage (a), Before Impact, e.g., t = 0.0 ms in Figure 2

Just before impact, the droplet possesses both kinetic energy, K_a, and total surface energy, S_a, but negligible potential energy [14]. Because the solid surface energy is the reference and is set to zero, the total surface energy consists of only the liquid-vapour surface energy, $S_{a,LV}$, i.e., the product of the droplet surface area, A, and the liquid-vapour surface tension, γ_{LV}. The total impact energy, E_{Impact}, is the sum of the kinetic energy and the surface energy:

$$E_{\mathrm{Im}\,pact} = K_a + S_a \tag{1}$$

$$K_a = \left(\frac{1}{2}pu^2\right)\left(\frac{1}{6}\pi D^3\right) \tag{2}$$

$$S_a = S_{aLV} = A\gamma_{LV} = \pi D^2 \gamma_{LV} \tag{3}$$

where ρ is the liquid density, u is the impact velocity, and D is the droplet diameter.

Stage (b), Maximum Spread, e.g., t = 3.5 ms in Figure 2

This is the point at which the liquid flow changes direction from spreading outwards to recoiling inwards. The surface energy of the droplet, S_b, is at a maximum while the kinetic energy, K_b, is zero. The potential energy is negligible compared with other forms of energy [14]. The total energy therefore is equal to the surface energy:

$$E_b = S_b. \tag{4}$$

A characteristic length which is proportional to the surface energy is the liquid-solid contact diameter, d_m, i.e., the maximum spread diameter, and is normalized with respect to the original droplet diameter, D, to give a maximum spreading factor, d_m/D.

The flattened droplet is assumed to be a thin circular disk with a top surface area of

$$\frac{\pi}{4}D_M^2$$

and a circumferential edge area of

$$\frac{2}{3}\pi\frac{D^3}{D_M}$$

The surface energy, S_b, is the sum of the liquid-vapour surface energy, $S_{b,LV}$, and the energy of the new solid-liquid surface, $S_{b,SL}$, minus the solid-vapour surface energy, $S_{b,SV}$, which has been lost in the process:

$$S_b = S_{b,LV} + S_{b,SL} - S_{b,SV} = \left(\frac{\pi}{4}d_m^2 + \frac{2}{3}\pi\frac{D^3}{d_m}\right)\gamma_{LV} \tag{5}$$

$$+\frac{\pi}{4}d_m^2(\gamma_{SL} - \gamma_{SV}),$$

where γ_{SL} is the solid-liquid surface tension and is the solid-vapour surface tension.

Because the values of γ_{SL} and γ_{SV} are not readily available, Young's equation [17] was used to relate their difference, γ_{SL} - γ_{SV}, to γ_{LV} and the static contact angle θ :

$$\gamma_{LV}\cos\theta = \gamma_{SL} - \gamma_{SV}. \tag{6}$$

Substituting Eq. 6 into Eq. 5 gives:

$$S_b = \left[\frac{\pi}{4}d_m^2(1-\cos\theta) + \frac{2}{3}\pi\frac{D^3}{d_m}\right]\gamma_{LV} \tag{7}$$

The energy conservation between stage (a) and stage (b) can now be established as:

$$K_a + S_a = S_b + Diss_{a-b,} \qquad (8)$$

where $Diss_{a-b}$ is the energy consumed by viscous dissipation of the droplet in going from stage (a) to stage (b).

As shown above, K_a, S_a and S_b have been defined with an unknown, d_m. Thus, Eq. 8 can be used to solve for d_m if $Diss_{a-b}$ is quantified.

An energy dissipation model is developed using the boundary layer theory of the stagnation point flow with empirical constants obtained by correlating the present experimental data to give the dissipation energy normalized with respect to the original droplet surface energy:

$$\frac{Diss_{a-b}}{\pi D^2 \gamma_{LV}} = 0.2 \frac{We^{0.83}}{\mathrm{Re}^{0.33}} \left(\frac{d_m}{D} \right)^2 . \qquad (9)$$

The resulting equation used to calculate dm/D is:

$$\left[\frac{1}{4}(1-\cos\theta) + 0.2 \frac{We^{0.83}}{\mathrm{Re}^{0.33}} \right] \left(\frac{d_m}{D} \right)^3 - \left(\frac{We}{12} + 1 \right) \left(\frac{d_m}{D} \right) + \frac{2}{3} = 0 \qquad (10)$$

where d_m/D has an exact solution, since for an equation of the form $x^3 + px + q = 0$, the real root is given by: .

$$x = \sqrt[3]{-\frac{q}{2} + \sqrt{\left(\frac{q}{2}\right)^2 + \left(\frac{p}{3}\right)^3}} + \sqrt[3]{-\frac{q}{2} - \sqrt{\left(\frac{q}{2}\right)^2 + \left(\frac{p}{3}\right)^3}}$$

Figure 6 compares the model predictions (Eq. 10) with the present experimental measurements, and the agreement is to within

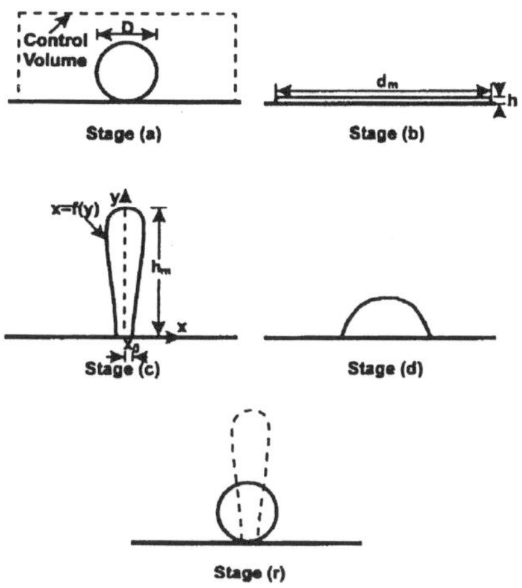

Figure 5. Four stages of the impact process and one rebound criterion.

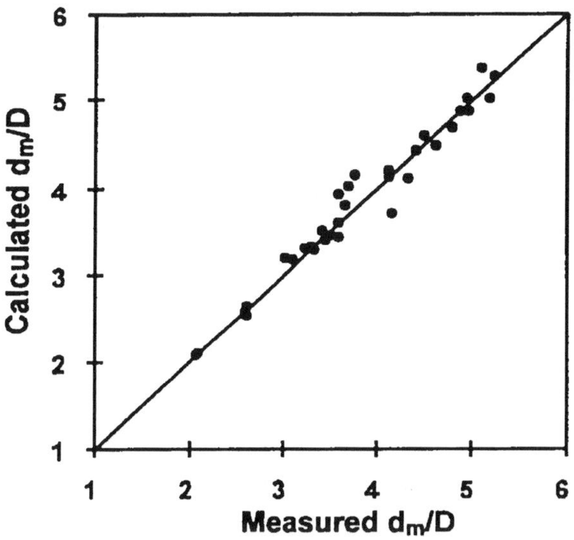

Figure 6. Comparison of the calculated maximum spread using the present model (Eq. 10) with the experimental data in this study.

10% over a range of impact velocities (1 - 6 m/s), viscosities (1 - 100 mPa ·s), static contact angles (30 - 120°) and droplet sizes (1.5 - 3.5 mm). Figure 7 compares the calculated maximum spread using the present model with the literature data over a wide range of impact velocities (1.4 - 28 m/s), viscosities (1 - 296 mPa ·s), surface tensions (25 - 73 mJ/m^2), static contact angles (34 - 100°) and droplet sizes (0.04 - 3.7 mm). The predictions agree with more than 90% of the available literature data to within 10%. Based on the predicted maximum spread value, the energy in stage (b), i.e., the maximum energy available for recoil and/or rebound, can be calculated using Eq. 7. This is the basis for the development of a rebound model.

Equation 10 shows that the maximum spread is a function of the Reynolds number, the Weber number, and the contact angle. The predicted effects of the Reynolds number and the Weber

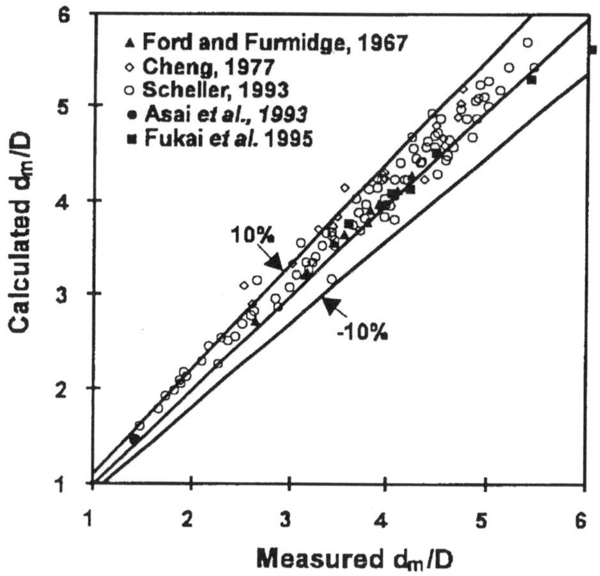

Figure 7. Comparison of the calculated maximum spread using the present model (Eq. 10) with the experimental data in the literature.

number using Eq. 10 are illustrated in Figure 8 for the contact angle of 90°. As the Reynolds number and the Weber number increase, the maximum spread increases. Since impact energy increases with increasing Reynolds number and Weber number, a higher impact energy must lead to a greater maximum spread.

Stage (c) Maximum Recoil/Rebound, e.g., t = 28.0 ms in Figure 2

At this moment, the droplet changes its direction of motion from up to down under the influence of gravity. This stage signifies the maximum extent to which a droplet recoils upwards, and the droplet possesses potential and surface energy but no kinetic energy.

Stage (c) defines the maximum extent to which a droplet recoils or bounces. At this moment, since the droplet is assumed to be at rest and the kinetic energy, K_c, is zero, only the potential energy, P_c, and the surface energy, S_c, are considered, so the total energy $E_c = P_c + S_c$.
It is assumed that the droplet in this stage is a body of revolution about the y-axis. Thus, the potential energy, P_c, and the surface energy, S_c, are:

$$P_c = \pi \rho g \int_o^{h_m} y \cdot x^2 dy \tag{11}$$

$$S_c = 2\pi \gamma_{LV} \int_o^{h_m} x \cdot \sqrt{1 + [f'(y)]^2}\, dy - \gamma_{LV} \pi x_o^2 \cos\theta \tag{12}$$

where , and is the first order derivative with respect to y, see Figure 5. However, the geometric configuration of the liquid body in stage (c) is not well defined, and thus there is no analytical expression for x = f(y). In this study, P_c and S_c are obtained by numerical integration of the digital profile of a droplet in stage (c) taken using the fast shutter-speed camera.

Stage (d), Equilibrium, e.g., t > 80.0 ms in Figure 2

The droplet possesses a minimum energy which is equal to the static surface energy.

Stage (r), Rebound Criterion

This is a fictitious configuration of stage (c) which is used as a measure of rebound. After the deformation and flattening of a droplet upon impact, it is assumed that the excess surface energy enables the droplet to recover its spherical shape and its original diameter. The droplet is just above the solid surface and is considered to be momentarily at rest. Therefore, an energy balance involves the potential energy, P_r, and the surface energy, S_r, as follows:

$$P_r = \frac{\pi}{12} \rho g D^4 \tag{13}$$

$$S_r = S_{r,LV} = \pi D^2 \gamma_{LV} \tag{14}$$

Since the potential energy as defined by Eq. 13 never exceeds 2% of the total energy within the broad parametric bounds of the experiments, it is neglected in the development of the rebound model. Therefore, the total energy in this stage reduces to:

$$E_r = S_r = \pi D^2 \gamma_{LV}. \tag{15}$$

If the energy possessed by a droplet in stage (c), E_c, is greater than in stage (r), E_r, the droplet will further pop up and be swept away by the strong shear force of the flue gas in the context of fouling on heat exchanger tubes by molten droplet impact, i.e., rebound occurs when $E_c > E_r$. Since the geometry of a droplet in stage (c) is complex, it is difficult to obtain E_c directly. But the energy in stage (b), E_b, can be calculated using Eq. 4 with the maximum spread prediction. Therefore, E_c can be related to E_b and $Diss_{b-c}$ from energy conservation considerations between stages (b) and (c):

$$E_c = E_b - Diss_{b-c}, \tag{16}$$

where $Diss_{b-c}$ is the energy dissipated between stages (b) and (c) as the droplet undergoes recoil and/or rebound. Consequently, a rebound criterion can be expressed as:

$$E^*_{ERE} = \frac{E_c - E_r}{E_r} = \frac{E_b - Diss_{b-c} - E_r}{E_r} > 0, \tag{17}$$

where E^*_{ERE} is defined as the excess rebound energy normalized with respect to the energy possessed in stage (r); and E^*_{ERE} describes the tendency of a droplet to rebound upon impact. A droplet remains on the surface when $E^*_{ERE} \geq 0$.

The following empirical correlation of $Diss_{b-c}$ as a function of the maximum spread d_m/D and the static contact angle was developed based on the experimental impact data:

$$\frac{Diss_{b-c}}{E_r} = \frac{E_b - E_c}{E_r} = 0.12\left(\frac{d_m}{D}\right)^{2.3} (1 - \cos\theta)^{0.63} \tag{18}$$

Substituting Eqs. 4, 7, 15 and 17 into Eq. 18, results in the proposed rebound model:

$$E^*_{ERE} = \frac{1}{4}\left(\frac{d_m}{D}\right)^2 (1 - \cos\theta) - 0.12\left(\frac{d_m}{D}\right)^{2.3} (1 - \cos\theta)^{0.63} + \frac{2}{3}\left(\frac{D}{d_m}\right) - 1 \tag{19}$$

Figure 9 shows the excess rebound energy, E^*_{ERE}, as a function of maximum spread, d_m/D, and contact angle, . When the contact angle is less than 90°, the excess rebound energy decreases

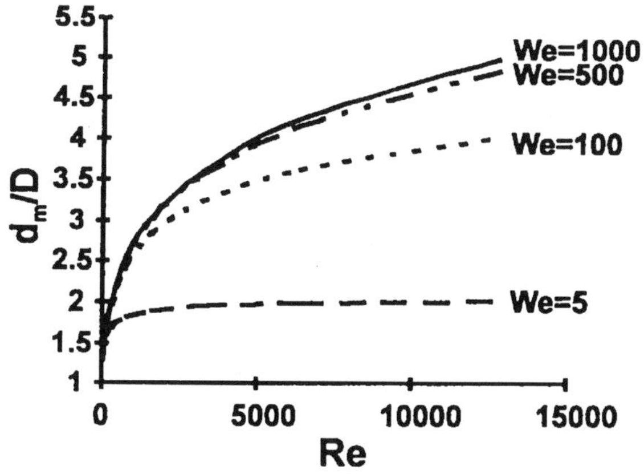

Figure 8. Variation of maximum spread with Re and We.

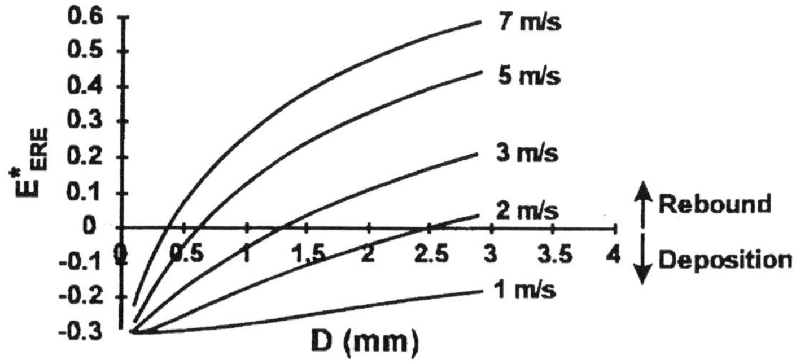

Figure 10. Prediction of the deposition and rebound of molten carryover.

with increasing maximum spread. The more a droplet spreads, the less is the tendency for it to rebound, because more energy is dissipated by viscous flow than is generated by excess surface as a result of droplet deformation. However, when the contact angle is greater than 90°, the tendency of droplet rebound increases as the maximum spread increases and the contact angle increases. This is confirmed by the current experimental observations.

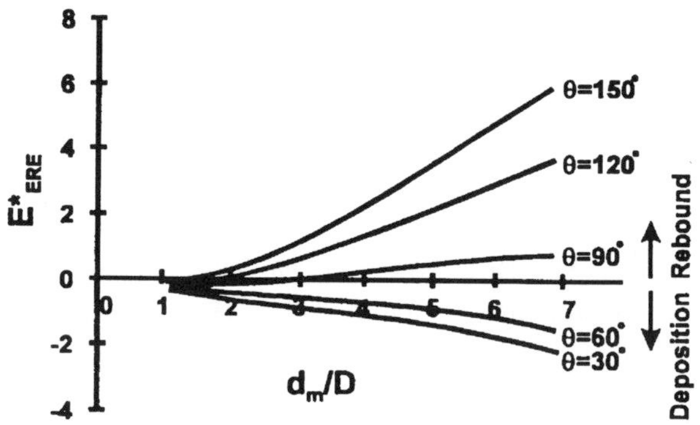

Figure 9. Excess rebound energyas a function of the maximum spread and the static contact angle, using Eq. 19.

Application

The tendency of a molten carryover droplet to either deposit or rebound can be predicted using the rebound model, as shown in Figure 10. Typical physical properties of molten carryover at the entrance to the superheater are used in the calculation: density is 2000 kg/m^3; viscosity is 4.0 mNs/m^2; surface tension is 160 mN/m; and the static contact angle is assumed to be 90°. In this specific case, the extent of rebound increases with increasing carryover diameter and impact velocity. The critical carryover diameter, above which carryover bounce off, decreases as impact velocity increases. For low impact velocity, e.g., 1 m/s, carryover particles stick on the surface when they hit. This indicates that smaller carryover particles with lower impact velocities are more likely to deposit on heat exchanger surfaces.

The rebound model provides a solid base for predicting the dynamic capture of molten carryover upon impact in kraft recovery boilers.

CONCLUSIONS

Parametric studies were conducted of liquid droplets impacting on solid surfaces; this process is analogous to the inertial impaction of molten carryover on heat exchanger surfaces in kraft recovery boilers.

- Four stages characterize the liquid droplet impact: (a) before impact, (b) maximum spread, (c) recoil/rebound, and (d) final spread.
- Liquid viscosity and impact velocity are the most important parameters which affect the maximum extent to which a droplet spreads. Viscosity and static contact angle play dominate roles in determining the tendency of rebound of a droplet upon impact.
- A semi-empirical model was developed to predict the maximum spread as a function of the Reynolds number, the Weber number and the static contact angle. The maximum spread increases with the Reynolds number and the Weber number, and the effect of the contact angle is secondary. Predicted results agree with experimental data to within 10%. This precision of calculating the maximum spread makes it possible to develop a rebound model since a droplet possesses the maximum surface energy for recoil and rebound at its maximum spread.
- A rebound model was formulated to predict the tendency of the droplet to bounce upon impact. This is a strong function of the maximum spread and the static contact angle.
- The rebound model can be used to predict the tendency of molten carryover to stick to tube surfaces upon impact in kraft recovery boilers.

ACKNOWLEDGMENTS

This work is a part of a research program on "Control of Recovery Boiler Fireside Deposits and Corrosion", jointly supported by ABB Combustion Division, Ahlstrom Corporation, Bacock & Wilcox Company, Bergemann USA, Champion International Corporation, Diamond Power Specialty Company, E.B.Eddy Forest Products Ltd, Georgia Pacific Corporation, International Paper Company, Irving Pulp & Paper Limited,

James River Corporation, Kværner Pulping Camp Corporation, Westvaco Corporation, Weyerhaeuser Company and Willamette Industries, Inc. and the National Sciences and Engineering Research Council of Canada.

LITERATURE CITED

1. **Taylor, G.I.**, "A Method of Constructing the Paths of Raindrops of Different Diameters Moving in the Neighbourhood of (1) a Circular Cylinder, (2) An Aerofoil, Placed in a Uniform Stream of Air; and a Determination of the Rate of Deposit of the Drops on the Surface and the Percentage of Drops Caught," *Aero. Res. Com.*, R. & M. 2025 (1940).

2. **Israel, R. and D. E. Rosner**, "Use of a Generalized Stokes Number to Determine the Aerodynamic Capture Efficiency of Non-Stokesian Particles from a Compressible Gas Flow," *Aerosol Science and Technology*, 2, p. 45 (1983).

3. **Wang, H.C.**, "Theoretical Adhesion Efficiency for Particles Impacting a Cylinder at high Reynolds Number," *Journal of Aerosol Science*, 17 (5), p. 827 (1986).

4. **Wessel, R.A., and J. Righi**, "Generalized Correlations for Inertial Impaction of Particles on a Circular Cylinder," Aerosol Science and Technology 9, p. 29 (1988).

5. **Dahneke, B.**, "The Capture of Aerosol Particles by Surfaces," *Journal of Colloid Interface Science*, 37, p. 342 (1971).

6. **Rogers, L.N. and J. Reed**, "The Adhesion of Particles Undergoing an Elastic-Plastic Impact with a Surface," *Journal of Physics, D: Applied Physics,* 17, p. 677 (1984).

7. **Davis, R.H.**, "The Rate of Coagulation of a Dilute Polydisperse System of Sedimenting Spheres," *Journal of Fluid Mechanics*, 145, p. 179 (1984).

8. **Wall, S., W. John, H.C. Wang, and S.L. Goren,** "Measurements of Kinetic Energy Loss for Particles Impacting Surfaces," Aerosol Science and Technology, 12, p. 926 (1990).

9. **Madejski, J.**, "Solidification of droplets on a cold surface," *Int. J. Heat Mass Trans.*, 19, 1009 (1976).

10. **Chandra, S., and C.T. Avedisian**, "On the Collision of a Droplet with a Solid Surface," Proc. R. Soc. London A, 432,.13 (1991).

11. **Bennett, T, and D. Poulikakos**, "Splat-quench Solidification: Estimating the Maximum Spreading of a Droplet Impacting a Solid Surface," *J. Mat. Sci.*, 28, 963 (1993).

12. **Asai, A., M. Shioya, S. Hirasawa, and T. Okazaki**, "Impact of an ink drop on paper," J. Imag. Sci. Tech., 37, 205 (1993).

13. **Cheng, L.**, "Dynamic Spreading of Drops Impacting onto a Solid Surface," *Ind. Eng. Chem. Process Des. Dev.,* 16 (2), p. 192 (1977).

14. **Ford, R.E., and C.G.L. Furmidge**, "Impact and Spreading of Spray Drops on Foliar Surfaces," Wetting, Soc. Chem. Industry Monograph, 25, 417 (1967).

15. **Fukai, J., Y. Shiiba, T. Yamamoto, O. Miyatake, D. Poulikakos, C.M. Megaridis, and Z. Zhao**, "Wetting Effects on the Spreading of a Liquid Droplet Colliding with a Flat Surface: Experiment and Modeling," Phys. Fluids, 7 (2), p. 236 (1995).

16. **Scheller, B.L.**, "Drop Formation and the Spreading Behavior of Newtonian and Shear-Thinning Liquid Drops Impacting a Solid Surface", PhD Thesis, Univ. Of Maine, Orono (1993).

17. **Young, T.**, "Miscellaneous Works Vol. I.", Peacock, I.G. ed., Murray, London, 418 (1855).

Novel Approaches to Deinking of Xerographic and Laser-Printed Paper

T.L. Moon, J. Li and R. Nagarajan
Chemical Engineering Department, The Pennsylvania State University
210 Fenske Laboratory, University Park, PA 16802

An important class of paper which presents challenges for ink removal is office waste because of the manner in which ink is applied in photocopiers and laser-printers. We describe some novel approaches to deinking such papers using a single block copolymer as the deinking chemical in a two-step process of pulping followed by flotation. These include: (a) Use of triblock copolymers composed of polyethylene oxide (PEO) and polypropylene oxide (PPO). (b) Use of diblock copolymers composed of PEO and a block which resembles the ink toner chemistry (either polystyrene, PS or polymethyl methacrylate, PMMA). (c) Inclusion of a second chemical, stearic acid, to aid as an agglomerating agent for flotation. (d) Integration of an enzyme treatment step using three commercially available enzymes (SP342, Pulpzyme HC and Gamanase) before flotation. The deinking performance is evaluated in terms of the amount of fiber lost in the operations, the brightness of handsheets formed from the deinked pulp, and the dirt count estimated using computerized image analysis. All four deinking approaches allow effective ink removal from the office waste. The novelty of the process developed lies in the extreme simplicity of the deinking chemical formulation, the low concentrations of chemicals employed, and the good biodegradability characteristics of the block copolymers.

INTRODUCTION

Governmental regulations coupled with a greater awareness by the public to reduce the volume of waste in landfills has necessitated research in the area of removing ink from paper for the purpose of recycling the fibers. A class of paper which presents unique challenges to deinking is office waste consisting of papers from office copiers and laser printers. The ink formulations used in these papers are difficult to remove because they are applied by a combination of pressure and heating. The copolymer vehicle (composed of a styrene-acrylate copolymer) which carries the pigment is liquified by the heating, which causes the particles to seep in and wrap around the paper fibers at the surface. They then dry to solid particles again [1].

Because the ink particles are effectively fused to the paper surface, their removal requires a combination of mechanical and chemical action. The mechanical action is usually supplied by a pulper in which the paper is beat into its respective fibers. The ink particles are detached from the fibers by a combination of factors including: hydrodynamic flow of the liquid phase in the pulper, swelling of the fibers, flexing and bending of the fibers, and abrasion of the fibers against each other [2]. Because mechanical action alone is not sufficient to bring about effective ink removal, chemicals are added in the pulper to increase the ink removal efficiency. In general, the type of deinking chemicals added depends on the type of ink, the nature of the paper stock, and the manner in which the ink has been applied in the printing process. The principal chemicals involved in the pulper may include sodium hydroxide, sodium silicate, chelating agents, hydrogen peroxide, surfactants, and collector chemicals [3].

Upon detachment of the ink from the paper fibers in the pulper, further steps are necessary to separate the ink and the fibers into two distinct phases. This is usually accomplished by washing, screening, flotation, or a combination of these operations. During flotation, selective adhesion of ink particles to the surface of air bubbles pumped into the pulp slurry occurs [4]. The ink particles float to the top of the slurry with the air bubbles and thus out of the paper pulp. Surfactants are added to the flotation cell to stabilize the air bubbles and promote attachment of the ink particles to the bubble surfaces. They are also used to form a froth or foam layer at the top of the flotation cell when the bubbles rise to the surface. Because the ink often comes off the paper as flat, flaky particles, they need to aggregate in order to be the correct size and shape for flotation. While this size range is usually reported to be 10-100 μm, some reports suggest an optimum of 30-60 μm [5].

OUR APPROACH TO DEINKING BASED ON SIMPLER STRATEGIES

Deinking chemical formulations with multiple components are quite complicated, since each chemical specie creates problems requiring the presence of another chemical to counteract them. The role of each individual chemical becomes ambiguous because of the unknown synergistic or antagonistic effects and the design of the formulation is largely empirical. More importantly, the total amount of deinking chemicals that should be added per unit mass of paper fiber is quite large, of the order of 2 to 3 weight percent based on the fiber mass. Finally, the water streams from the deinking process are contaminated by the numerous chemicals employed and will have to be treated. To overcome these economic and environmental concerns, we explore novel approaches which utilize one or two deinking chemicals in place of the conventional multicomponent mixture formulations.

Block Copolymer Surfactants

The selection of block copolymer surfactants is motivated by the fact that these molecules exhibit a multitude of surface chemical characteristics that are essential for the deinking application. One group of block copolymers employed are triblock copolymers of the type, polyethylene oxide (PEO)- polypropylene oxide (PPO) - polyethylene oxide (PEO). In these, hydrophilic PEO blocks are attached to each end of a hydrophobic PPO block. These are known as poloxamers and one of the commercial products is called Pluronics. By changing the molecular weights of the blocks, the overall composition and total molecular weight of the block copolymer can be altered. This permits many different physical properties and surface chemical functions within this family of block copolymers. Further, these triblock copolymers possess good biodegradability properties (some are even approved by the FDA as food additives).

A second group of block copolymers tested are the diblock copolymers, polymethyl methacrylate (PMMA)-PEO and polystyrene (PS)-PEO. They are chosen because of the chemical similarity of their hydrophobic blocks to the styrene-acrylate copolymers present in ink toner formulations.

Agglomerating Agent in the Flotation Step

Another approach involves the addition of an agglomerating or collecting chemical to the flotation stage. Thus two chemicals (block copolymer and agglomerating agent) are involved in this deinking formulation. The agglomerating agent studied for this purpose is stearic acid. Because of the ink-fusing action of a photocopier, xerographic inks come off the paper mainly as large, flat plate-like particles [3]. It is speculated that the stearic acid would agglomerate some of the smaller ink particles into larger, more spherical complexes which are more easily attached to and carried on the air bubbles.

Addition of Enzymes

Enzymes which specifically alter the chemistry of the cellulose fibers making up paper have been under investigation in recent years. Enzymes have successively been used in newspaper deinking [6]. Their application for fiber modification of non-impact printed paper [7-10] has been investigated. Most recently, Jeffries et al. have compared enzymatic deinking with conventional deinking on mixed office waste [7,11] and found favorable results. The pilot plant trials later confirmed that enhanced toner removal was achieved with primary flotation and limited washing and cleaning [12]. In this study, we examine three commercially available enzyme cultures. These include SP342 (an enzyme mix of *Humicola insolents*), Pulpzyme HC (xylanase from *Bacillus pimilus*) and Gamanase (mannanase of *Aspergillus niger*). SP342 is a mixture of cellulases and hemicellulases, Pulpzyme HC is a xylanase preparation, and Gamanase is a hemicellulase. A selected PEO-PPO-PEO triblock copolymer was used as surfactant in combination with the enzymes.

EXPERIMENTAL METHODS

Deinking experiments were performed on papers which wer xerographically printed on site. All contained the same image (15.24 cm diameter black circle on an 21.6 cm x 28 cm sheet) an thus the same ink content. The percent consistency desired in th flotation cell (defined as weight in grams of oven dry fiber in 10 g of pulp-water mixture [13]) determined the amount of pape used in a deinking run. Deinking chemicals added to the proces are specified as a percentage of the amount of paper added. I most of our studies discussed below, the copolymer was added a 0.1 % of the paper weight.

Pulping

Thirty grams of paper were added to a Hamilton Beach 14 spee Blendmaster blender which served as the pulper to repulp at 3% consistency. The copolymer was weighed out and also placed i the blender. Water preheated to 43°C was poured into the blende and the contents were pulped together for a total of 4 min. Th water temperature did not appreciably change during pulping After pulping, the pH of the pulp was noted; for nearly all th block copolymers used (the triblock poloxamers and th diblocks), the pH of the pulp was approximately 8.5.

Flotation

The pulp was transferred to the flotation cell (now diluted with water a 43°C to give a consistency of 1%) and mixed for approximately on minute before checking the pH. After noting the pH, flotation was car ried out for a total of 6 min. Depending on the particular block copoly mer used, the froth formed quickly or slowly, was thick or thin, and ha much ink or very little ink at the top. This froth was scraped off the to of the flotation cell by hand using a small paddle. Care was taken t not scrape too deeply into the froth, as this raises the risk of collectin paper fibers along with the ink. Figure 1 shows the two steps of th deinking procedure as employed in this study. For the experiments wit stearic acid, 0.1% of stearic acid (based on paper weight) was added t the flotation cell at the same time as the paper-surfactant pulp and th same procedure as above was followed.

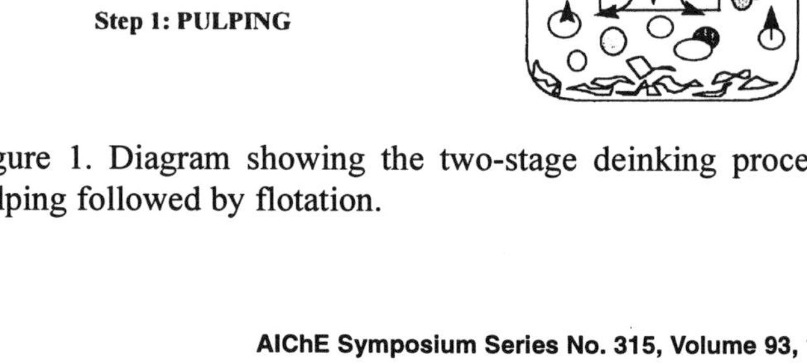

Figure 1. Diagram showing the two-stage deinking process c pulping followed by flotation.

Procedures for Deinking Analysis

Handsheets or brightness pads were prepared from the deinked pulp, to facilitate the evaluation of deinking performance [14]. The procedure used to make the handsheets was TAPPI Test Method T218 om-91, "Forming Handsheets for Reflectance Testing of Pulp (Büchner Funnel Procedure)" [15]. For a typical deinking experiment, a total of six test sheets was prepared.

Because deinking is at heart a recycling process, an understanding of how much paper fiber is lost during each run is desirable. The amount of paper before deinking was weighed without ink printed on it, the assumption being that the deinked paper would have the majority of ink removed and would therefore most closely resemble the clean paper. The percent fiber loss was simply determined by subtracting the amount of deinked paper from the amount of original clean paper, dividing this by the amount of clean paper and multiplying by 100.

Brightness tests are common paper industry tests used to numerically measure the degree of reflectance of a paper sample when exposed to blue light at a wavelength of 457 nm. The procedure employed was TAPPI Test Method T452 om-92, "Brightness of Pulp, Paper, and Paperboard (Directional Reflectance at 457 nm)." [16]. A total of five measurements were made at different points on each handsheet. The measurements were then averaged to get an average brightness value for the experiment.

A computerized image analysis method was employed for determining the amount of ink on the sheets both before and after deinking. A Genius ColorPage-I™ flatbed scanner hooked to a Gateway 4DX2-66 Computer was used to capture the images of the handsheets and a software package known as PhotoFinish™ was used to process the images. Mocha™ image analyzing software was utilized to count the ink specks. The scanning resolution was set at 1200 dpi (dots per inch). For each handsheet five sample regions (each with 2.54 cm X 2.54 cm area) were scanned. Each pixel (the smallest element in an image) has an area of 448 μm^2, and hence, ink specks with diameters as low as 23.9 μm can be detected. The intensity of the image ranged from 0 (darkest) to 256 (brightest) on a gray scale. The cut-off intensity for ink specks was set to be 160 and the deinking efficiency was calculated based on ink spots having intensities between 0 and 160. For each handsheet, the total dirt count from the five spots with a total area of 32.3 cm^2 (11% area of a handsheet) was determined. The result for each experiment was based on the average value of all handsheets (typically six) made from that particular experiment.

Method for Enzymatic Treatment

Pluronic P65 was selected as the surfactant in the enzyme runs. Following the same pulping procedure used in all runs, the stock was pulped at 3% consistency with 0.1% of the copolymer for 4 min. The pulp from the blender was poured into a Büchner fun-

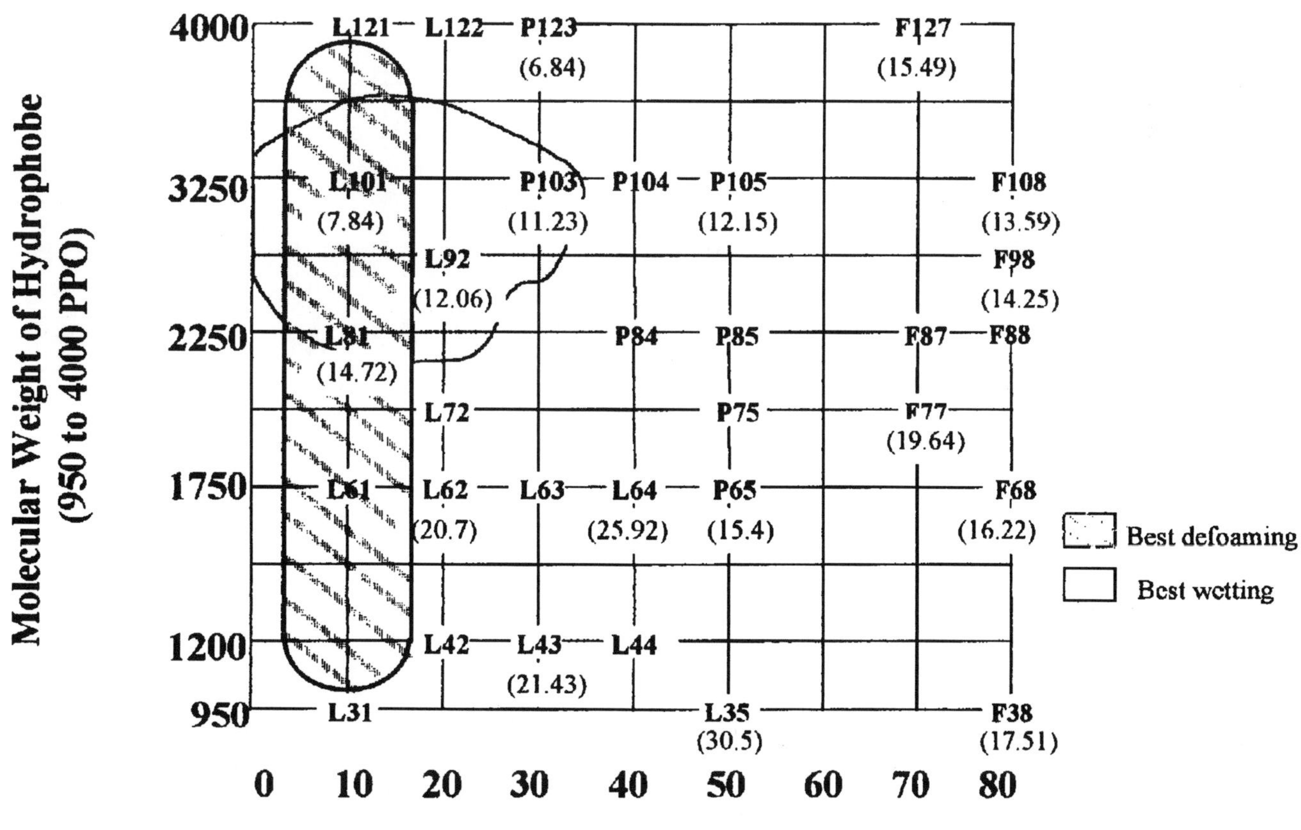

Figure 2. Pluronic grid showing percent fiber losses (numerical values listed belowthe name of the Pluronics). The shaded areas identify the Pluronics for best defoaming and wetting.

nel, and a calculated amount of water was drained out to give a 6% consistency to the pulp. (Some ink specks may have been washed out during this step, so the water drained out was reserved and poured back into the flotation cell. This step insured that the washing effect was minimized for the final handsheets, and that the consistency in the flotation cell remained identical at 1% for all runs with or without enzyme treatment.) The pulp at 6% consistency was transferred to the pulper for enzymatic treatment. Increased pulp consistency would cause more mechanical action between fibers and hence aid toner removal. We chose to treat the pulp at 6% rather than at a higher consistency because of concern about poor mixing at high consistencies and the limitation of experimental facilities to circumvent this problem. The enzyme culture was added to the pulper at a loading rate of 1 ml enzyme per kilogram of paper. To ensure even distribution of enzyme, the original enzyme culture was diluted to 1:100 solution before being added to the pulper. The enzyme cultures are most active at neutral to slightly acidic range and at a temperature of approximately 50 °C. In some runs, the pH and the temperature were adjusted for maximum enzyme activity. Dilute sulfuric acid (H_2SO_4) was added to lower the pH of the pulp; the pulper was placed in a water bath to reach the optimum temperature for the enzyme. In other runs, no pH or temperature adjustment was done. The physical action of enzyme treatment and the initial pulping are identical, except the enzyme treatment took 20 min. Friction at higher consistency causes heat generation, so the temperature of the pulp generally goes up during this process. In order to keep the operating conditions constant, the exterior of the pulper was cooled by an ice pad.

RESULTS AND DISCUSSION

Studies with Triblock Copolymers

Seventeen Pluronics with diverse physical properties were tested for their effectiveness in deinking. The fiber loss and brightness results are presented on grids which provide a graphic representation of the relationship between the copolymer molecular weight, composition and the resulting surfactant characteristics [17]. The numbers preceded by letters within the grid refer to designations for the block copolymers. Superimposed on the grid in Figure 2 are the regions of copolymer composition and molecular weight where they are effective for defoaming and wetting. Similarly, in Figure 3, the copolymers effective for foaming, wetting and detergency are identified.

Results on fiber loss. The percentage of fiber lost is indicated on the grid in Figure 2. The results range from a low 6.84 weight percent fiber loss with P123 to a very large loss of 30.5 weight percent with L35. Deinking processes described in the literature for various kinds of papers and inks suggest that fiber loss should be below about 15 weight percent for a reasonable deinking operation. The copolymers at the top left-hand corner of the grid have low amounts of PEO and high PPO molecular weights. This area corresponds to the lowest fiber losses. Thus, it appears that the more hydrophobic the copolymer is, the lower the fiber losses. As one moves to the most hydrophilic area on the grid corresponding to low PPO molecular weights and high percentages of PEO (i.e., the lower right-hand corner), the fiber losses increase.A probable

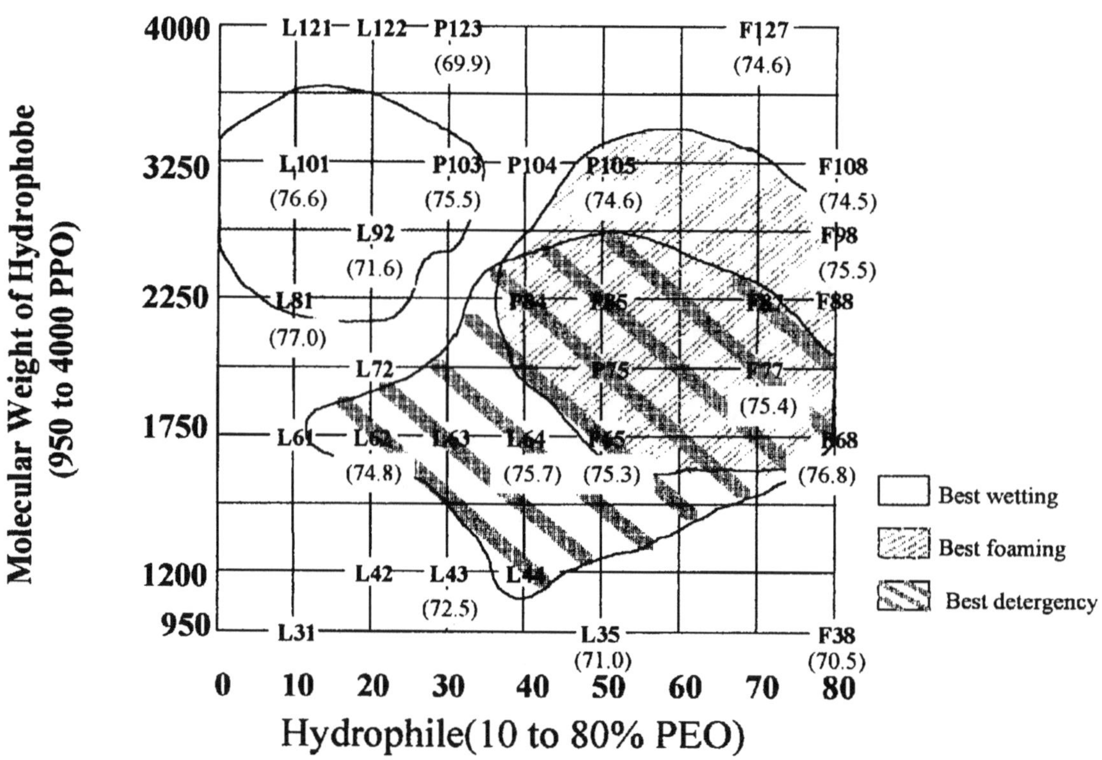

Figure 3. Pluronic grid showing the brightness values (numerical values listed below the name of the Pluronics). The shaded areas indicate the best Pluronics for detergency, foaming, and wetting.

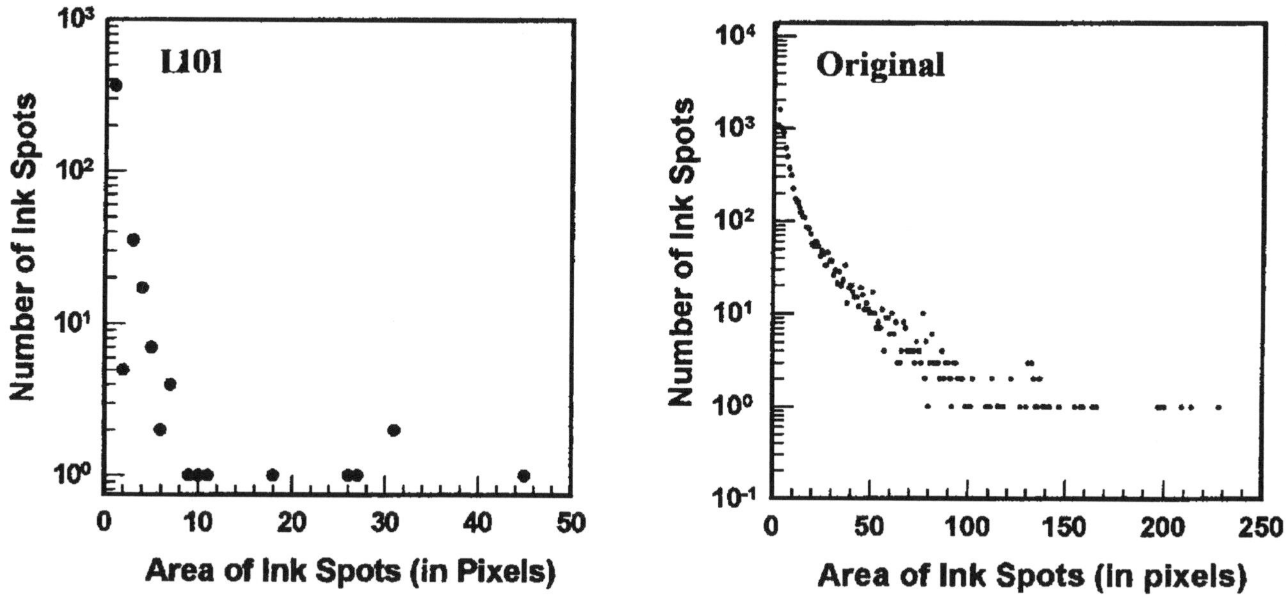

Figure 4. Comparison of the ink specks size distribution for paper deinked with Pluronic L101 and the original test paper with no ink removed. 1 pixel corresponds to an area of 448 μm^2.

explanation is that chemical interactions such as hydrogen bonding between the hydrophilic copolymer and the cellulose fiber may make it possible to carry the paper fiber into the foam layer at the top of the flotation cell. Other physical properties of the block copolymer may also be responsible for the higher fiber losses. Copolymers known as good foamers have some of the higher fiber losses. This is reasonable since a thicker foam generally means that there is a greater likelihood of trapping paper fibers and collecting them with the ink, thus resulting in higher fiber losses. Copolymers considered the best for wetting have some of the lowest fiber losses, but this may also be attributed to the fact that the best wetters are also some of the best defoamers.

Results on brightness. The brightness results are also shown on a grid (Figure 3). A clean paper containing no ink has a brightness of 83 on this scale (upper limit) while the repulped test paper prior to ink removal has a brightness value of 64.9 (lower limit). The grid shows that large brightness values are generally found in the middle of the grid, corresponding to mid-range PPO molecular weights (from approximately 1800 to 2300). As the molecular weight of the PPO becomes either too low or too high, the brightness values drop dramatically. The region of large brightness on the grid corresponds to block copolymers displaying effective detergency, foaming, and wetting.

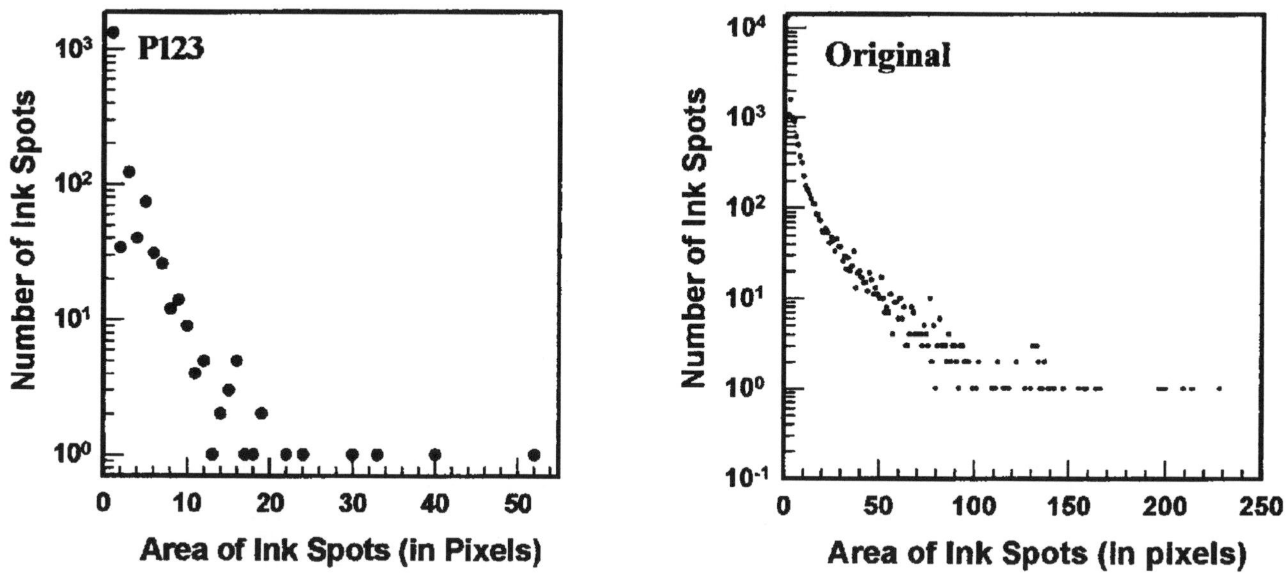

Figure 5. Comparison of the ink specks size distribution for the paper deinked with Pluronic P123 and the original test paper with no ink removed. 1 pixel corresponds to an area of 448 μm^2.

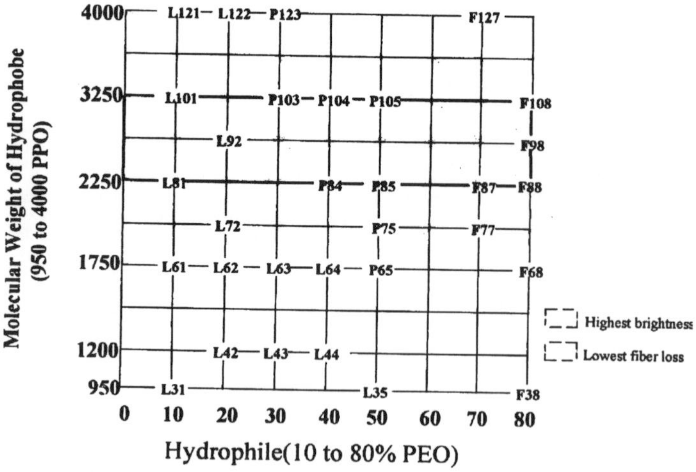

Figure 6. Pluronic grid showing the areas of highest brightness and lowest fiber loss.

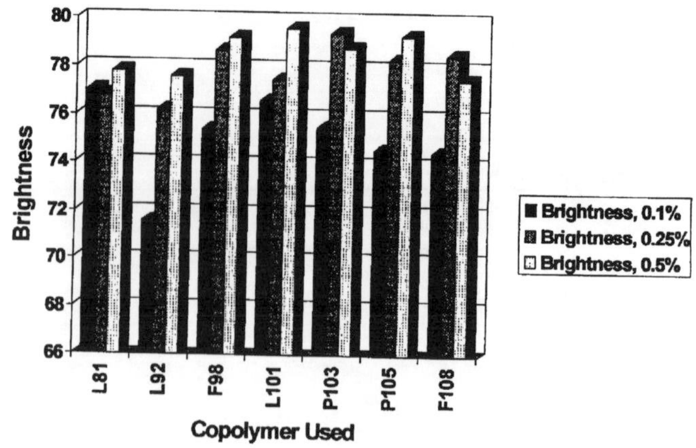

Figure 8. Effect of changing the Pluronic concentration on the brightness values obtained from Pluronics in the 80 series through the 100 series.

Results on ink counts. To determine how much ink was removed from the paper, the number of ink particles on reformed sheets was counted and compared to that on an original test sheet. For illustrative purposes, the ink count measurements done for deinking experiments with Pluronic L101 (which gave a large brightness value) and Pluronic P123 (which gave the smallest brightness value) are presented here. From the ink specks size distributions shown on Figures 4 and 5, one can observe that the number of ink specks decreased upon treatment with the copolymers when compared to the number of ink specks on an original master sheet in all the size ranges. Pluronic L101, the more hydrophobic copolymer, removed more ink specks than P123. The copolymer which gives one of the higher brightness values also leads to a lower ink specks count.

Effect of copolymer concentration. The results on fiber loss and brightness have been grouped together onto the same grid to determine the optimum range of copolymers for deinking (Figure 6). The data show that there is an overlap of lowest fiber losses and highest brightness values between the Pluronic 80 series and the 100 series. These copolymers were thus studied at higher concentrations (0.25% and 0.5% by weight of paper) and the results are shown in Figures 7 and 8. As the copolymer concentration increases, the fiber loss generally increases as well (Figure 7). These copolymers produced a larger quantity of foam in the flotation cell when present at higher concentrations, and therefore the possiblity of trapping more paper in the froth increased. It should be noted that those copolymers which are known as good foamers (those with a higher percentage of PEO) demonstrate the largest increases in fiber loss with increasing concentration. Brightness results for the experiments with higher copolymer concentrations are reported in Figure 8. As the concentration increases from 0.1% to 0.25% and then to 0.5% the brightness of the deinked sheets increases as well.

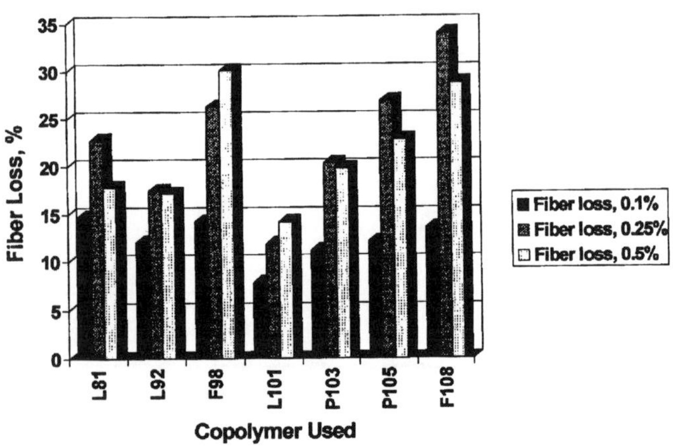

Figure 7. Effect of changing the Pluronic concentration on the percent fiber losses obtained from Pluronics in the 80 series through the 100 series.

Results Based on Conventional Deinking Chemicals

Some copolymers were tested as part of multicomponent chemical formulations with conventional deinking chemicals. For example, a deinking formulation tested included 2.0% NaOH (a fiber swelling agent), 0.8% H_2O_2 (a bleaching agent) and 0.2% EDTA (ethylene diamine tetra-acetic acid, a chelating agent) besides 0.1% triblock copolymer. The brightness values obtained for the deinked sheets using this formulation were on average at least one brightness point higher, with some brightness values as high as 80. This appears to be an upper limit of brightness for this formulation at these concentrations. The percent fiber losses were all higher, some as much as 10% higher, compared to the fiber losses obtained while using the copolymer alone. These findings provide the realization that the addition of more chemicals may raise the brightness of the pulp, but at the expense of losing significantly more fiber.

Studies with Diblock Copolymers

Five diblock copolymers, two of the PMMA-PEO type and three of the PS-PEO variety, were also examined for their deinking abil-

Table 1. Fiber loss and brightness results for the diblock copolymers added at 0.1% by weight of paper.

Diblock Copolymer Used	%Fiber Loss	Brightness
PMMA(1000)PEO(2000)	22.44	79.6
PMMA(1000)PEO(3000)	27.26	81.5
PS(1000)PEO(1000)	20.74	79.5
PS(1000)PEO(3000)	26.30	80.2
PS(3000)PEO(1000)-SO_3NH_4	25.74	81.1

ty at 0.1% concentration. Table 1 lists the five molecules indicating also the molecular weight of each block. One of the PS-PEO copolymers contained a SO_3NH_4 functional group on the end of the PEO block.

Fiber loss results. The percentage of fiber lost is indicated in Table 1. The fiber loss range from 20.74 weight percent in the case of PS(1000)-PEO(1000) to 27.26 weight percent in the case of PMMA(1000)-PEO(3000). As the hydrophilic character (PEO) of the copolymer increases relative to the other block, the percent fiber loss increases as well, similar to what was seen in the case of the triblock copolymers. Even though there is a higher proportion of PS than PEO in PS(3000)-PEO(1000), the presence of the SO_3NH_4 group attached to the end of the PEO block keeps the copolymer very hydrophilic and the fiber losses are higher. The most hydrophobic copolymers, PS(1000)-PEO(1000) and PMMA(1000)-PEO(2000), would be the best copolymers to use from the point of view of keeping the fiber losses low.

Brightness results. The brightness results are also presented in Table 1. It is seen that all the diblock copolymers give large brightness values that are about 5 points higher than those obtained using triblock copolymers. The brightness values increased as the hydrophilic content increased in the molecules. Considering the results for fiber loss and brightness together, one can observe again a trade-off between high sheet brightness and low fiber loss.

Studies with Stearic Acid

The PEO-PPO-PEO triblock copolymers at 0.1% were added with the paper at the beginning of the pulping stage whereas stearic acid at 0.1% was added at the beginning of the flotation stage. After the contents were allowed to mix, the air flow into the cell was activated and flotation proceeded.

Table 2. Fiber loss and brightness results for experiments using just the Pluronics and the Pluronics with stearic acid added to the flotation stage.

Surfactant Used	0.1% Pluronic Alone: % Fiber Loss	0.1% Pluronic Alone: Brightness	0.1% Pluronic and 0.1% Stearic Acid: % Fiber Loss	0.1% Pluronic and 0.1% Stearic Acid: Brightness
L81	14.72	77.0	11.14	81.7
L92	12.06	71.6	12.54	80.0
F98	14.25	75.4	17.92	81.4
L101	7.84	76.6	10.72	76.0
P103	11.23	75.5	18.76	77.8
P105	12.15	74.6	11.17	78.5
F108	13.59	74.5	13.41	81.0

Fiber loss results. The results on fiber lost for the stearic acid experiments are indicated in Table 2. The results show that either the fiber loss remained roughly the same or increased on the addition of stearic acid. This is somewhat surprising, because the foaming ability of every copolymer decreased when stearic acid was present in the flotation cell. The bubbles in the froth were larger and less stable, and the froth itself was thinner and did not cover the entire top surface of the pulp underneath it. One would therefore anticipate that fiber losses on the addition of stearic acid should be lower than that with the copolymer alone. There is a strong possibility that the thinner foam was overcompensated for by placing the paddle deeper into the pulp, thus lifting out a larger portion of paper fibers with the ink.

Brightness results. The brightness results are also presented in Table 2. In every experiment except one, the brightness of the deinked sheet was higher when stearic acid was also added. In fact, there was an increase of almost eight points in brightness for Pluronic L92. The only copolymer which did not fit this trend was Pluronic L101. The addition of stearic acid also demonstrates the trade-off between low fiber losses and high brightness.

One may attribute the observed increase in brightness to the role of stearic acid in agglomerating smaller ink particles and thereby enhancing their removal via flotation. It has been shown earlier that a deinking formulation with a high percentage of stearic acid gave very good results [18] but this was attributed to the effectiveness of stearic acid in ink removal from the fiber. Therefore, stearic acid may aid in removing undetached ink from the fiber in the flotation cell instead of being an agglomerator, collecting the already detached ink. To test the role of stearic acid, ink counts were performed on sheets deinked using Pluronic L92, with and without stearic acid (Table 3). If stearic acid is collecting very small ink particles, then one should observe a decrease in the number of the smaller ink particles. The data in Table 3 show that the number of small particles indeed dropped significantly in the presence of stearic acid, but the number of larger ink particles also dropped even more dramatically. This suggests that multiple mechanisms including agglomeration are operative in the presence of stearic acid in the flotation cell.

The ability of stearic acid to provide significantly better deinking results by itself without any triblock copolymers was tested in two ways. In the first test, stearic acid at 0.1% was introduced at the beginning of the pulping stage, followed by flotation, without

Table 3. Comparison of number of ink specks in various size ranges on sheets deinked with Pluronic L92 alone and on sheets deinked with Pluronic L92 and stearic acid (32.3 cm^2 area examined).

	Particle Size (µm)					
	<30	30-60	60-90	90-120	120-150	>150
L92 + Stearic Acid	262	142	45	14	2	4
L92 alone	903	503	178	39	36	35

adding any other chemicals. It was found that the fiber loss increased to nearly 22.4%, as much as 10-15% higher than the fiber losses obtained using the copolymers alone. However, the brightness value of 75.8 achieved with stearic acid is in the same range as those obtained using the copolymers alone. In the second test, the pulping was done with no chemicals added and stearic acid was added at the beginning of the flotation stage. While the fiber loss of 12.4% was comparable, the brightness value of 71.8 was much below that achieved with the best copolymers. Therefore, to achieve higher brightness values, stearic acid may be used in conjunction with the copolymer. The use of stearic acid alone is not an attractive alternative over the use of copolymers, because it does not significantly improve brightness values while at the same time it dramatically increases the fiber losses. However, if one wants to use stearic acid in place of a copolymer, then (from a brightness point of view) it should be added at the beginning of the pulping stage rather than at the beginning of the flotation stage.

Studies with Enzymes

The three enzymes, SP342, Pulpzyme HC and Gamanase, studied have maximum or near maximum activities in the temperatures range of 45°C to 60°C. Since the enzyme activity is not sensitive to temperature within this range, it is not an important variable for the deinking studies. The optimum pH for activity of the three enzyme cultures fell between 6-8, 6.5-9, and 3-6, respectively [19]. The typical pH value of the sludge after the pulping stage was 8.5.

Deinking experiments were performed with all three enzymes without optimizing the pH to achieve maximum enzyme activity. The deinking results are compared against those based on the block copolymer alone in Table 4. All three enzymes improved the deinking efficiency compared to that based on Pluronic P65 alone, without sacrificing fiber yield. The total ink area in ppm after deinking dropped from 125 for the block copolymer alone to 67 or less when the enzyme treatment was included. The use of Pulpzyme HC resulted in a lower fiber loss as well as higher brightness, and therefore may be considered more efficient. It is

Table 4. Comparison of deinking results for Pluronic P65 alone and with enzymes.

pH adjusted

		P65 alone	P65 and Enzymes		
	Original	No Enzyme	SP 342	Pulp. HC	Gamanase
Total Ink (ppm)	14159	125	41	35	34
Brightness	64.9	75.3	75.9	76.9	76.8
Fiber Loss		15.4%	14.5%	12.8%	17.5%
Ink Removal Efficiency		99.1%	99.7%	99.8%	99.8%

Without pH adjustment

		P65 alone	P65 and Enzymes		
	Original	No Enzyme	SP 342	Pulp. HC	Gamanase
Total Ink (ppm)	14159	125	52	35	67
Brightness	64.9	75.3	75.0	76.9	76.2
Fiber Loss		15.4%	11.5%	12.8%	12.05%
Ink Removal Efficiency		99.1%	99.6%	99.8%	99.5%

no surprise that Pulpzyme HC out-performed the other two enzymes, since its activity has a maximum near the pH of the sludge from the pulper. SP342 gave no improvement in brightness values compared to that for P65 alone, even though the total dirt area was much lower. Possibly, the presence of the enzyme resulted in the dispersion of ink particles as fines. They could not be removed from the pulp and were not detected as spots, but their presence lowered the brightness values.

When the pH was adjusted to a value of 4.5, to maximize Gamanase activity, the ink removal efficiency increased. The dirt count was similar to that in the case of Pulpzyme HC but the fiber loss was higher. The latter may be due to sulfuric acid used for pH adjustment, causing cellulose fibers to weaken and thus producing more fines that were lost in the flotation step. In the case of SP 342, no improvement was seen with the pH adjusted to a value of 7.

CONCLUSIONS

A two-step deinking process based on pulping followed by flotation, utilizing a block copolymer in *very low* concentrations, is developed here for xerox and laser-printed papers. The PEO-PPO-PEO triblock copolymers were found to be effective for deinking. The copolymers which gave the lowest fiber losses were those with the most hydrophobic character; they also had excellent wetting and defoaming characteristics. The copolymers which provided the brightest deinked pulp generally had mid-range PPO molecular weights. These were noted as best for detergency, foaming, and wetting. Some diblock copolymers (PS-PEO and PMMA-PEO) which more closely resemble the chemistry of the ink toner particles were also studied. They yielded better brightness values but somewhat larger fiber losses compared to the triblock copolymers. A trade-off between low fiber loss and high brightness was observed with both the triblock and the diblock copolymers.

The addition of stearic acid at the beginning of the flotation stage enhanced the deinking ability of a triblock copolymer. While the fiber losses were not significantly changed by the stearic acid, the brightness values rose considerably. The ink count results suggest that stearic acid may be drawing the plate-like ink particles together and thereby promoting greater ink removal. At the same time it is probably removing some of the undetached particles from the fibers. It was also found that stearic acid by itself would not provide higher brightness levels without the copolymers included. The use of enzymes was shown to decrease the number of ink specks and improve the ink removal efficiency compared to the use of a copolymer alone. The results also show that two of the enzymes are most effective when the pH corresponds to their optimal activity while for the third enzyme, the pH optimization led to very little improvement.

Considering the quantities of chemicals involved, the deinking chemical formulation based on the block copolymer would be much more economical compared to conventional deinking formulations. Also, the good biodegradability properties of the block copolymer can be taken advantage of. From *both economic and environmental points of view*, the process outlined in this study appears promising.

LITERATURE CITED

1. **Shrinath, A., J. T. Szewczak, and I. J. Bowen,** "A Review of Ink-Removal Techniques in Current Deinking Technology," *Tappi Journal*, **74** (7), pp. 85-93 (July 1991).

2. **Borchardt, J.K.,** "Possible Deinking Mechanisms and Potential Analogies to Laundering," *Prog. in Paper Recycling,* **2** (2), pp. 47-57 (Feb. 1993).

3. **Ferguson, L.,** "Deinking Chemistry: Part 1," *Tappi Journal,* **75** (7), pp. 75-83 (July 1992).

4. **Crozier, R.D.,** "Flotation: Theory, Reagents, and Ore Testing," 1st ed., Pergamon Press, Oxford (1992).

5. **Schriver, K.E.,** "Mill Chemistry Must be Considered Before Making Deink Line Decision," *Pulp and Paper*, **64** (3), pp. 76-79 (March 1990).

6. **Eom, T.J. and S. S. Ow,** Korean Patent 3934772 (Nov. 22, 1990).

7. **Jeffries, T., J. H. Klungness, M. Sykes, and K. Corpsey,** "Comparison of Enzyme-Enhanced With Conventional Deinking of Xerographic and Laser-Printed Paper," *Tappi Journal,* **77** (4), pp. 173-179 (Apr. 1994).

8. **Jackson, L.S., J. Heitmann, and T. W. Joyce,** "Enzymatic Modification of Secondary Fiber," *Tappi Journal,* **76** (3), pp. 147-154 (March 1993).

9. **Borchardt, J.K.,** "A Primer for Surfactants Used in Deinking," *Progress in Paper Recycling* , **2** (1), p. 55 (Jan. 1992).

10. **Bhat, G.R., J. A. Heitmann, and T. W. Joyce,** "Novel Techniques for Enhancing the Strength of Secondary Fiber," *Tappi Journal*, **74** (9), pp. 151-157 (Sept. 1991).

11. **Sykes, M., J. Klungness, T. Jeffries, K. Cropsey, and S. Abubakr,** "Enzymatic Deinking of Sorted Mixed Office Waste: Recommendations for Scale-Up," Proceedings of the Technical Association of the Pulp and Paper Industry's 1995 Recycling Symposium, New Orleans, LA, Paper No. 2-5 (Feb. 1995).

12. **Heise, O. U., J. P. Unwin, J. H. Klungness, W. G. Fineran Jr., M. Sykes, and S. Abubakr,** "Industrial Scaleup of Enzyme-Enhanced Deinking of Nonimpact Printed Toners," *Tappi Journal,* **79** (3), pp. 207-212 (March 1996).

13. **TAPPI Test Method T 240 om- 81,** "Consistency (Concentration) of Pulp Suspensions," TAPPI Press, Atlanta, GA (1981)

14. **McKinney, R.W.J.,** "Evaluation of Deinking Performance: A Review of Test Methods," *Tappi Journal*, **71** (1), pp. 129-131 (Jan. 1988).

15. **TAPPI Test Method T 218 om-91,** "Forming Handsheets for Reflectance Testing of Pulp (Büchner Funnel Procedure)," TAPPI Press, Atlanta, GA (1991).

16. **TAPPI Test Method T 452 om-92,** "Brightness of Pulp, Paper, and Paperboard (Directional Reflectance at 457 nm)," TAPPI Press, Atlanta, GA (1992).

17. "Pluronic and Tetronic Block Copolymer Surfactants," Brochure from BASF Corporation, Speciality Products, Parsippany, NJ.

18. **Read, B.R.,** Proceedings of the Technical Association of the Pulp and Paper Industry's 1991 Pulping Conference, available from TAPPI Press, Atlanta, GA, p. 851 (1991).

19. Product Information, courtesy of Novo Nordisk Enzyme Process Division, Denmark.

ACKNOWLEDGMENTS

This work was supported by funds provided by the TAPPI Foundation, International Paper Company, and the GAANN Fellowship. The authors also wish to thank BASF Corp. for their donation of the various Pluronics used in this study, as well as the Lock Haven, PA site of International Paper for the use of their brightness meter. Valuable discussion on general deinking procedures was provided by Dr. John K. Borchardt of Shell Chemical Company and by Dr. Ross T. Gray and Tina O'Lear, both of ECC International Inc., Pittsburgh, PA. Dennis Kalberg of International Paper in Erie, PA was also invaluable in the initial stages of the process.

Bleaching of Colored Broke by Elemental Chlorine Free (ECF) and Totally Chlorine Free (TCF) Processes

V.R. Parthasarathy
Fluor Daniel Forest Products (Technology), Greenville, SC 29607

*Six different colored broke samples were bleached by Elemental Chlorine Free (ECF) and Totally Chlorine Free (TCF) bleaching processes. Broke samples were bleached either individually or in blends. Some of the colored brokes were mixed with brownstock and bleached. CIE's L*a*b* numbers were used to rank the efficiency of various bleaching processes. Single-stage TCF bleaching with hydrogen peroxide, oxone or formamidine sulfinic acid (FAS) did not provide bright pulp suitable for use in premium grade papers like Laser Jet, etc. Chlorine dioxide is very effective in bleaching colored brokes and their blends, as is hypochlorite. Single stage bleaching of broke in many instances provided Lightness (L*) value of 90.0 but failed to meet the "chroma" specification of*

$$\sqrt{a^{*2}+b^{*2}} \leq 10.0$$

that is important to use in premium grade papers. Multistage bleaching of colored broke, broke blends and broke blends mixed with brown stock resulted in pulps of ≥ 85% ISO brightness, ≥ 90.0 lightness, and chromacity of

$$\sqrt{a^{*2}+b^{*2}} \leq 10.0$$

. These pulps are suitable for use in white, near-white and colored paper grades. Goldenrod, cherry and pink colored broke are difficult to bleach. Blue and green colored broke were bleached easily either individually or blended among themselves or with brown stock.

INTRODUCTION

Colored broke is a valuable fiber source for the mill and its recovery has enormous economical benefits. Many options are available for the recovery of colored broke in a paper mill. One is storing and using it later during the same color run on the paper machines. Color stripping of broke by either single or multi-stage bleaching is another option. A third option is to mix and bleach the colored broke with brown stock in the mill's bleach plant. In the bleaching of colored broke, chlorine offers the best of color stripping and brightness. However, with the recent awareness on the environmental impact of using elemental chlorine [1-3] in bleaching, it cannot be used for color stripping of broke. Therefore, interest is growing in Elemental Chlorine Free (ECF) and Totally Chlorine Free (TCF) bleaching of broke.

Among the bleaching chemicals, chlorine dioxide is one of the most efficient in the color stripping of broke [4]. Oxygen in combination with hydrogen peroxide and ozone is also very effective [5, 6]. Oxone[1], a triple salt of potassium peroxymono, peroxydi, and persulfate had been explored for the delignification of chemical pulps [7]. With its strong oxidizing power, oxone makes an ideal candidate for the color stripping of paper. Sodium hydrosulfite and formamidine sulfinic acid (FAS) are reducing chemicals that had been used in the bleaching of colored broke [8, 9]. The former can provide only moderate brightness gains but is cheaper than FAS. Sequential bleaching of colored broke with an oxidant and a reducing chemical is a common two-stage sequence to achieve optimum color stripping [10].

[1]"Oxone" 2KHSO5. KHSO4.K2SO4 is the registered trade mark of E.I. DuPont.

COLOR, BRIGHTNESS AND LIGHTNESS

Color is defined by "value, hue, and chroma" [11]. Brightness is a measure of reflectance of blue light at 457 nm [12]. While the brightness of achromatic paper is purely a function of Lightness (L*), the brightness of colored paper is a function of both L* and chromacity. Chromacity is defined by CIELAB's a* (Green-Red) and b* (Blue-Yellow) [12] values. For white and near-white papers, brightness is a single measurement. That does not apply for colored papers. In the bleaching of colored broke, the brightness, whiteness (L*) and chroma (a* and b*) assumes importance only after a significant amount of color is stripped. Additionally, they have to meet the criteria set by TAPPI standard (T 524 om-86) for white or near-white paper [13]; T 524 om-86 defines the white or near white pulp or paper as having L* ≥ 84.0 and chromacity value

$$\left(\sqrt{a^{*2}+b^{*2}}\right) \leq 10.0$$

[13]. This specification of L*a*b* is not adequate to define

pulps for use in premium white and color grades since they need certain minimum brightness. In this study, a deviation to T 524 om-86 is applied to define the efficiency of color stripping of a given bleaching chemical or process; bleached pulp having $L^* \geq$ 90.0 and

$$\sqrt{a^{*2} + b^{*2}} \leq 10.0$$

at a minimum brightness of 85% ISO is considered white or near-white pulps suitable for use in premium grade papers.

COLOR INDEX (CI)

A unified index called "Color Index (CI)" [11] defines the changes in color and is the magnitude of color difference as measured by the difference in 'value' and "chroma." Color Index (CI) does take into account the difference in the lightness, darkness and the chromatic differences between the sample and the standard and is computed as

$$\Delta E^* \sqrt{\left[(\Delta L^*)^2 + \Delta(a^*)^2 + \Delta(b^*)^2\right]}$$

Sharpe and Lowe [14] and Rangamannar [15] used the E* value to evaluate the efficacy of a particular color stripping or bleaching process to provide white or near-white pulps. However, in this paper both TAPPI's definitions of white and near-white (TAPPI test methods T524 om-86) and the color Index (CI) value is used to evaluate the efficiency of different bleaching processes.

PROCESS DESCRIPTION

A fine paper mill, using bleached hardwood pulps with bleached softwood kraft pulp for the manufacture of premium white and color paper grades, is the source of colored broke used in this study. One of the objectives of this study is to evaluate different chlorine free bleaching sequence(s) for the color stripping and bleaching of broke. The other objective is to explore the appropriate broke blends that can be subjected to bleaching with brownstock in the mill's bleach plant. Both laboratory and mill bleaching data are presented in this report. A common practice at this mill is to segregate and use colored broke during the same color run on the paper machine. Many times, the colored broke is bleached individually or in combination with chemical pulp at the final D-stage of the bleach plant. At the mill, the colored broke is made-down in a continuous hydrapulper, stored in a mixing chest before being sent to either paper machine or bleach plant. The retention time at the storage chest can be varied between 3 to 4 hours. The broke can be bleached at the hydrapulper or at the storage chest or at both places. Chlorine dioxide, sodium hypochlorite and sodium hydrosulfite can be used for bleaching only in the storage chest. They cannot be used in the hydrapulper due to release of toxic fumes, limitations on reaction time, air-stock mixing, etc. Bleaching with hydrogen peroxide, FAS, etc., can be carried out in hydrapulper and/or in storage chest.

The bleach plant is s short sequence (C/D E(OP)D) plant with conventional vacuum washers for pulp washing between stages. The first-stage has a upflow tower, and is a low consistency stage. The

Table 1. Initial Values of Brightness, L*a*b*, Color Index (CI) - Colored Broke, White Paper, and Bleached Hardwood Pulp

	Brightness, %ISO	L*	a*	b*	(a*^2 + b*^2)^0.5
COLORED BROKE					
Goldenrod	27.18	82.7	5.18	34.95	35.33
Blue 1	72.24	78.23	-7.81	-9.10	11.99
Green 1	55.72	83.24	-12.84	12.06	17.62
Pink	62.94	82.36	18.44	4.49	18.98
Mixed Broke*	41.90	77.83	-6.03	17.28	18.30
Blue 2	77.70	82.5	-11.85	-7.31	13.92
Green 2	59.90	85.9	14.98	12.28	19.37
Canary	43.50	91.1	-10.25	31.02	32.67
Cherry	29.80	57.4	45.33	3.30	45.45
Blue 2/Green 2**	67.20	84.7	-13.14	4.65	13.94
Canary/Cherry**	38.80	68.3	31.71	8.35	32.79
HARDWOOD PULP	88.40	96.05	-0.61	3.35	3.41
WHITE PAPER					
a) Xerox Copy	84.00	93.20	0.99	-0.09	0.99
b) Ink Jet****	90.00	93.30	3.60	-5.40	6.49

** A blend of golden rod, blue 1, green 1 and pink in equal proportion*
*** Blend of blue 2/green 2 and blend of canary/cherry (blended in equal proportion)*
****Bleached hardwood pulp*
***** Compensated for the fluorescent component of the paper*

alkaline extraction and chlorine dioxide bleaching stages have both upflow tubes and downflow towers and are medium consistency bleaching stages. In the present bleaching arrangement at the mill, the colored broke can be bleached with the chemical pulp in three ways. 1) They can be mixed with brown stock before chlorine and chlorine dioxide injection and subjected to C/D or DE(OP)D bleaching. 2) They can be blended with the C/D pulp at the upflow tower laundermate, washed and thickened before alkaline extraction. 3) By adding them at C/D washer repulper or at E(OP) washer repulper before the steam mixer, the colored broke can be bleached by two-stage E(OP)D- or by single-stage D-bleaching. Such arrangements allow much flexibility in bleaching depending on the ease or difficulty of a colored broke to reach certain final pulp brightness.

RESULTS AND DISCUSSION

The initial values of brightness, whiteness (L*) and

$$\sqrt{a^{*2}+b^{*2}}$$

for colored broke and broke blends are listed in Table 1 and are compared with the values of bleached hardwood pulp and white papers manufactured at the mill. None of the colored broke met the TAPPI specification of L*90.0 and chromacity of

$$\sqrt{a^{*2}+b^{*2}} \leq 10.0$$

Among the colored broke, blue, green and blue/green combinations had the lowest CI values at a given brightness and lightness; yellow colored papers as canary and goldenrod had the lowest brightness but highest CI value. The initial CI value at a given brightness is important because it provides the indication as to the ease or difficulty of bleaching a colored broke to a given brightness. High brightness of bleached colored broke is critical to use these pulps in premium color grades and high brightness papers.

Single- and Two-Stage Bleaching of Colored Broke and Broke Blends (Laboratory Results)

Laboratory results of single and two-stage ECF and TCF bleaching of colored broke and broke blends are detailed elsewhere [16]. Changes in brightness, L*a*b*, Color Index (CI), and

$$\sqrt{a^{*2}+b^{*2}}$$

("chromacity') of blue, green and goldenrod colored broke subjected to various bleaching processes are depicted in Figures 1 to 6.

The relationship between CI and

$$\sqrt{a^{*2}+b^{*2}}$$

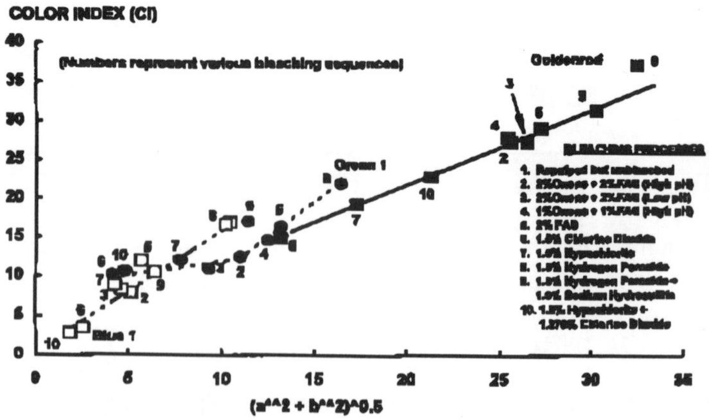

Figure 1. Relationship Between Color Index (CI) and (a*^2 + b*^2)^0.5 for the Bleached Pulp From Different Colored Broke by Various Bleaching Processes.

for blue, green and goldenrod colored brokes subjected to single and two-stage TCF and ECF bleaching is linear (Figure 1). Among the different bleaching sequences, single-stage chlorine dioxide bleaching is the best providing the lowest CI and

$$\sqrt{a^{*2}+b^{*2}}$$

value at highest brightness. Single-stage bleaching with 1.5 % chlorine dioxide charge is as good as two-stage bleaching with hypochlorite (1.5%) and chlorine dioxide (1.275%) for color stripping of broke and that applies to even difficult to bleach goldenrod colored broke (Figure 1). The high CI and chromacity

$$\left(\sqrt{a^{*2}+b^{*2}}\right)$$

values for bleached pulp from goldenrod indicate that it is preferable to exclude this broke from other broke blends when bleaching.

The changes in lightness (L*) of bleached blue, green and goldenrod colored broke to changes in brightness is logarithmic (Figure 2). Except for blue colored broke, after certain brightness, the changes in lightness of pulps become negligible and

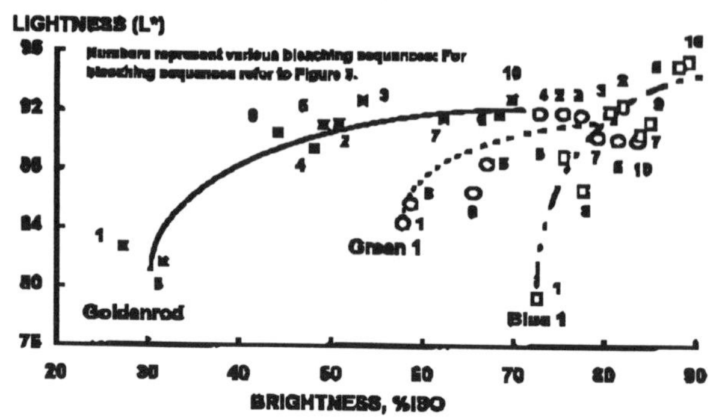

Figure 2. Relationship Between brightness and Lightness of Goldenrod, Blue and Green Colored Broke Bleached by single- and Two-Stage ECF and TCF Bleaching Sequences.

asymptotic to increases in brightness. More so than the L*, brightness gain is dependent on the individual bleaching process and on the initial color of the broke. For goldenrod and green colored broke, the L* remained constant, not influenced by the individual bleaching processes. However, the influences of individual bleaching processes on the lightness (L*) of bleached pulp from blue colored broke was significant. The bleached pulp from goldenrod or green colored broke reached a maximum lightness of about 92.0 at 60 %ISO and 70 %ISO brightness respectively and then changes little with changes in brightness. The bleached blue colored broke reached a lightness of 94.0 at about 90 %ISO brightness before becoming asymptotic to increase in brightness. Goldenrod colored broke while reaching very high lightness did not bleach to high brightness even when using an efficient bleaching chemical like chlorine dioxide This is a strong indication that goldenrod should not be blended and bleached with other colored broke since it would hinder the brightness development of pulp. Neither a single-stage nor two-stage bleaching provided high enough brightness from goldenrod colored broke to use the bleached pulp in white or near-white paper grades. Again, TCF bleaching is a not an option for goldenrod colored broke. Any bleaching sequence for goldenrod colored broke should be a multi-stages and should include both hypochlorite and chlorine dioxide in the bleaching sequence. Green and blue colored broke are easy to bleach brokes. They bleached easily, reaching 80.0%+ ISO brightness. A colored broke blend containing equal amounts of blue, green, canary, and pink colored brokes but no goldenrod, bleached easily by single-stage hypochlorite bleaching. The bleached pulp from such a broke blend did meet the conditions of brightness (~ 85.0% ISO), lightness (L* ≥ 85.0) and chromacity

$$\left(\sqrt{a^{*2}+b^{*2}} \leq 10.0\right)$$

for use in premium grade papers like Laser Jet, etc.

Single-stage TCF bleaching of colored broke or their blends did not provide L* and "chroma" values to classify them as white and near white pulp under TAPPI Standards T 524 om-86 and true for all the

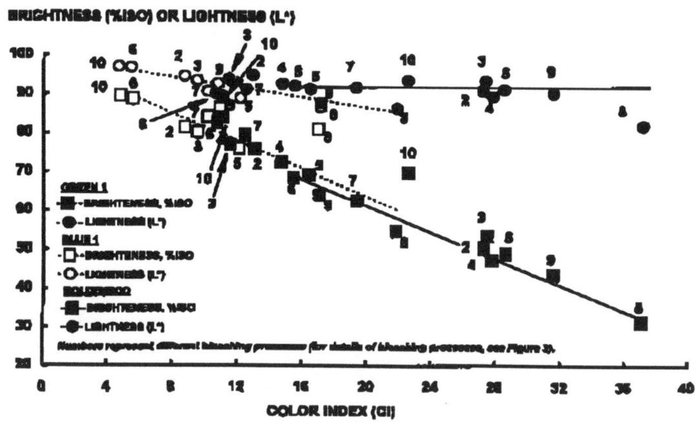

Figure 3. Relationship Between Color Index (CI) and Brightness and Lightness of Bleached Goldenrod, Blue, and Green Colored Broke by Different Bleaching Processes.

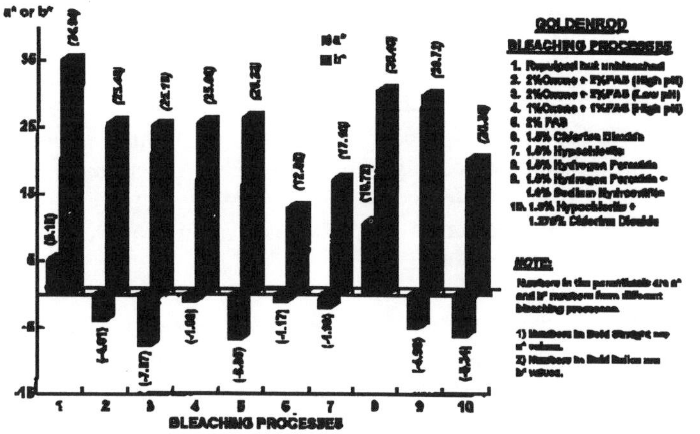

Figure 4. Changes in a* and b* Values of Bleached Goldenrod Colored Broke by Different Bleaching Sequences.

colored broke. The maximum pulp brightness obtained by TCF bleaching is 80 %ISO, below the brightness requirements of premium grade white papers. Single stage hydrogen peroxide bleaching is the least efficient process in bleaching colored broke as seen from CI versus brightness/whiteness relationship (Figure 3).

Among the single-stage elemental chlorine free (ECF) bleaching, chlorine dioxide bleaching was the best followed by hypochlorite bleaching as reflected in the a* and b* values of bleached pulp (Figure 4). Single stage FAS bleaching was only moderately better than hydrogen peroxide, despite using large amounts of FAS [2% on pulp basis](Figures 3, 5, and 6). This is true for even easier to bleach green and blue colored broke. This conclusion is different from the one reached by Kronis [17] who compared the efficiency of FAS, hypochlorite and hydrogen peroxide in bleaching different colored broke and concluded that FAS is better than both hydrogen peroxide and hypochlorite.

Single stage hydrogen peroxide and chlorine dioxide bleaching of blue/green, and canary/cherry (mixed in equal proportion) blends is described elsewhere [16]. Single stage

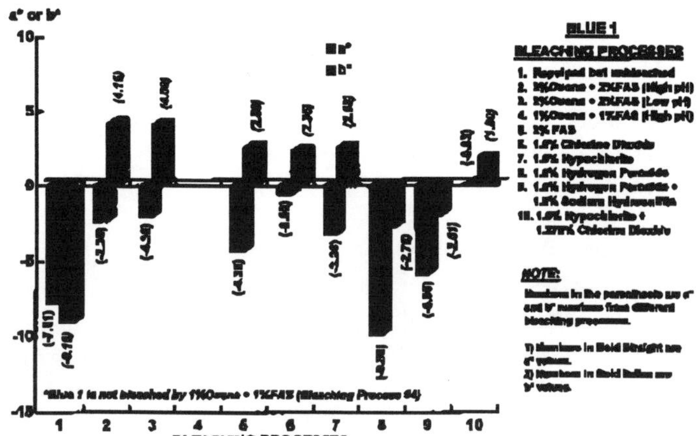

Figure 5. Changes in a* and b* values of Bleached Green Colored Broke Bleached by Different Sequences.

chlorine dioxide bleaching of broke blends (canary/cherry and blue 2/green 2) did provide pulps meeting the brightness, whiteness (L*), a* and b* requirements of premium grade white and color papers. Compared to blue/green colored broke blend, canary/cherry blend required more chlorine dioxide and longer reaction time to reach 86.0% ISO brightness. While single-stage chlorine dioxide bleaching of both broke blends and individual broke provided greater than 82% ISO brightness, the maximum brightness obtained from single stage bleaching with hydrogen peroxide was only 80% ISO.

Among the two-stage bleaching processes, hypochlorite/chlorine dioxide bleaching provided the best results. While hydrogen peroxide and sodium hydrosulfite bleaching are significant next to hypochlorite and chlorine dioxide bleaching for providing high brightness for the blue colored broke, it is last among the bleaching sequences to provide comparable brightness in bleaching either goldenrod or green colored broke. The bleached pulp from two-stage TCF bleaching of blue and green colored broke by oxone and FAS, met the L*a*b* requirement for white and near white paper (Figures 3,5 and 6). However, the bleached pulp had failed to reach the minimum pulp brightness (% ISO) required to use it in grades such as Laser jet, etc. Among the TCF bleaching sequences, only the combination of hydrogen peroxide and hydrosulfite provided pulp brightness close to 85.0% ISO, that too for blue colored broke only.

Maintaining a low pH at the first-stage oxone bleaching is important to reach values of

$$\sqrt{a^{*2}+b^{*2}} \leq 10.0$$

when bleaching by two-stage oxone/FAS combination. For blue and green colored broke, two-stage oxone and FAS bleaching are better than hydrogen peroxide and sodium hydrosulfite bleaching. The green colored broke could be bleached by the two-stage oxone and FAS bleaching to white or near white pulps

$$\left(L^{*}+\geq 90.0 and \sqrt{a^{*2}+b^{*2}} \leq 10.0\right)$$

but to lower brightness (73-77 %ISO). High chemical application (2% oxone and 2% FAS) and low pH at first-stage (oxone) is important to gain pulps of greater than 75% ISO brightness.

The goldenrod colored broke was the most difficult color to bleach by either single or multistage ECF and TCF bleaching. Both the oxone/FAS and peroxide/hydrosulfite bleaching sequences failed to decolorize goldenrod to acceptable brightness. Even with 2% oxone and 2% FAS application and low pH first-stage, goldenrod could be bleached only to a maximum brightness of 54 %ISO, too low to use as a bleached fiber source in any type of writing and printing white or color grades. Only the ECF bleaching with hypochlorite and chlorine dioxide combination (Figures 3 and 4), could provide high brightness; the maximum brightness obtained with this bleaching sequence was only 70.0% ISO. Single stage chlorine dioxide bleaching of goldenrod using 1.5% chlorine dioxide charge was as efficient as

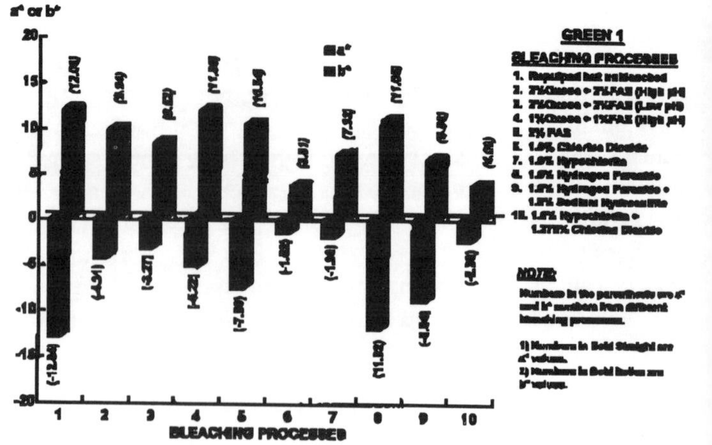

Figure 6. Changes in a* and b* Values from the Bleaching of Green Colored Broke by Different Bleaching Sequences.

two-stage hypochlorite/chlorine dioxide bleaching using 1.275% chlorine dioxide in the final D-stage.

<u>Bleaching of Colored Broke Blends with Brown Stock (Laboratory Results)</u>

The summary of the laboratory results obtained from the multi-stage C/D or DE(OP)D, two-stage E(OP)D and single D-stage bleaching of broke blends mixed with the brown stock is detailed elsewhere [16]. The E(OP)D bleaching of colored broke after blending with C/D or D100 delignified brown stock responded as well as the brown stock/broke blends subjected to the full bleaching sequence of C/DE(OP)D or DE(OP)D. The pulp easily reached final brightness of 86% ISO with a chlorine dioxide application of 0.6% (on oven dry. pulp basis). The pulp obtained met the chromacity value defined by TAPPI for white and near white papers

$$\left(L^{*}+\geq 90.0 and \sqrt{a^{*2}+b^{*2}} \leq 10.0\right)$$

The single D-stage bleaching of colored broke blends mixed with C/D or DE(OP)-delignified soda-AQ pulps provided pulps of equivalent brightness and chromacity values as the brown-stock and broke blends subjected to complete three-stage bleaching. Instead of the present procedure of bleaching the broke blend along with the brown stock by the complete C/DE(OP)D bleaching, this allows the bleach plant to receive the repulped individual or blends of broke from the storage chest and feed it directly to the E(OP)-stage repulper before injecting the chlorine dioxide.

<u>Bleaching of Colored Broke Blends with Brown Stock (Mill Trial)</u>

Mill trial conducted on bleaching the broke blends with the brownstock confirmed the above findings (16). To get bleached pulp brightness of 85% ISO and from bleaching colored broke, it is important that both goldenrod and cherry broke be excluded from the broke blend. If there is occasional contam ination of the blend from these two brokes, the blend should be subjected to the entire C/DE(OP)D bleaching. Otherwise, both brightness and L*a*b* values of the final bleached would be affected. The broke blend of blue/green/canary/pink is easily bleached to a final brightness of 86% ISO and

$$\sqrt{a^{*2}+b^{*2}} \leq 10.0$$

after mixing with chlorinated brown stock and bleached by E(OP)D bleaching provided the broke blend to brown stock does not exceed 25%. If the broke blends are fed to the E(OP) repulper for single D-stage bleaching with extracted soda-AQ pulps, their amount should not exceed 10% of the total furnish.

Tighter consistency control of the blended broke is essential before sending it to the bleach plant to avoid "blow-backs" and upsets to the retention times at the stages. The biggest advantage of bleaching the broke by multi-stage C/D or DE(OP)D is obtaining high brightness pulps with appropriate chromacity values even when the bleach plant experiences certain process upsets. Bleaching of broke with the brownstock did not increase the bleach chemical consumption.

CONCLUSIONS

Bleaching of colored broke (excluding goldenrod colored broke) by single and multi-stage ECF bleaching provided pulps of high brightness and appropriate L*a*b* to use in high quality paper grades like Laser Jet, etc. Goldenrod colored broke is the most difficult to bleach and can be bleached to a maximum brightness of 70% ISO that too using combination of chlorine dioxide and hypochlorite. Chlorine dioxide bleaching is the most efficient, providing pulps of brightness 85.0% ISO and

$$\sqrt{a^{*2}+b^{*2}} \leq 10.0$$

Chlorine dioxide can successfully replace hypochlorite in the bleaching of colored broke to provide pulps having high brightness and low color index (CI). Among the bleach chemicals studied, the performance of single stage bleaching in providing high brightness and appropriate L*a*b* values can be ranked as follows: Chlorine dioxide > hypochlorite > FAS > hydrogen peroxide. This ranking applies to both individual and broke blends. The performance of the two-stage bleaching processes depended on the nature of the colored broke and their initial CI and brightness values. Based on the performance, the two-stage bleaching processes can be ranked as Hypochlorite/chlorine dioxide > 2% Oxone/2% FAS (Low pH first-stage) > 2% Oxone/2% FAS (high pH first-stage) > 1% Oxone/1% FAS (High pH first-stage) > 1.5% hydrogen peroxide/1.0% sodium hydrosulfite. Broke blends (excluding goldenrod and cherry) mixed with brown stock when bleached by multistage bleaching [C/D or DE(OP)D] resulted in pulps of brightness $\geq$ 85,0% ISO, L* $\geq$ 95.0, CI and chromacity values equal to that of TAPPI's white and near-white paper

$$\left(\sqrt{a^{*2}+b^{*2}} \leq 10.0\right)$$

This pulp can be easily used in high brightness paper grades such as Laser Jet, etc., Mixing the broke blends with alkaline extracted chemical pulps and bleaching by single D-stage provided pulps equivalent to C/D or DE(OP)D bleached pulps.

Table 2. Bleaching of Broke Blends* Mixed with Pulp by C/D or D)E(OP)D, E(OP)D and D-Sequence (s).
Note: Broke blend is the blend of blue, green, canary and pink colored brokes in equal portion.

	BLEACHING SEQUENCE									
	C/D or DE(OP)D**		C/DE(OP)D***		D100E(OP)D***		E(OP)D***		D***	
%Substitution in the First-Stage	30/100	30/100	30	30	100	100	100	100	100	100
Kappa Factor at the First-Stage	0.30	0.30	0.30	0.30	0.30	0.30	0.30	0.30	0.30	0.30
%ClO2 Charge in Final D-Stage	0.6	1.2	0.6	1.2	0.6	1.2	0.6	1.2	0.6	1.2
Final D-Stage pH	3.5/4.5	3.5/4.5	3.5/4.5	3.5/4.5	3.5/4.5	3.5/4.5	3.5/4.5	3.5/4.5	3.5/4.5	3.5/4.5
ClO2 Consumed, % of Addition	96.3/95.8	93.8/94.5	96.3	95.8	95.8	94.3	91.4	90.5	97.8	98.3
Brightness	88.1/87.3	88.9/88.5	86.4	88.2	86.1	87.9	86.4	88.0	87.5	88.1
L*	96.0/97.1	97.2/97.3	95.6	96.2	96.0	96.1	95.3	95.6	95.4	95.8
a*	-0.61/-0.55	-0.58/-0.43	-0.72	-0.50	-0.55	-0.41	-0.77	0.49	-0.48	-0.42
b*	3.61/3.04	3.21/2.98	3.75	2.95	3.23	3.01	3.00	3.12	2.76	2.54
(a*^2+b*^2)^0.5	3.66/3.09	3.26/3.01	3.82	2.99	3.28	3.04	3.10	3.16	2.80	2.57

** In the full C/D or D100E(OP)D bleaching and E(OP)D bleaching, the ratio of soda-AQ pulp to broke blends is 75:25 and in the single-stage D-bleaching it is 90:10*
*** Soda-AQ pulp. No broke blend added. Base line data*
**** Hardwood broke blend - Bleaching conditions arelisted in "Experimental" section.*
Two-stage E(OP)D and Single stage D- bleaching of broke is with hardwoodpulp

EXPERIMENTAL

SINGLE-STAGE BLEACHING

Conditions:

	Chlorine Dioxide (D) - Bleaching*	FAS - Bleaching**	Hypochlorite (H) - Bleaching**	Hydrogen Peroxide (P) - Bleaching
Pulp Consistency,%	3.0-5.5	3	3	5.5
Temperature, °C	60	60	60	60
Time, minutes	60 to 240	60	60	60 to 240
Chemical Charge, %@	1.0 -1.5	2	1.0 to 1.5	1.0 to 1.5@@
Final Pulp pH	3.5-5.5	10.5	>8.5	10.0-10.5

* *pH adjustment is by the addition of 4N sulfuric acid to the pulp.*

** *Note: pH adjustment is by NaOH.*

@ *Chemical charges are on as such basis except for NaOCl which is on active Cl_2 basis*

@@ *Other chemicals added during hydrogen peroxide bleaching include % DTPA = 0.3, % Magnesium Sulfate (as MgO) = 0.05, % Sodium Silicate = 3.5, % Sodium Hydroxide = 0.5, % Hydrogen Peroxide (as % H_2O_2) = 1.5.*

TWO-STAGE BLEACHING:

Conditions:

Unless otherwise indicated, pulp is washed between stages.

Oxone (PAA) - Formamidine Sulfinic Acid (FAS) Bleaching

Sequence: **Stage 1 (Oxone) + Stage 2 (FAS) + Final Washing**

Stage 1

Conditions: Pulp Consistency, % = 3.0, Temperature,°C = 60, Time, minutes = 60, pH = 4.0, Oxone charge, % = 1.0 or 2.0 (as oxone)

Stage 2

Conditions: Same as single-stage bleaching except FAS charge = 1.0% or 2.0% as FAS

Sodium Hypochlorite (H) - Chlorine Dioxide Bleaching (D)

Sequence: **Stage 1 (Sodium Hypochlorite) + Stage 2 (Chlorine Dioxide) + Final Washing**

Stage 1

Conditions: Same as single-stage bleaching except NaOCl charge = 1.5% (as Active Cl_2).

Stage 2

Conditions: Same as single-stage bleaching except ClO_2 charge = 1.275%.

Hydrogen Peroxide (P) - Sodium Hydrosulfite (Y) Bleaching

Sequence: **Stage 1 (Hydrogen Peroxide) + Stage 2 (Sodium Hydrosulfite) + Final Washing**

Stage 1

Conditions: Bleaching conditions are same as in single-stage hydrogen peroxide bleaching.
Note: At the end of hydrogen peroxide bleaching, sodium sulfite (0.1% on O.D. Fiber) was added to the pulp to terminate the bleaching and scavenge the residual hydrogen peroxide that otherwise would decompose and destroy the sodium hydrosulfite.

Stage 2

Conditions: Pulp Consistency, % = 6.0, Temperature,°C = 60, Time, minutes = 120, Final pulp pH = 5.0-6.0, % Sodium Hydrosulfite (as % Na_2HSO_3) = 1.275.

Three-Stage Bleaching Conditions:

Sequence(s): **C/D(or D)E(OP)D or E(OP)D or D**
Note: In two-stage E(OP)D bleaching and in the single-stage D bleaching , the ratio of pulp to colored broke (blend) is 90:10]

C/D Stage
(or D-Stage): Pulp Consistency, % = 3.0, Temperature, °C = 45, Time, minutes = 45, Chemical Charge (on O.D. Fiber Basis): % Substitution in C/D-Stage = 30.0/100.0, Kappa Factor = 0.30, Note: Applied Kappa Factor in the first-stage is same for both C/D and D treatments and is based on the Kappa number of the brown stock.

E(OP)-Stage: Pulp Consistency, % = 12.0, Temperature, °C = 75, Time, minutes = 120, pH = 10.8-11.5 Chemical Charge: % Sodium Hydroxide = 2.25, % Magnesium Sulfate as MgO = 0.05, % Hydrogen Peroxide as H_2O_2 = 0.2, Oxygen Pressure, bars = 3.2 to 0 (at bottom of the retention tube to top of the retention tube).

LITERATURE CITED

1. **Hise, R.**, "Process Changes to Lower the Formation of Dioxins and Chlorinated Organic Compounds During the Chemical Pulp Bleaching". in 1991 Bleach Plant Operations Short Course, TAPPI PRESS, Atlanta, GA., pp. 105-118 (June 1991).

2. **Larson, L.E., and H. de V. Partridge**, "Bleaching with Hypochlorites" in "The Bleaching of Pulp", R.P., Singh.Ed., TAPPI PRESS. Atlanta, GA., p. 106 (1979).

3. **Axegard, P., et al.**, "The Matrix for Softwood Kraft Pulp - Pulp Quality and Effluent Load". in 1991 Bleach Plant Operations Short Course, TAPPI PRESS, Atlanta, GA., pp.21-22 (June 1991).

4.**Quinnett, P.E., and L.R. Ward.**, "Chlorine Dioxide for Processing Mixed Paper Waste"., Proceedings of 1995 TAPPI Pulping Conference, TAPPI PRESS, Atlanta, GA., pp. 695-706 (October 1995).

5. **Naddeo, R., et al.**, "Oxidative Cleaning and Bleaching of Secondary Fibers" in Waste Paper II Conference., Chicago, IL (May 1991).

6. **van Lierop, B., and N. Liebergott**, "Bleaching of Secondary Fiber Pulps" in CPPA 1st Forum on Recycling Proceedings, Toronto, ON. pp. 175-182 (October 1991).

7. **Minor, J.L., and E.L. Springer,** "Delignification of Wood Fibers with Peroxymonosufate- A Potential New Pulping Process". Proceedings of 1989 International Symposium on Wood and Pulping Chemistry, TAPPI PRESS, Atlanta, GA., pp. 133-138 (May 1989).

8.**Hache, M.J.A., et al.**, Pulp. Paper Can., 95 (12) : T 511-T 514 (1994).

9. **Darlington, B., et al.**, " Secondary Fiber Color Stripping - Evaluation of Alternatives," Proceedings of 1992 TAPPI Pulping Conference, TAPPI PRESS, Atlanta, GA., pp. 67-74 (November 1994).

10. **Forsberg, P., and J.M., Genco**, "Bleaching Mixed Office Waste to High Brightness," Proceedings of 1993 TAPPI Pulping Conference, TAPPI PRESS, Atlanta, GA., pp. 1219-1229 (November 1993).

11. **Judd, G.B., and G.Wyszeci**, "Color in Science, Business and Industry," Wiley-Interscience, New York, NY. (1975).

12. **Popson, S.J., and D.D. Malthouse**, "Measurement and Control of the Optical Properties of Paper". Technidyne Corporation, New Albany, IN. 4th Publication (1990).

13."L,a,b,45°, 0° Colorimetery of White and Near-White Paper and Paper Board". T 524 om-86, TAPPI Test Methods 1991, Vol. 1. TAPPI Press, Technology Park/Atlanta, GA (1992).

14. **Sharpe, P.E., and R.W. Lowe**, "The Bleaching of Colored Recycled Fibers," Proceedings of 1993 TAPPI Pulping Conference, TAPPI PRESS, Atlanta, GA., pp. 1205-1217 (November 1993).

15.**Rangamannar, G.**, "Influence of Colored Papers on the Brightness of Ledger Pulps," Proceedings of 1993 TAPPI Pulping Conference, TAPPI PRESS, Atlanta, GA., pp. 1231-1236 (November 1993).

16.**Parthasarathy, V.R.**, "Elemental Chlorine Free (ECF) and Totally Chlorine Free (TCF) Single and Multi-Stage Bleaching of Colored Broke" in 1996 TAPPI/AIChE Conference (October 1996). To be published in July 1997 TAPPI J.

17.**Kronis, J.D.**, "Adding Some Color to Your Wastepaper Furnish," Proceedings of 1992 TAPPI Pulping Conference, Atlanta, TAPPI PRESS, Atlanta, GA., pp. 223-231 (November 1992).

18."Diffuse Brightness of Pulp (d/00)" T 525 om-86, TAPPI Test Methods 1991, Vol. 1. TAPPI Press, Technology Park/Atlanta, GA (1992).

19."Forming Hand Sheets for the Reflectance Tests of Pulp". T 218 om-83, TAPPI Test Methods 1991, Vol. 1. TAPPI Press, Technology Park/Atlanta, GA (1992).

20."Color of Paper and Paperboard in CIE Y; x,y or Y Dominant Wavelength and Excitation Purity".T 527 su-72, TAPPI Test Methods 1991, Vol. 1. TAPPI Press, Technology Park/Atlanta, GA (1992).

Repulping of Wet-strength Poly-coated Cartons Using Non-chlorine Chemicals

R. de Jong, and J. Rogers
James River Corporation, Neenah Technical Center, Neenah, WI 54957

S. Aziz and H.S. Dugal
Integrated Paper Services, 101 West Edison Avenue, Appleton, WI 54912

The objective of this study was to repulp high wet-strength and poly-coated milk cartons economically without the use of chlorine chemicals. Four types of poly-coated waste as well as caustic, hydrogen peroxide, and oxone treatments were evaluated in this study. The experimental plan was formulated to investigate the effect of soaking, size, and chemicals on fiber separation. This study was conducted in the laboratory and at two pilot plants.

In the laboratory experiments a sample size of 7.5 x 7.5 cm was found to be optimum. Soaking time had little influence on fiber separation. The use of chemicals (caustic, alkaline peroxide, oxone) increased the fiber yield at temperatures of 60°C and a reaction time of one hour. Shredding and presoaking with non-chlorine chemicals was not effective.

Experiments using aerobic and anaerobic microorganisms in the Quantum mixer at different consistencies showed the importance of consistency; a higher consistency in the Quantum mixer led to better fiber separation. Increasing the intensity of mixing resulted in higher yields but also resulted in shredding of the poly into small particles which could not be separated from the fiber by commercial screening techniques.

Pilot plant studies at Lamort (France) were partially successful. It was observed that adding water and mixing for half an hour prior to adding chemicals improved the fiber separation. However, the best result obtained still had a high flake content and poor fiber yield after 60 minutes of pulping.

Studies at Voith-Sulzer (Appleton) using the above technique decreased the flake content to 5% but required an extended pulping time (120 minutes) with caustic and peroxide.

The conclusion from this study is that chemicals are still required to achieve adequate fiber/poly separation and that they should be added in the proper sequence (i.e. caustic followed by peroxide after the cartons have been repulped at high consistency for 30 minutes). Attempts to reduce the high cost of chemicals by shredding, presoaking and using microorganisms had little effect on yield.

As a result, the systems investigated in this study were not economical for implementation at the Ashland mill of James River Corp. Further studies need to be conducted looking at newer concepts in fiber delamination including pulper design.

1. OBJECTIVE

This research was designed to investigate the possibility of repulping high wet-strength and poly-coated milk cartons economically without using chlorine containing compounds. Chemicals used in this study were caustic, hydrogen peroxide, and oxone.

2. BACKGROUND

The Ashland mill of James River Corp. processes post consumer milk cartons in their medium consistency pulper by adding 12.5 to 17.5 kg calcium hypochlorite per ton of post consumer milk cartons. This high shear pulper requires a long beating time which results in a considerable amount of shredding of the poly on the poly-coated milk cartons. The small poly particles are very difficult to remove from the stock.

The mill would like to be able to use high wet-strength, post-consumer milk cartons in their high consistency Voith pulper without adding any chlorine-containing chemicals. The mill would also like to restrict the pulping time to less than 90 minutes and keep the temperature below 60°C and still obtain a flake content of less than 5% (where flake content is defined as the amount of non-poly material retained on a 0.15 mm cut screen after pulping).

A possible approach to improve the repulping of post consumer milk cartons would be to pre-soak the shredded material in chemicals without chlorine like caustic. As a result, the following procedure was proposed for handling high wet-strength milk cartons in the lab.

- Shred cartons by tearing to avoid compression of the shredded edges.
- Impregnate shredded cartons with hot caustic plus per oxide.
- Store shredded, impregnated cartons in a heated hop per next to the pulper for 1-4 pulper cycles.
- Discharge hopper contents into (high consistency) pulper as 10, 25, or 100% of the furnish depending

upon how effective the shredding, impregnation system works and how long a pre-soaking time is required.

3. DESCRIPTION OF WORK DONE

Four types of poly-coated waste were used in the study:

Poly Waste #1) Medium wet-strength, unprinted poly-coated board from a bale of Tetrapak waste.

Poly Waste #2) High wet-strength, unprinted milk carton board from one bale, which turned out to contain varying amounts of wet-strength as determined by the nitrogen content of the poly-coated board samples (i.e., 0.045 to 0.09%).

Poly Waste #3) Flat, printed Tropicana juice cartons with poly on both sides, obtained from two pallets of unused juice cartons, which had been rejected for being out of date. These cartons contained 0.07% nitrogen.

Poly Waste #4) Unshredded, post consumer milk cartons with glued bottoms, which could be more difficult to repulp than the unglued, flat juice cartons.

3.1 Work Done at Integrated Paper Services (IPS)

The experiments carried out at the IPS laboratory to simulate mixing were performed using a high-shear mixer from Quantum Technology. The high-shear mixer is equipped with a central control unit which controls the mixing frequency, chemical addition and heat-up time, and temperature. In most of the experiments, the frequency of mixing was kept at 20 hertz for 1 second every 40 seconds. For the experiments using enriched mill process water, it was necessary to decrease the frequency to 10 hertz with continuous mixing , inorder to obtain measurable repulping without shredding the poly.

Three hundred grams of shredded poly-coated wet strength cartons (Poly Waste #1 and #2) of different sizes (1.25 x 5, 2.5 x 2.5, 5 x 5, and 7.5 x 7.5 cm), with the required amount of water and chemicals, were introduced into the mixer. In some cases, the shredded cartons were pre-soaked for 3-24 hours before mixing in the Quantum mixer. After mixing, the pulp slurry was removed, diluted to 1%, and screened on a Valley flat screen (0.2 mm slot width). In some experiments, the slurry was soaked overnight before screening. The accepts and rejects were collected and yield of the fiber was determined.

The yield for the treatment process was based on any material that passed through the 0.2 mm screen (i.e. 300 grams minus the weight of rejects) divided by 300 grams of original sample. This method includes fines, ash, and solubles as part of the yield and therefore gives a high yield.

3.2 Work Done at Neenah Technical Center

The first six experiments used a modified Waring Blender. All experiments were evaluated for percent yield using a 0.15 mm flat screen.

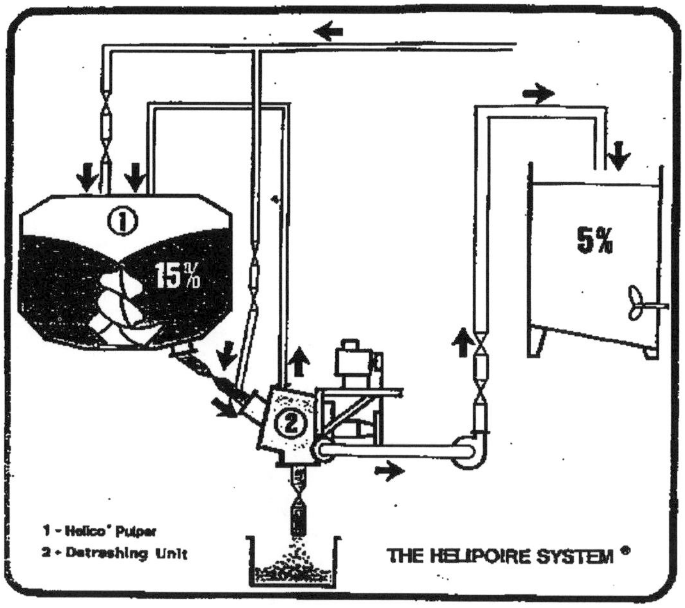

Figure 1. Schematic of Lamort Pulper/Scavenger Arrangement in Pilot Plant.

Since the Waring Blender experiments were unsatisfactory, the Fiber Lab continued the experiments with the Lamort high consistency laboratory pulper. The following conditions were applied:

- •Temperature: 60°C
- •Pulping time: 45 minutes
- •Consistency: 10%
- •No soaking

An indication of the amount of wet-strength in the samples was obtained by determining the nitrogen content using a Kjeldahl analysis for nitrogen.

3.3 Work Done at Pilot Plant Trials

After determining the wet-strength content, two pallets of Poly Waste #3 were shipped to France for testing in the Lamort pilot plant in their 230 kg high consistency pulper.

The trials were limited to pulping the furnish with different chemicals. Paper and water were first introduced, followed by 30 minutes pulping without chemicals in order to increase the shear forces during the initial stages. Only then were the chemicals added. Pulping was then continued for 60 to 75 minutes of total slushing time.
Testing and analyses were done on:

- •stock sampled during pulping to determine wet strength flake content on a Sommerville screen with 5 mm holes, after removing the poly by hand.
- • stock after dumping through the scavenger with 6 mm holes, using Sommerville with 0.15 mm slots.

• rejects from the scavenger for estimating the percent of fibers within the rejects.

The pilot plant pulper/scavenger arrangement is shown in Figure 1.

Another pilot plant trial was run at Voith, Appleton, using post consumer milk cartons (Poly Waste #4). The same sequence of adding water and chemicals was used during the trial at Voith as was used in the smaller Lamort pulper.

4. RESULTS AT IPS

4.1Tetrapak Medium Wet-Strength Cartons (Poly Waste #1)

The first six experiments were performed on 1.25 x 5 cm strips of medium wet-strength Tetrapak at neutral pH (7-8) with 3-24 hours soak time and 50-90°C mixing temperature. The fiber yield obtained from these trials was (37-77%). Adding a cellulase enzyme to the system did not improve the yield.

An additional sixteen experiments were performed on 7.5 x 7.5 cm square Tetrapak. In these experiments, the pH was varied from 2 to 12. The best results (yields of 80% or higher) were obtained in the pH range between 10 and 12.

4.2 High Wet-Strength Cartons (Poly Waste #2)

The next set of experiments (17) was carried out on high wet-strength milk carton board (7.5 x 7.5 cm squares). The pulp

Table 4.2.1: Effect of Caustic, Peroxide and Soaking on Yield of Poly Waste #2

Exp. #	Chemicals	pH	Yield %	Comments
23+24	None	7	16	Control
25+26+27	0.15% Caustic	10	6-14	3-24 hrs presoak
30	1.1% Caustic	12	25.4	
29+31	1.1% Caustic + 1% Peroxide	12	49-52	
32	" "	12	68.6	18 hrs post soak
28+33+34	" "	12	74-76	3 hrs presoak + 18 hrs post soak
35+40	" "	11-12	79.6	Mix at 80 + 90°C 18 hrs post soak
36+37	" "	12	81	3-20 hrs post soak
38+39	" "	12	78-81	1 hr post soak with + without surfactant

yield increased from 16% without any chemicals to 25% with the addition of caustic and up to 81% with the addition of peroxide and caustic (Table 4.2.1). Three experiments were conducted on Poly Waste #2 to determine the effect of soaking. The data indicated that presoaking was not necessary.
The next step was to look at other chemicals which might help

Table 4.2.2: Effect of Chemicals Without Soaking on Yield of Poly Waste #2

Exp. #	Chemicals	pH	Yield %	Comments
44	1.1% Caustic + 0.5% Peroxide	12.5	50.6	No soaking
43	1.1% Caustic + 1% Peroxide	12	61.9	No soaking + 0.3% non-ionic surface active agent
41	" "	12.5	73.9	No soaking
42	" "	12	77.7	No soaking + 0.1% silicate
59	" "	12.5	80.4	No soaking + 0.3% nonionic surface active agent
58	2% Caustic + 1% Oxone	12	75	No soaking
48	5% Caustic +2% Oxone	13	84.8	No soaking

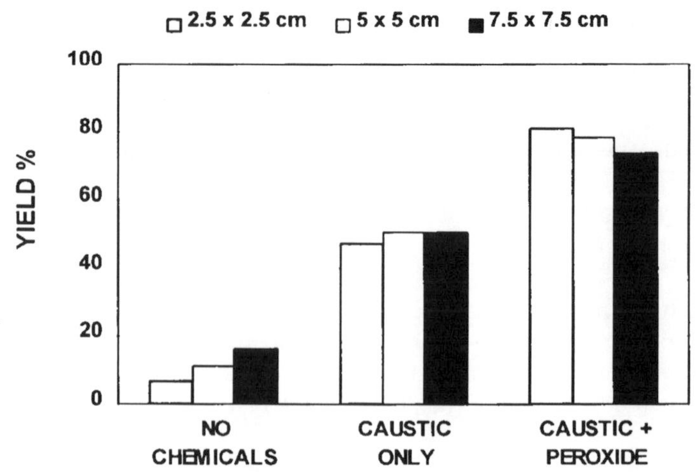

Figure 2. Effect of Chemicals and Size on % Yield of Poly Waste #2.

in the separation process. Twenty experiments were performed on high wet-strength milk carton board (Poly Waste #2). The results are shown in Table 4.2.2.

It was found that surfactants can increase the yield of the fibers by approximately 6%. Oxone was effective in separating the fibers but required higher charges of sodium hydroxide than the peroxide experiments.

Next, we explored the effect of sample size. Three different sizes were chosen (2.5 x 2.5, 5 x 5, 7.5 x 7.5 cm) as shown in Figure 2. The data showed that chemicals were more important than size.

5.RESULTS AT NEENAH TECHNICAL CENTER

5.1 Waring Blender Results on Poly Waste #2

Six experiments were run in the Waring Blender using 7.5 x 7.5 cm squares of high wet-strength, poly-coated milk carton board (Poly Waste #2) using tap water and different chemicals and amounts. It was not possible to obtain a yield over 35% as shown in Table 5.1.
Five experiments were run in the Lamort high consistency lab

Table 5.1 - Experiments in Waring Blender on Poly Waste #2
(Using 7.5 x 7.5 cm squares, without soaking, in tap water at 60°C, 2% consistency, and mixing for 30 seconds.)

Exp. #	Chemicals	% Yield
1	None	15.4
2	1% Caustic	30.0
3	1% Caustic + 1% Peroxide	24.2
4	1% Hypo	29.4
5	6% Caustic	33.8
6	6% Caustic + 6% Peroxide	24.9

pulper using 7.5 x 7.5 cm squares of high wet-strength, poly-coated milk carton board (Poly Waste #2) and adding Neenah tap water, Ashland mill water, and different chemicals. A yield of more than 65% was obtained regardless of the chemicals used as shown in Table 5.2.

Table 5.2 Lab Pulper Results on Poly Waste #2
(Using 7.5 x 7.5 cm squares, without soaking, in tap or mill water at 60°C, 10% consistency and pulping for 45 minutes.)

Exp. #	Chemicals	Water	% Yield
1	None	Tap	16.0
2	1% Caustic	Tap	65.2
3	1% Caustic + 1% Peroxide	Tap	65.2
4	1% Hypo	Tap	65.6
5	1% Hypo	Mill	66.2

5.3 Lab Pulper Results on Poly Waste #4

The flat, high wet-strength juice cartons (Poly Waste #3), which were tested in France, did not contain any glued bottoms. The Ashland mill reported that the glued bottoms were more difficult to repulp. As a result, seven additional experiments were run in the Lamort high consistency lab pulper using post consumer milk cartons with glued bottoms (Poly Waste #4). The influence of aerobic and anaerobic microorganisms was also evaluated and compared with tap water plus 1% caustic. The results are summarized in Table 5.3.

The best results (52.4% and 44.1% yield) were obtained with 1% caustic at 60oC, and 40oC, respectively. The yield was lower than that obtained with the juice cartons. The yields with microorganisms were less than 35%.

Table 5.3 - Experiments in Lab Pulper with Poly Waste #4
(Using tap water or microorganisms, 0 to 1% caustic, at 40-60°C, 10% consistency and pulping for 45 minutes.)

Exp. #	Chemicals	Water	Temp. °C	% Yield
1	None	Tap	60	21.0
2	1% Caustic	Tap	60	52.4
3	1% Caustic	Tap	40	44.1
4	0.3% Nonionic surfactant	Tap	40	30.6
5	" "	Microorganisms (aerobic & anaerobic)	40	28.1
6	" "	Microorganisms (aerobic)	40	27.9
7	" "	Microorganisms (anaerobic)	60	26.1

6. RESULTS FROM PILOT PLANT TRIALS

6.1. Results from Lamort in France

The results from the seven trials conducted at the pilot plant of Lamort are shown in Table 6.1 and in Figure 3.

All the pulpers were successfully discharged through the detrashing unit, except for the pulper in trial E with 1% hypo, when the detrasher became plugged. (The Ashland target of less than 5% flakes to obtain a fiber yield of more than 95% was, however, not achieved after pulping 60 minutes.)

Table 6.1: Pilot Plant Results with Lamort Pulper in France on Poly Waste #3
(Showing fiber yield after 60 minutes of pulping, where fiber yield is defined as 100% - the % wet-strength flakes retained on a screen with 5 mm holes)

Trial #	Temperature °C	Chemicals added	% FiberYield after 60 minutes of pulping
A	60	1% Caustic	85
B	60	0.5% Caustic	85
C	60	0.5% Caustic + 1% Peroxide	87
D	38	0.5% Caustic	72 (poor)
E	60	1% Hypo	59 (very poor)
F	60	0.5% Caustic + 1% Oxone	93 (best)
G	41	0.5% Caustic + 1% Peroxide	69 (poor, but good with 90% yield after 75 minutes pulping)

The composition of the rejects from the detrasher unit was also determined. Trial F and G had the best composition with 3 and 6% fiber in the detrasher rejects, respectively. Trial E had the worst composition with 55% fiber in the rejects. The high amount of wet-strength flakes in the detrasher unit caused the 6 mm holes in the detrasher unit to plug with fiber and plastic during Trial E.

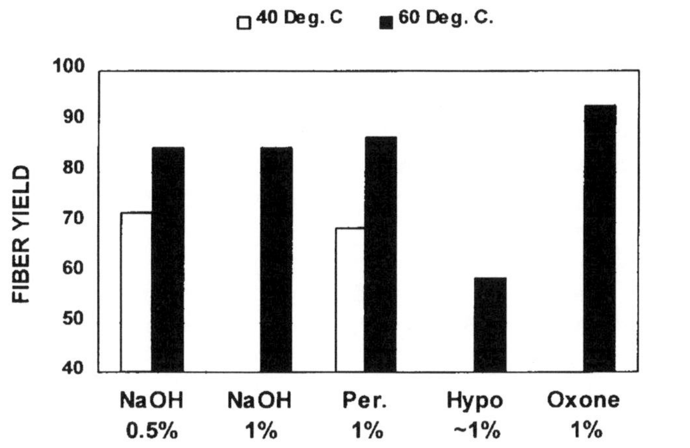

Figure 3. Effect of Chemicals and Temperature on % Fiber Yield During Lamort Trials.

6.2. Results from Voith-Sulzer in Appleton on Poly Waste #4

Full scale trials at Voith-Sulzer in Appleton, Wisconsin, using unshredded post-consumer milk cartons with glued bottoms were only partly successful, in spite of using the staggered approach for adding water and chemicals applied in July, 1995 at Lamort. It was possible to obtain a flake content of less than 5%, but only when applying 1% caustic plus 1% peroxide at 60°C to the pulper and after slushing in the pulper for 120 minutes. This would be an uneconomical proposition for the Ashland mill due to high chemical costs and high energy consumption during the extended beating time.

7. DISCUSSION

The original objective of the project was to repulp high wet-strength and poly-coated milk cartons without the use of chlorine chemicals or other high (chemical) costs. It was envisaged that shredding and soaking the milk cartons would improve repulping. As a result, the following parameters were investigated:

- influence of size and soaking
- effect of microorganisms
- effect of consistency
- effect of mixing intensity

7.1 Influence of Size and Soaking on Fiber Separation

The first experiments were done to observe the effect of size and soaking on fiber separation. Three different sizes were chosen (2.5 x 2.5, 5 x 5, 7.5 x 7.5 cm) as shown in figure 1. The data showed that the chemicals were more imprtant than size and that the 7.5 x 7.5 cm samples gave optimal results. We also found that soaking had a minimal effect on fiber separation.

These results were discussed with mill personnel. During the discussion with the mill personnel, it was suggested that we should do more experiments using a lower pH. The high pH would be hard to implement without many process modifications.

In order to reduce the pH we decided to explore (a) increased mixing, and (b) the use of micro-organisms. It had been reported by the Netherlands Organization for Applied Scientific Research, TNO, that delamination can be achieved by using mill process water [2]. According to them, the process water contains sufficient quantities of microorganisms which interact with the wet strength and help in delamination. Eight experiments were conducted using process water from a Wisconsin paper mill. These results produced disappointing yields in the range of 3 to 15%. At this point we decided to get advice on understanding the use of microorganisms from Dr. Rouf .

7.2 Effect of Microorganism Enriched Mill Process Water

After consulting with Dr. Rouf, it was determined that the past experiments with mill process water may have been unsuccessful due to very low microorganism activity. Dr. Rouf enriched the mill process water with microorganisms at his laboratory and provided the enriched water to IPS for use in the experiments conducted in the Quantum mixer. A total of eighteen experiments were conducted. The mixing frequency for these experiments was reduced to 10 hertz, and the mixer was run continuously inorder to improve the yield with out shredding the poly. A control sample with water was run which produced a yield of 55%. The results of these experiments are shown in Table 7.2. As can be seen, the pulp yield improved to a maximum of 75% for the juice cartons. These results were very encouraging. We achieved the mill's desire to operate at a lower pH and still obtained a high pulp yield. It seemed that the enriched process water was the reason for this excellent separation. One thing which surprised us was that the pulp yield was not affected by the type of microorganisms introduced; e.g., aerobic water versus anaerobic water. This made us analyze the data more carefully.

7.3 Effect of Consistency

Table 7.2 Effect of Microorganisms on Poly Waste #3

Exp. #	Water Type	Consistency	% Yield
89, 104, 103,105	Distilled Distilled	15 23	55.3-55.6 71-71.6
91, 92, 97 98 90, 102, 108	Microorganisms (aerobic and anaerobic) " "	23-30 10-16	74.6-75.5 43.5-66.6
94, 106 93, 96	Microorganisms (aerobic)	10-15 23	66.7-67.3 58-73.4
95, 107	Microorganisms (anaerobic)	15	63.1-66.8

On further analyses of the data, we noticed a strong consistency effect. During some of the experiments (which had given us the higher yields), the first phase (30 minutes) was at higher consistency. Figure 4 is a plot of the average yield obtained at 15 and 23% consistencies for distilled and mill enriched waters. This figure clearly shows the influence of consistency on fiber separation. For example, increasing the consistency from 15 to 23% for distilled water mixing resulted in improvement of fiber yield from 55 to 71%.

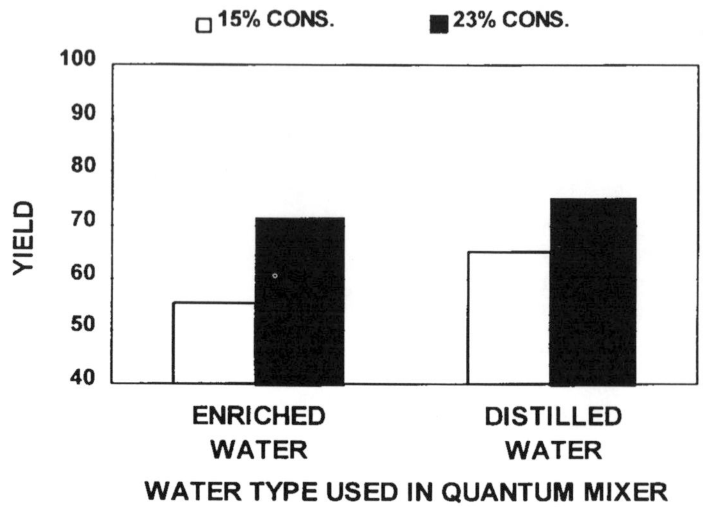

Figure 4. Effect of Consistency on % Yield Using Enriched and Distilled Water.

7.4 Effect of Mixing Intensity

The shear forces produced during pulping depend on a lot of factors, such as:

- consistency
- viscosity (chemicals/temperature)
- circulation speed (rotor design/rpm)
- type of pulping equipment

The importance of these factors was demonstrated by the different results on Poly Waste #1 to #4 using different pulpers with different rotors at different consistencies. Table 7.4 compares these results for 1% caustic at 60°C.

Table 7.4 Influence of Pulping Equipment on Repulping
(Using 1% caustic at 60°C and 3 furnishes.)

Type of Pulping Equipment	Furnish Type	Consistency %	Pulping	Results	
			Delamination*	% Yield	% Flakes
Waring Blender	#2	2	Poor	30	
Lab Pulper	#2 #4	10 10	Good Good	65 52	
Quantum Mixer	#1 #2	15 15	Good Good	80 25-51	
Lamort pulper after 60 min pulping	#3	15	Good		15
Voith pulper after 60 min pulping	#4	15	Poor		>15

* Poor is little separation of poly/fiber and good indicates poly/fiber is separated.

The furnish, the type of pulping equipment, and the consistency influence the mixing conditions and the test results. If the pulping time is too long, or the shear force intensity during pulping is too great, then the poly breaks up into little pieces which are too difficult to remove.

During the pilot trials, a better poly/fiber separation was attrib-

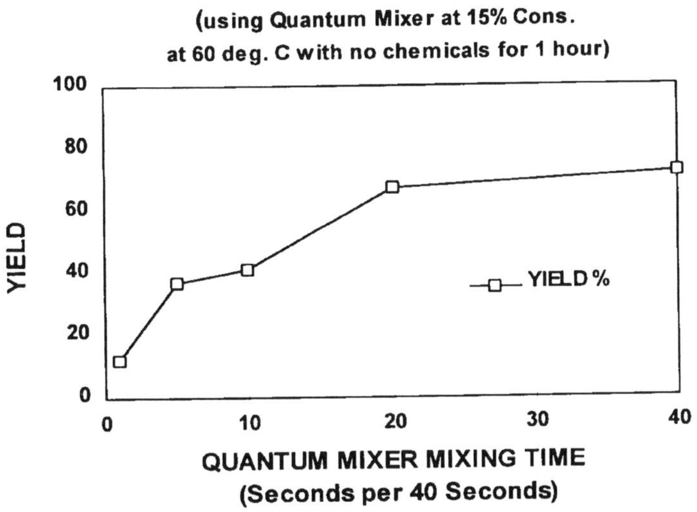

Figure 5. Effect of Mixing Intensity on % Yield of Poly Waste #3

uted to better mixing. We decided to explore this as a way of possibly obtaining better separation at lower pH. A number of experiments were performed in the Quantum mixer (Figure 5) during which the mixing time was increased. The results show that the higher the energy put into the sample, the better the pulp yield. For example, pulp yields of 72% at pH of 7 were obtained. As the mixing was increased, we noticed that the poly was being shredded into smaller particles. The small particles of poly are undesirable since in mill operations one cannot effectively separate these particles from the pulp. Hence the product obtained becomes contaminated.

8. CONCLUSIONS

8.1 The Quantum Mixer, Waring Blender, NTC Lab Pulper and pilot plant pulper all gave different results on similar furnishes.

8.2 High consistency and good mixing are key variables in obtaining a high yield of fiber.

8.3 Chemicals are still required to achieve adequate fiber/poly separation and deflaking at 60°C. The best results were obtained with 1.1% caustic plus 1% peroxide or oxone at that temperature added after an initial delamination stage without chemicals.

8.4 Attempts to reduce the high cost of chemicals by shredding, presoaking and using microorganisms had negligible effect on the yield.

8.5 All the processes tried were uneconomical for the Ashland mill.

9. RECOMMENDATIONS

9.1 As a consequence of the above conclusions, the results of this study can not be applied to the Voith pulper at the Ashland mill.

9.2 A way to predict the behavior of the commercial pulper in the laboratory should be determined.

9.3 New ways of improving delamination during pulping should be investigated.

10. LITERATURE CITED

1.**Lamort**, "Technical Bulletin," nr.84/1051, (Sept. 1991).

2. **Hesselink, P.G.M., Doddema H.J., and van der Meulen Bosma, F.O.J.**, "Biotechnological Delamination and Pulping," Netherlands Organization for Applied Scientific Research TNO, Delft, the Netherlands, Report nr. 90/3549/4403-80102/DN, (Aug. 1990).

11. ACKNOWLEDGEMENTS

The financial support for this study was provided by the Wisconsin Department of Development Contract #RTA FY94-0265. The laboratory experiments were performed by J. Albers at Integrated Paper Services and S. Seidl under the direction of A. Button at James River. The assistance of Dr.MA.Rouf (Chairman,Microbiology Dept.,University of Wisconsin-Oshkosh) with the preparation of the microorganisms is also greatly appreciated.

Corrosion of Composite Tube Air Ports in Kraft Recovery Boiler - Cr_2O_3 Solubility in NaOH

Matthew J. Estes
Proctor & Gamble, Winton Hill Technical Center, 6100 Center Hill Avenue, Cincinnati, OH 45224

Alan W. Rudie
Institute for Paper Science and Technology, 500 10th St., N.W., Atlanta, GA 30318

Jeffrey A. Colwell
Battelle Inc., 505 King Avenue, Columbus, OH 43201

Composite tubes in kraft recovery boiler air ports can experience severe corrosion. One possibility is the corrosion is caused by molten NaOH. The corrosion of stainless steel in molten salts is dependent upon maintaining the protective Cr_2O_3 film formed on the surface. The goal of this project was to measure the solubility of $Cr_2 O_3$ as a function of the basicity of NaOH. Plots of the chromium ion concentration indicate a minimum solubility at -log $a_{(Na20)}$ of 8.2. The data support two mechanisms for dissolution of the Cr_2O_3. The basic dissolution had a slope of -0.988, consistent with the formation of Na_2CrO_4. The acidic dissolution had a slope of 1.48, consistent with the formation of Cr^{3+}. The results from weight loss experiments on chromium metal were then compared to the solubility curves for Cr_2O_3 to confirm that metal oxide solubility was the rate limiting step in the corrosion mechanism. The weight loss experiments on chromium metal also confirmed that the corrosion rate was a function of the basicity of molten NaOH. Understanding the mechanisms and the environment in which minimum solubility occurs for chromium oxide allows a solution to be proposed to lessen the composite tube air port corrosion of kraft recovery boilers.

Introduction

In recent years, the frequency of corrosion in recovery boiler air ports has risen dramatically throughout the pulp and paper industry (1-5). The increase coincides with the growing use of composite tubes, which are fabricated with a stainless steel outer shell and carbon steel inner body. A unique characteristic of air port corrosion is that the stainless steel is rapidly attacked, exposing the carbon steel underneath. Stainless steel corrosion rates up to 35-150 mils per year (mpy) have been measured (1-5). After completely consuming the stainless steel, corrosion of the underlying carbon steel occurs at a rate of 5-15 mpy (1-5). However, there is some concern that the carbon steel can suffer sulfidation corrosion, which was the reason for using composite tubes in air ports.

Molten sodium hydroxide has been postulated as a possible cause for corrosion at air ports (1-3, 5). The severity and uniformity of corrosion are consistent with a molten salt reaction. It has been suggested that sodium hydroxide vaporizing from the smelt bed can migrate to the air port (1). The walls of the air port at 275 to 500°C are relatively cool (4-6). At these temperatures, sodium hydroxide vapors would condense on the waterwall tube surface as a liquid.

There is some evidence for NaOH being present at kraft recovery boiler air ports. Colwell and Fonder collected deposit samples on a cooled probe and measured the melting point *in situ* inside a kraft recovery boiler. Analysis of a limited deposit showed a Na:K ratio of approximately 6:1. The melting points measured yielded a value close to the theoretical melting point of NaOH:KOH of 6 to 1 ratio (3).

Laboratory experimentation has shown that molten sodium hydroxide corrodes stainless steel faster than carbon steel (3). Other corrosive mechanisms in addition to molten NaOH may take place as suggested by Falat (6), but they do not explain why stainless steel corrodes faster than carbon steel.

Theory of molten salt reactions

The stability of an oxide layer on a metal is a function of the oxide's thermodynamics, growth kinetics, and the dissolution of that oxide in the environment. Solubility of an oxide can be defined as the maximum weight of metal ions in solution divided by the weight of the solution. Solubility of a nominally protective oxide is often a function of the salt basicity; where solubility is the defining mechanism, the corrosion rate should also be a function of basicity.

Basicity of a molten salt is defined on the basis of the dissociation of the compound, in this case, NaOH:

$$2\ NaOH = Na_2O + H_2O \qquad (1)$$

For sodium hydroxide, sodium oxide is defined as the Brönsted base because it accepts a proton, and water is defined as the Brönsted acid because it donates a proton. Basicity, for NaOH, can also be defined by -log $a_{(Na20)}$, analogous to the definition of pH.

A considerable volume of work has been reported in the literature showing the solubility of different metal ions plotted as function of basicity as shown in Figure 1 (7-15).

Figure 1: Measured oxide solubilities in Na_2SO_4 at 1200°K (12).

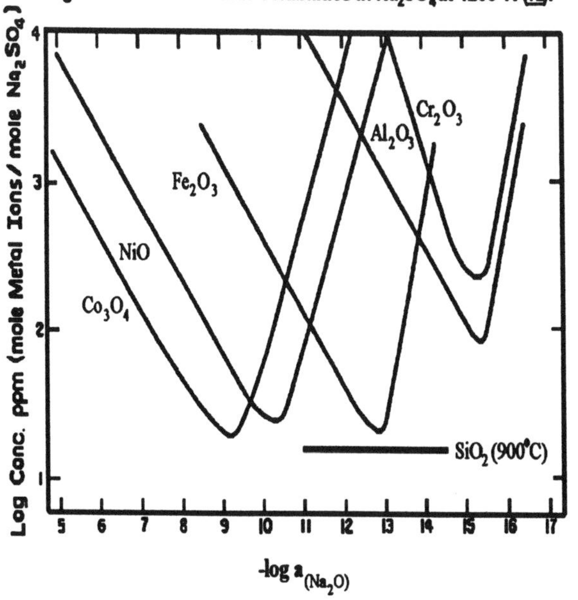

The plots have basicity values along the x-axis and the solubility limit of the metal oxide along the y-axis. These plots show the basicity conditions that give the minimum solubility for a particular metal oxide. If the basicity conditions for a metal oxide is located to the left side of the minimum, the material undergoes a basic dissolution mechanism such as:

$$M_xO_Y + O^{2-} = M_xO^{2-}_{Y+1} \qquad (2)$$

where M = metal O = oxygen

If the basicity conditions for a metal oxide is located to the right side of the minimum, the material undergoes an acidic dissolution mechanism, which is:

$$M_xO_Y = X\,M^{a+} + Y\,O^{2-} \qquad (3)$$

Thus, the location on the curve provides information about which type of dissolution is occurring.

In addition, based on various assumed soluble ions, the dependence of the solubility of a particular ion on basicity can be related to the experimental slope of the curve. An example of acidic dissolution would be:

$$\left(\frac{-\log M^{a+}}{-\log O^{2-}}\right) = \frac{Y}{X} \qquad (4)$$

Agreement between the experimental slopes and theoretical slopes based on stoichiometric equations can be used to confirm the identity of the soluble ion (2).

Equipment

A schematic cross section of the furnace, the crucibles holding the sample, and the electrodes to measure basicity is shown in Figure 2.

Figure 2: Equipment setup

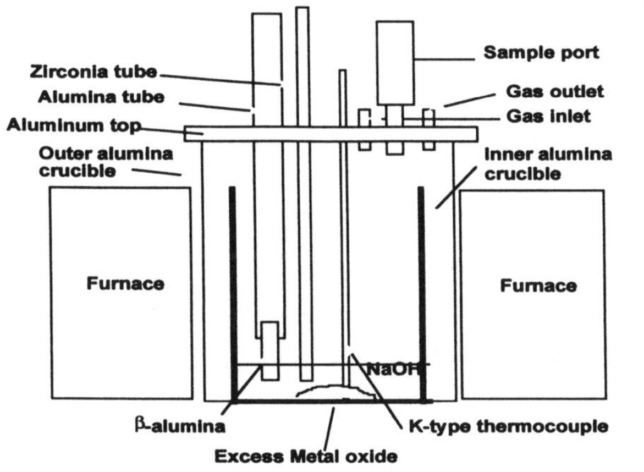

The ceramic crucibles were surrounded by a cylindrical radiant furnace that provided uniform heating. Temperature was measured with a calibrated K-type thermocouple protected by an α-alumina tube and was controlled by a CN3201 Omega® controller. An aluminum top and a silicone O-ring encapsulating in Teflon®, sealed the top of the crucible. Port holes fitted with Cajon® vacuum fittings with Viton® O-rings provided access to the reactor for the thermocouples, gas inlet and outlets, and electrodes.

Electrodes were inserted into the melt to measure the basicity (16-17). The electrode system consisted of a combination of a β-alumina probe and a zirconia probes. β-Alumina ($Al_2O_3 * 5\,Na_2O$) conducts sodium ions exclusively. The ß-alumina tubes were purchased from Polyceram and attached using a Durabond 989™ adhesive to an α-alumina tube to minimize the expense. The inside of the tube was filled with sodium metal, in contact with a platinum wire that was routed outside and sealed. A potential across the ceramic tube wall results from a difference in sodium activity.

Zirconia fully stabilized with yttria ($ZrO_2 * Y_2O_3$) conducts oxygen ions exclusively. The inside end of a closed-end tube was painted with three coats of platinum paste and attached to a platinum wire. The tube was open to the air and thus uses the oxygen in air as a reference. A potential across the ceramic tube wall results from a difference in the oxygen activity.

The half-cell reaction at each electrode is as follows:

$$\text{Oxidation: } 2\ [Na \rightarrow Na^{+} + e^{-}] \quad (5)$$

$$\text{Reduction: } \tfrac{1}{2}\ O_2 + 2e^{-} \rightarrow O^{2-} \quad (6)$$

Using the Nerst equation, it follows that

$$E = E° - \frac{RT}{ZF} \ln a_{(i)} \quad (7)$$

where Z represents the moles of electrons participating in the reaction. For our case with NaOH, equation (7) becomes

$$E = E° - \frac{RT}{2F} \ln a + \frac{RT}{F} \ln (Na) + \frac{RT}{4F} \ln P(O_2) \quad (8)$$

Using pure Na metal as the reference, in the standard state (Na) is unity and ln (Na)=0.

$$E = E° - \frac{RT}{2F} \ln a_{(Na_2O)} + \frac{RT}{4F} \ln P(O_2) \quad (9)$$

R is equal to 8.314 J * mol^{-1} * K^{-1}, and F is equal to 96484.6 C/mol. The temperature, T, was held constant at $773^{o}K$ for the experiments. Because atmospheric air was the reference for the zirconia tube, the $P(O_2)$ was equal to 0.21 atm. By measuring the voltage, E, the resultant activity of sodium oxide was used to determine the basicity of the sodium hydroxide melt.

The E° value in equation (9) was experimentally determined as follows: first, the sodium hydroxide was completely dried at 500°C under nitrogen gas for three days to remove excess water and contaminants. Then, the basicity was changed by adding sodium oxide to the sodium hydroxide in different concentrations, and the corresponding voltage was measured. A plot of the mole fraction of sodium oxide versus the measured voltage is shown in Figure 3.

Figure 3: Value for E'.

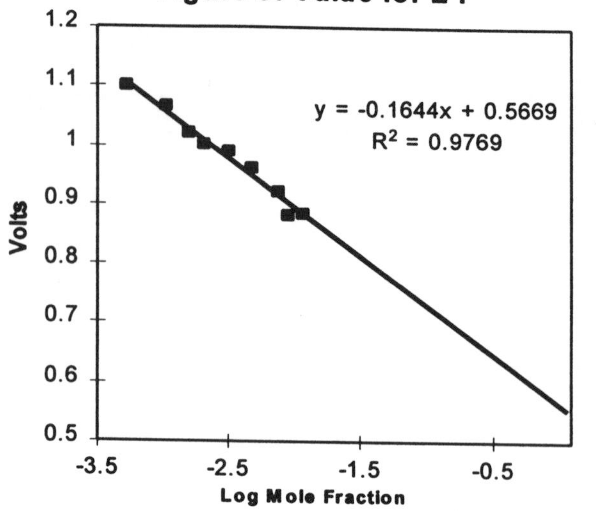

Extrapolating the line to zero sodium oxide content gives E° = 0.567.

For the experiments, sodium oxide was added to make the NaOH more basic. Sodium oxide pellets were purchased from Aldrich®.

To make NaOH more acidic water vapor was added by bubbling the carrier gas through a water column fitted with a polyurethane frit. By fixing the water column temperature and assuming saturation, which was tested by Schwerdtfeger, the vapor pressure was determined (18).

Samples for chemical analysis were collected by inserting a ceramic rod through a ball valve on the aluminum top. The depth of rod insertion was calibrated so that samples of approximately the same size and nearly the same area were obtained. The cold rod caused the sample to freeze to the rod as it was withdrawn from the melt. The sample was then quickly scraped off with ceramic tweezers into a volumetric flask partially filled with deionized water. The weight change of the flask was recorded. The flask was then filled to the mark with deionized water and stored until analyzed. Four replicate salt samples were taken at each basicity.

Chromium concentration was measured using a Perkin Elmer Optima 3000 Dual View Inductively Coupled Plasma (ICP) Spectrometer. Standards were prepared, values loaded into the computer, and then compared with unknown samples using EPA's guidelines for solid waste (19). The optimal conditions for the ICP had been found to be a pressure of 70 psi for argon and wavelength 267 nm for chromium. Chromium gave a linear response from 0.04 ppm to 100 ppm. Each unknown sample was analyzed three times, and the average value was reported.

Results and Discussion

Results for the chromium concentration vs. basicity are summarized in Figure 4 below.

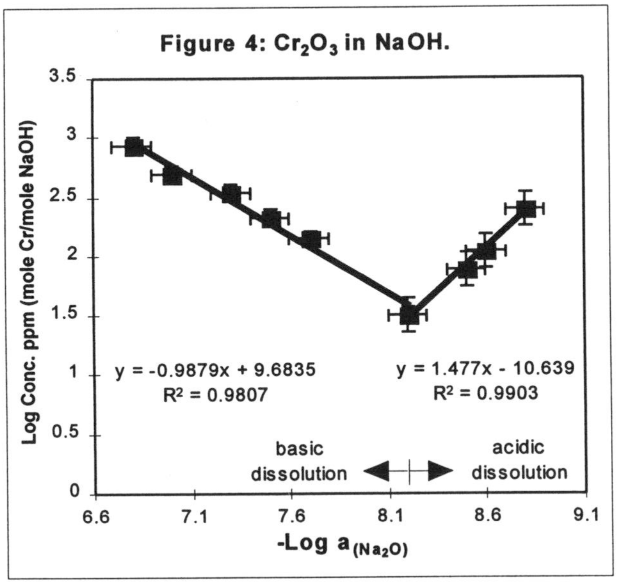

Each basicity condition shown in Figure 4 is a mean of three values. Equilibrium was ensured at a given basicity by measuring until the concentration did not change over three days. Chromium oxide was found to have a minimum solubility at a basicity of 8.2. The solid line represents the least squares best fit of the data. Basicity error bars were estimated at -log $a_{(Na_2O)}$ ± 0.1 based upon fluctuations in the voltage readings. Concentration of chromium error bars show ± one standard deviation.

If the dissolution reaction for basic conditions is assumed to be:

$$Cr_2O_3 + 2\,O^{2-} + \frac{3}{2}O_2 = 2\,CrO_4^{2-} \tag{10}$$

At constant $P(O_2)$,

the theoretical slope for the reaction is given by:

$$\left(\frac{-\log CrO_4^{2-}}{-\log O^{2-}}\right) = -1 \tag{11}$$

The measured value of -0.9879 is in very good agreement with theory. Also, the melt was yellow in color, which is consistent with the formation of chromate (20).

If the dissolution reaction for acidic conditions is assumed to be:

$$Cr_2O_3 = 2\,Cr^{3+} + 3\,O^{2-} \tag{12}$$

The theoretical slope for the reaction is given by:

$$\left(\frac{-\log Cr^{3+}}{-\log O^{2-}}\right) = \frac{3}{2} \tag{13}$$

The measured value of 1.477 is in very good agreement with theory. Also, the melt was green in color, which is consistent with the formation of chromium ions (20).

To further confirm that the mechanism for dissolution had changed, the oxidation state of the chromium ion was determined. A chemical method described by Cranston and Murray was used (21). Chromium (III) was scavenged by reacting the solution with 0.01 M iron(III) hydroxide at pH 8. The chromium was precipitated and collected on a 0.45 µm Nucleapore™ filter precleaned with 12 M HCl. The chromium complex was then dissolved in 6 M HCl and analyzed by ICP. Chromium (VI) and chromium(III) are both precipitated by reacting the solution with iron(II) hydroxide at pH 8. The iron(II) was oxidized to iron(III), and the chromium(VI) was reduced. The precipitated chromium was dissolved in 6 M HCl and analyzed as before. Chromium (VI) was determined by the difference.

Results indicate that the mechanism for dissolution had changed under different basicity conditions on either side of the minimum. Cr(III) was the dominant species at -log $a_{(Na_2O)}$ of 8.8, and Cr(VI) was the dominant species at -log $a_{(Na_2O)}$ of 6.8. The change in the oxidation state of chromium confirms that the mechanism changed from basic to acidic dissolution.

Weight loss experiments

Weight loss experiments were performed using chromium metal in molten NaOH at basicity values of 6.8, 8.2, and 8.8. The three basicities selected represented basic dissolution, the solubility minimum, and an acidic dissolution, respectively. Figure 5 shows the results of the weight loss measurements performed at different basicities.

Figure 5: Chromium weight loss in NaOH at 500°C.

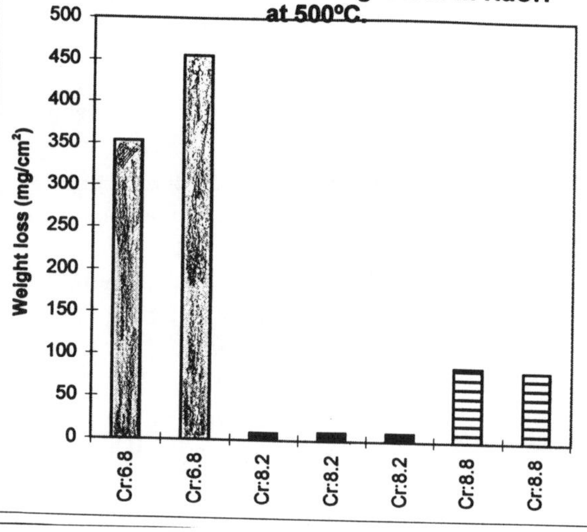

Figure 6: Weight loss as a function of time for Cr in NaOH at a basicity value of 8.2 at 500°C.

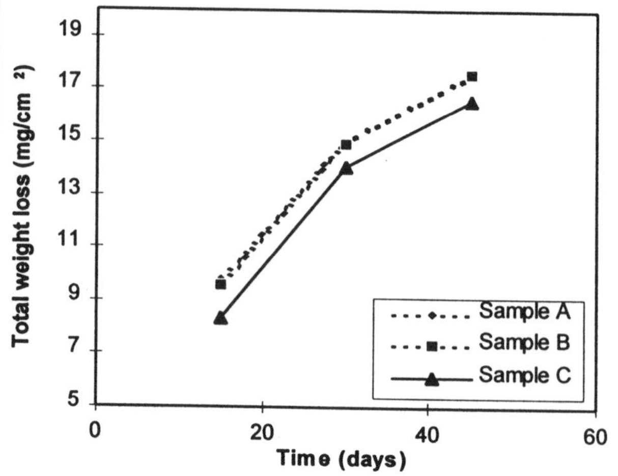

The weight loss variation can be predicted from the basicity and solubility curves. The chromium oxide solubility results in Figure 4 show that the solubility was at a minimum at a basicity value of 8.2. The weight loss of chromium metal is also low at a basicity value of 8.2.

At basicities of 6.8 and 8.8, the solubilities were much higher, which should lead to higher weight loss. Furthermore, the weight loss at a basicity value of 6.8 should be higher than at 8.8 based upon the solubility curve in Figure 4. Weight loss experiments with chromium metal at basicities of 6.8 and 8.8 in Figure 5 showed the predicted increase in weight loss. Therefore, basicity, which determines the solubility, does affect the weight loss of chromium metal.

In addition, three samples of chromium metal were placed in NaOH at a basicity value of 8.2. The minimum of the chromium oxide solubility was chosen to determine the effect of time on weight loss. After 15 days, the samples were removed and cleaned with water to remove excess NaOH. The samples were then ultrasonically cleaned for 5 minutes in acetone and dried in a desiccator prior to weighing. The same samples were replaced in the furnace with fresh NaOH for an additional 15 days. In figure 6, the results were graphed showing the weight changes of three 15-day trials.

Samples taken from the melt were analyzed by the ICP. The samples were yellow, indicating the presence of chromate, so the basicity was slightly on the basic side of the minimum. The average results are summarized in Table 1.

Table 1: Analysis for chromium by ICP

	Log chromium (ppm)
1st 15 days	1.70
2nd 15 days	1.18
3rd 15 days	0

The results show that the weight loss decreased each time a new trial began. The chromium ions in the melt also identically decreased in each trial until no chromium was detectable.

Visually, a dark green, fairly compact scale formed on the test pieces that was strongly adherent to the metal. Because the chromium oxide was not very soluble in NaOH under the tested conditions, the chromium oxide was able to form a passive layer, which reduced the rate of corrosion. Because the melt was replaced at each 15-day interval, the drop in corrosion rate was not due to the melt becoming saturated.

Conclusion

The solubility of Cr_2O_3 was determined as a function of basicity. The results clearly show a minimum solubility at -log $a_{(Na_2O)}$ of 8.2. The data suggest two mechanisms of dissolution of chromium oxide in molten NaOH. Basic dissolution (basicity values < 8.2) formed chromate, and acidic dissolution (basicity values > 8.2) formed chromium ions. Analysis of the oxidation state of chromium confirmed the products.

The solubility curves successfully predicted the variation in the weight loss of chromium metal. Chromium coupons at -log $a_{(Na_2O)}$ of 8.2, which is the minimum solubility of chromium oxide in NaOH, developed a green film that successfully lessened the corrosion.

Controlling conditions in the kraft recovery boiler air ports that establish a minimum solubility condition should lead to a lower corrosion rate of composite tubes. The basicity of a NaOH deposit in a kraft recovery boiler could be controlled by the amount of water vapor in the incoming air to reduce the corrosion rate and provide significant cost savings. The practicality of such an approach would have to be demonstrated. Addition factors, such as the presence of and effect of other metal oxides and chemical species should be investigated further.

Acknowledgments

The authors thank Drs. Miroslav Marek, and Preet Singh for valuable discussions and constructive suggestions. This research was supported by the Institute of Paper Science and Technology and its Member Companies. Portions of this work were used by M.J. Estes as partial fulfillment for the Ph.D. degree at the Institute of Paper Science and Technology.

References

1. **Bruno, F.,** "Primary air register corrosion in kraft recovery boilers," *Pulp and Paper International Corrosion Problem*, **4,** pp. 68-75 (1983)

2. **Colwell, J.A.,** "On the mechanism of corrosion of composite tubes at ports in kraft recovery boilers," *7th International symposium on corrosion in the Pulp and Paper Industry Proceedings,"* TAPPI press, Atlanta, GA p. 231 (1992)

3. **Colwell, J.A. and G. Fonder,** "Status report to the corrosion control project advisory committee," Institute of Paper Science and Technology press, Atlanta, GA (1994)

4. **Lunn, M.A., W.B.A. Sharp, J.D. Andrews, H.N. Tran, and D. Barham,** "Corrosion of composite tubes at recovery boiler air ports - A case history," *6th International symposium on corrosion in the pulp and paper industry,* Finnish Pulp and Paper Research Institute, Helsinki, **6**, pp. 151-162 (1989)

5. **Wensley, D.A.,** "Corrosion of recovery boiler waterwall composite tubes," *Material Performance*, **26** (11), pp. 53-55 (1987)

6. **Falat, L.,** "Corrosion of air ports," *Tappi Journal,* **79** (2), pp. 175-185 (1996)

7. **Deanhardt, M.L. and K.H. Stern,** "Solubility of yttrium oxide in Na_2SO_4 and NaCl melts," *Journal of electrochemical society: Electrochemical science and technology,* **29** (10), pp. 2228-2232 (1982)

8. **Gupta, D.K. and R.A. Rapp,** "The solubilities of NiO, Co_3O_4, and ternary oxides in fused Na_2SO_4 at 1200°K," *Journal of electrochemical society: solid-state science and technology,* **127** (10), pp. 2194-2656 (1980)

9. **Hwang, Y.S. and R.A. Rapp,** "Synergistic dissolution of oxides in molten sodium sulfate," *Journal of electrochemical society,* **137** (4), pp. 1276-1280 (1990)

10. **Jose, P.D., D.K. Gupta, and R.A. Rapp,** "Solubility of α-Al_2O_3 in fused Na_2SO_4 at 1200°K," *Journal of electrochemical society: Solid-state science and technology,* **132** (3), pp. 735-737 (1985)

11. **Leblanc, P. and R.A. Rapp,** "Solubility of Cr_2O_3 in Na_2SO_4-K_2SO_4-$Fe_2(SO_4)_3$," *Progress report submitted to E.P.R.I.,* (11), pp. 1-8 (1991)

12. **Leblanc, P. and R.A. Rapp,** "Solubility of Cr_2O_3 in Na_2SO_4-K_2SO_4-$Fe_2(SO_4)_3$. Behavior of the SO_3 sensor in more reduced environments," *Progress report submitted to E.P.R.I.,* (3), pp. 1-4 (1992)

13. **Rapp, R.A. and K.S. Goto,** "The hot corrosion of metals by molten salts, in molten salts," J. Braunstein and J.R. Selman eds., Electrochemical society, Pennington, NH, pp. 81 (1981)

14. **Zhang, Y.S.,** "Solubilities of Cr_2O_3 in fused Na_2SO_4 at 1200°K," *Journal of electrochemical society: Solid-state science and technology,* **133** (3), pp. 655-657 (1986)

15. **Zhang, Y.S. and R.A. Rapp,** "Solubility of α-Fe_2O_3 in Fused Na_2SO_4 at 1200°K," *Journal of electrochemical society, Solid-state science and technology,* **13**, pp. 734-735. (1985)

16. **Ito, Y.,** "Electrochemical studies on molten sodium hydroxide," *Journal of electroanalytical chemistry,* **124**, pp. 229-235 (1981)

17. **Petric, A.,** "Development of cation substituted beta alumina solid electrolytes and EMF measurement of the thermodynamic properties of ionic alloys," Ph.D. Thesis. University of Montreal., (1987)

18. **Schwerdtfeger, K. and E.T. Turkdogan,** "Equilibria and transport phenomena. Physicochemical measurements in metals research," R.A. Rapp, eds., Interscience Publishers of John Wiley & Sons Inc., New York, NY, pp. 321-355 (1970)

19. **US Environmental Protection Agency,** . "Method for Physical and Chemical Analysis of Solid Waste," SW-846 Method 6010A, (1972)

20. **Pourbaix, M.,** "Atlas of electrochemical equilibria in aqueous solutions," NACE, Houston, TX, pp. 257-271; 307-321; 330-342 (1974)

21. **Cranston, R.E. and J.W. Murray,** "The determination of chromium species in natural waters," *Analytica Chimica Acta,* **99**, pp. 276-282 (1978)

Acoustic Enhancement of Dry HCl Scrubbing With Calcium Oxide

J.R. Boerner and D.I. Orloff
Institute of Paper Science and Technology, 500 10th St. NW, Atlanta, GA 30318

Dry scrubbing of acidic gases is a common technique employed to clean flue gas resulting from combustion processes. It is capable of removing compounds such as SO_2 and HCl. A current disadvantage of dry scrubbing is incomplete conversion of the sorbent material, usually calcined limestone. Any improvement in reaction conversion between the sorbent and the acidic gas may lessen sorbent quantity requirements or decrease the size of the solid/gas contacting equipment, both of which are of high economic value.

Several examples in the literature show that processes conducted in an intense acoustic field can substantially improve chemical conversion in solid/gas reactions. Calcining and hazardous waste reduction are areas where acoustic field application has been shown to improve chemical conversion. This paper examines the effects of reactor temperature and CaO particle size on chemical conversion of HCl to $CaCl_2$ with and without an acoustic field.

Results of our entrained flow reactor studies show that reaction conversion increases with reactor temperature in the range 125 to 350 ^{o}C and with decreasing CaO particle size for three different size fractions possessing mass mean particle sizes of 90, 56, and 4μm. For the mid-size CaO particles, chloride ion capture was improved up to 86% in an acoustic field of 163 dB resonating between 600-750 Hz at 200 ^{o}C reactor temperature. No improvement in reaction conversion was found for the largest CaO particle fraction and only marginal improvement was noted in the cases involving the smallest CaO particles. Conversion enhancement may be due to greater concentration gradient of HCl around the CaO particles caused by enhanced mixing of the reactants resulting from the acoustic field.

1. INTRODUCTION

Scrubbing technology is used to remove acidic gases from combustion process stack gases. Scrubbers are typically an entrained flow configuration, but various embodiments of the scrubbing process are possible with the two main classifications being wet and dry methods. These scrubbing methods remove acidic gases such as SO2 and HCl with calcium based sorbent compounds [1,2,3,4]. Wet scrubbing is accomplished by contacting a countercurrent flow of flue gas with a slaked lime/ground limestone slurry in an open or packed tower. The acidic gas is transferred from the gas to the liquid. The salt formed in the scrubbing reaction must be precipitated in a separate reaction vessel. This system is common among utility boilers and has the advantage of proven design and high removal efficiencies of the acidic gas component.2 Disadvantages include high water consumption, corrosion potential, scale and plugging problems and sludge disposal difficulties[2].

Dry scrubbing utilizes a slurry of lime which is atomized into a fine mist that dries almost immediately upon injection into the flue gas where the acidic component reacts with the calcium compound to form a dry product [2,3]. In another embodiment of dry scrubbing the sorbent is injected directly into different sections of the furnace in powder form where it can react with the acidic gas [4]. Dry scrubbing possesses the advantages of a simpler process design, low pressure drop in the scrubbing system, and a dry reaction product composed of the salt of the acid. The dry product can be more easily disposed compared with a sludge resulting from wet scrubbing methods [2]. Even with these advantages dry scrubbing does not provide as high a removal rate of the acidic gas, a reason wet scrubbing methods are used more than dry scrubbing methods. An enhancement of gas scrubbing efficiency may improve the feasibility of use of dry scrubbing technology in order to use the advantages that the dry configurations offer.

Intense acoustic fields over 160 dB have been shown to improve some industrial-related reactions. Drying, incineration of municipal waste, coal combustion and calcination are examples of operations which can benefit from acoustic fields in the reaction vessel [5,6]. Laboratory and pilot plant scale experimentation show significant process improvements due to an acoustic field of approximately 160 dB.5,6 This work began with the idea that a 160 dB acoustic field might also improve the efficiency of dry scrubbing and thus make it more advantageous in the removal of acidic gases from combustion processes because this sound pressure level was found to be effective in the literature [5,6].

In this investigation, we are interested in a specific example of a type of reaction associated with dry scrubbers. The CaO/HCl reaction is of interest due to the recent work on this reaction to characterize temperature and particle size effects of the reaction [7,8,9,10]. The reaction proceeds as follows:

$2HCl(g) + CaO(s)$ æÆ $CaCl_2(s) + H_2O(g)$ (1)

However, the CaO/HCl reaction has been examined only with fixed bed systems requiring sample sizes on the order of a few grams. This past experience has uncovered some general reaction behavior between CaO and acidic gases such as reaction conversion increases with temperature and a decrease in sorbent particle diameter [8,9]. We will examine an entrained flow reaction so results will have applicability to a general embodiment of dry scrubbing.

2. EXPERIMENTAL APPARATUS

2.1 Entrained Flow Reactor

An entrained flow reactor (EFR), Figure 1, was designed to investigate reaction conversion of two-phase reactions. The EFR consists of a 3.8 cm diameter quartz reactor, a fluidized bed particle feeder, ammonia reaction quench, a computer interfaced data acquisition system and reaction product collection screen. Nitrogen carrier gas, introduced at the top of the reactor, was preheated with an electric gas circulation heater. Reaction temperature was maintained by three zones of radiant heaters controlled independently. Sorbent was injected into the carrier gas at the top of the reactor where it traveled through a 61 cm injection tube to achieve reactor temperature before contact with the gaseous reagent. The gas/solid mixture was quenched rapidly by an injection of NH_3 gas at the bottom section of the reactor called the decoupler. The reaction product passed through the decoupler and was accumulated in the collection cylinder.

The reactor tube is L-shaped allowing for introduction of the reactants and collection of the reaction products while maintaining a standing sound wave with a sound pressure maximum at approximately the tube center and pressure nodes approximately every 30.5 cm. A portion of the reactor was fitted with a series of quartz optical viewports for entry of a laser beam for conducting gas and particle velocity measurements. Thermocouple and pressure transducer ports are located on the vertical tube section for collecting temperature and sound pressure data. The reactor

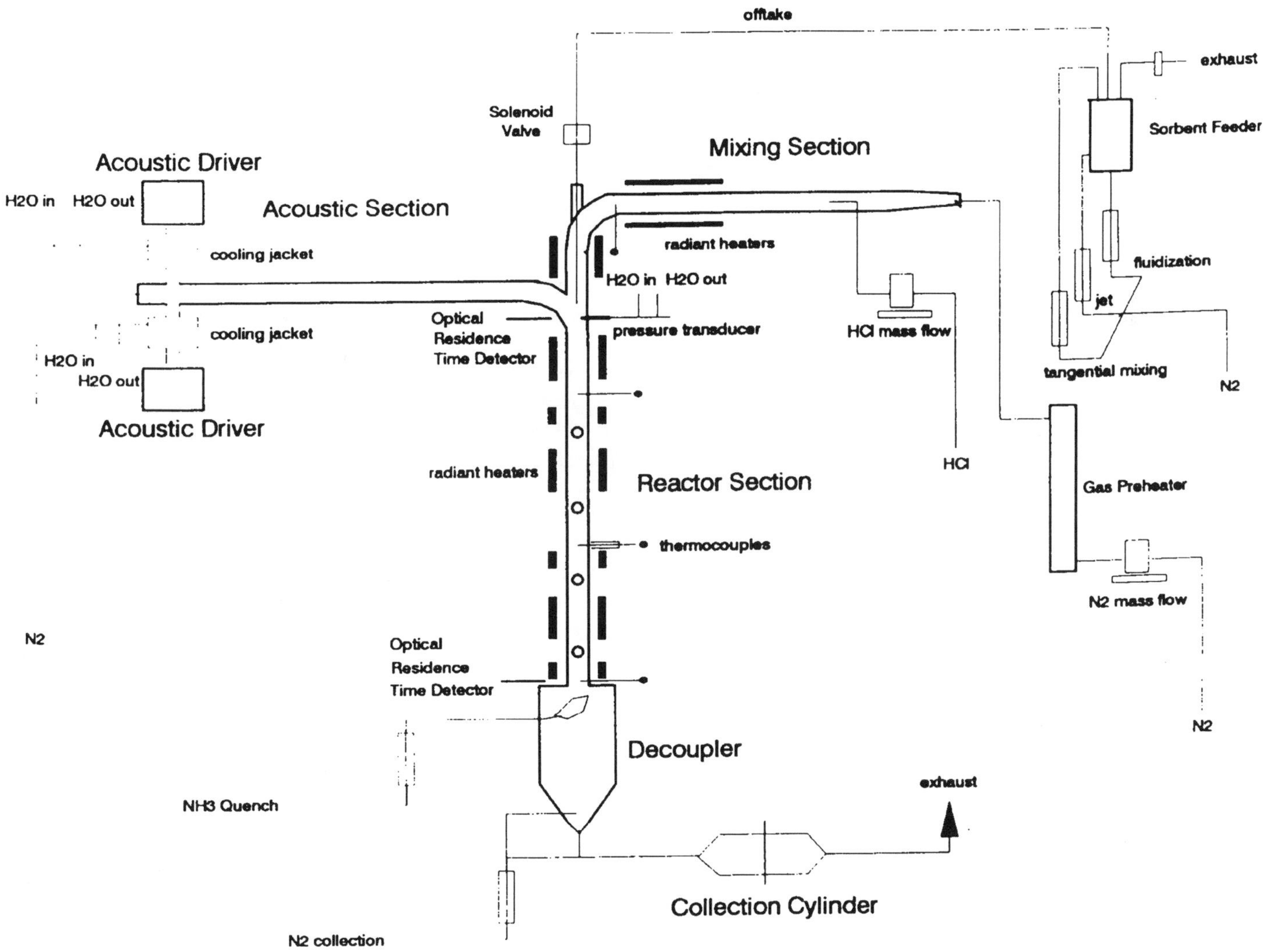

Figure 1. Experimental system schematic (EFR).

Table 1. CaO composition.

Compound	Composition, %
CaO	96.5
$CaCO_3$	3.0
MgO	0.5

tubes are categorized into three sections; the acoustic section, mixing section and reactor section.

This EFR differs from conventional entrained flow reactors, as it was fitted with compression drivers which were attached on the tube walls to generate an intense resonant sound wave within the reactor tube. Additionally, the decoupler at the bottom of the reactor section dampens the sound pressure oscillations from the carrier gas and provides the necessary retention time to quench any unreacted HCl. The entire reactor tube was encased in a steel shell to channel leaks to the exhaust hoods to prevent gas leaks from invading the surrounding environment. The reactor may operate in a batch or continuous manner. Continuous operation was the mode in which the present experiments were conducted.

A fluidized bed feeder was used to inject the desired quantity of CaO into the reactor. The sorbent feeder was operated in a continuous mode for the reaction conversion experiments in both steady flow and with acoustics. Fluidization gas, tangential mixing gas and jet gas were monitored with rotameters to achieve a consistent feed rate of CaO. It is important to maintain consistent particle density in the reactor to allow comparison of steady flow reaction results to those obtained in an acoustic field. Particle density is defined as the number of CaO particles per unit volume of the reactor tube.

3. MATERIALS AND METHODS

3.1 Hydrogen Chloride Supply and Carrier Gas

Dry hydrogen chloride gas and nitrogen from pressure regulated cylinders were monitored through mass flow meters. Nitrogen of 99.998% purity was used as the carrier gas in all experiments. A VLSI electronics grade hydrogen chloride reagent, 99.99% pure, ensures minimization of contaminants. Anhydrous ammonia of 99.99% purity was used to quench unreacted HCl in the reactor decoupler.

3.2 CaO Reactant

CaO is the sorbent compound used to react with the HCl gas in the reactor. The CaO reagent contained only trace amounts of magnesium and other metals. Analysis by Inductively Coupled Plasma (ICP) over several samples yielded the following composition of the CaO reagent as presented in Table 1. CaO density is 3.17 g/cm3.

Calcium oxide pellets of up to 4 mm in diameter were crushed in a ball mill and sifted in a shaker to retain the desired fractions on 125, 75, and 20 mm mesh screens. These three size fractions will be referred to throughout the paper as the Large, Medium and Small fractions, respectively. Specific surface area was measured by nitrogen adsorption on the sample according to the BET technique. Mass mean particle size and distribution were determined for each CaO size fraction as shown in Table 2 by a sedimentation method using a Micromeritics Sedigraph 5100. The distribution range indicates the particle size range of 80% of the particle mass composing a particular CaO fraction. The remaining 20% of mass is below the lowest size in the range given in Table 2. The size distribution is given as an illustration of the composition of the CaO fractions.

Table 2. CaO specific surface area and mean particle size

Mesh Screen Size/ Fraction Name	Specific Surface Area, m^2/g	Distribution Range	Mass Mean Particle Size, μm
20μm / Small	14.7	40 μm - 2 μm	3.95
75μm / Medium	11.8	110 μm - 40 μm	55.8
125μm / Large	11.4	150 μm - 80 μm	90.8

3.3 Ion Analysis

Calcium, chloride and ammonium ions were quantitatively determined for the reaction product. The amount of chloride in the product describes the extent of reaction conversion. Ion analysis was performed twice on each reaction product to account for conversion differences resulting from varying size distribution and extent of reaction within a given sample.

Calcium ion was determined by atomic absorption spectroscopy (AA). Chloride and ammonium ions were both determined with an ion selective electrode (ISE). No interfering ions were present in the calcium chloride and ammonium system, thus ISE's were an efficient, accurate method for quantifying chloride and ammonium ion content of a dilute solution. Calcium could be measured with an ISE, but ammonium is an interference to the calcium electrode, thus AA was utilized. Both AA and ISE results for the Cl and Ca ions were expressed in parts per million (ppm). These concentration results were used to calculate the percentage of Cl ion compared to the Ca ions present in the reaction product. The percentage of Cl ion is reported in the reaction conversion figures of this paper.

4. EXPERIMENTAL PROCEDURE

CaO fractions, once sifted, were baked in a muffle furnace at 900°C for 1 hour to evaporate bound water to achieve constant weight. Approximately 9g of sample was placed in ceramic crucibles, with no top. Once the sample was dried, it was placed into the particle feeder where the sample was not accessible by water vapor.

Table 3 shows the experimental parameters investigated in this study. The CaO particle size fractions were chosen to cover a wide range of CaO particle sizes. The Small fraction is representative of the size distribution utilized for dry scrubbing appli-

Table 3. Design of two-phase reaction experiments.

CaO Fractions	SPL, dB	HCl Concentration, ppm	Temperature, °C
Small	0/160	614	125
Medium	0/160	614	200
Large	0/160	614	350

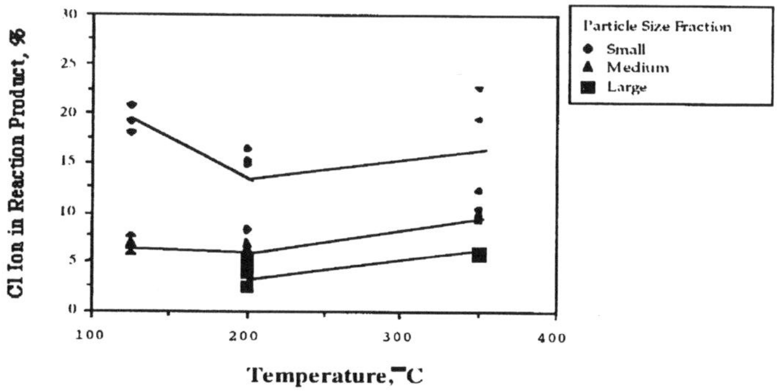

Figure 2. Reaction conversion with respect to sorbent particle size, no acoustics.

cations. The Large and Medium fractions were chosen to discover how they might be affected by an acoustic field. The HCl concentration was chosen to be in a range other investigations have used[7,9,10]. The temperatures chosen represent the range that is optimum for dry scrubbing. The optimum temperature window upper limit for this reaction is 550 -600°C [8,9,10].

At least three replicate experiments were conducted for each case in the experimental matrix. Each EFR experiment was continuously run for 15 minutes and the reaction product collected from the decoupler and collection cylinder. The particle residence time was approximately 0.90 s for each experiment. The flow rate of carrier gas through the reactor was 75 lpm and the concentration of HCl in the carrier gas 614 ppm. Sound pressure level (SPL) was maintained at about 163 dB unless otherwise noted. Variability in flow rates, HCl concentration and SPL was less than 2%. Temperature gradient was less than 3% in the reactor section.

Ion analysis was performed on the reaction product and the precipitate on the collection cylinder screen. No ammonium chloride (NH_4Cl) was found in the reaction product, only calcium and chloride ions. This was due to the physical nature of the CaO particles. The density is sufficiently high that the particles settle at the bottom of the decoupler. The particulates captured in the collection cylinder screen contain chloride and ammonium ions but only a trace of calcium ions. The NH_4Cl precipitate

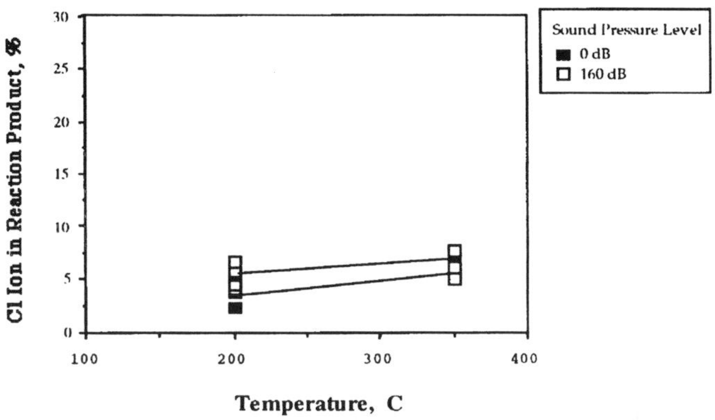

Figure 3. Reaction conversion for Large CaO fraction with and without acoustics.

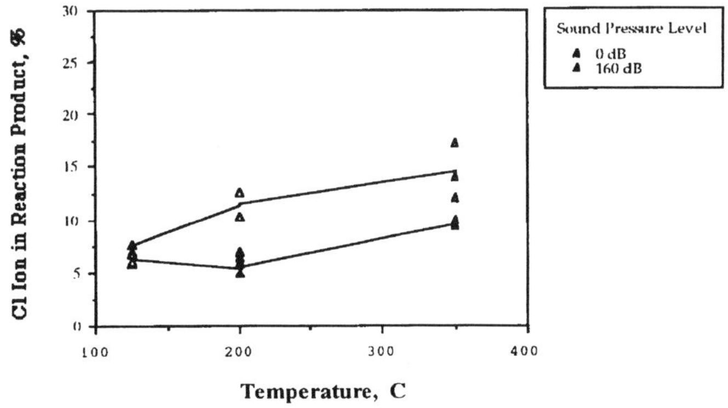

Figure 4. Reaction conversion for Medium CaO fraction with and without acoustics.

formed upon quenching the unreacted HCl remains entrained in the flow and was captured on the collection cylinder screen. Only the Cl found in the $CaCl_2$ is reported here.

5. RESULTS AND DISCUSSION

5.1 Product Conversion Results

A series of experiments was first run to compare the EFR results to other investigations and to establish reproducibility of the data. This was accomplished by running experiments of various CaO size fractions and reactor temperatures. Figure 2 illustrates the behavior of reaction conversion with CaO size fraction, reactor temperature and no sound pressure. Reaction conversion expressed as percent Cl ion in the reaction product is shown over a range of reactor temperature. Chloride ion in the reaction product is calculated as the percentage of chloride present in the product compared to the quantity of calcium in the product. Figure 2 Reaction conversion with respect to sorbent particle size, no acoustics.

Conversion is observed to increase with decreasing CaO particle size and increasing reactor temperature. The Large CaO fraction exhibits little change in conversion with reactor temperature. The Small and Medium CaO fractions do show an increase in reaction conversion with increasing temperature, with a maximum conver-

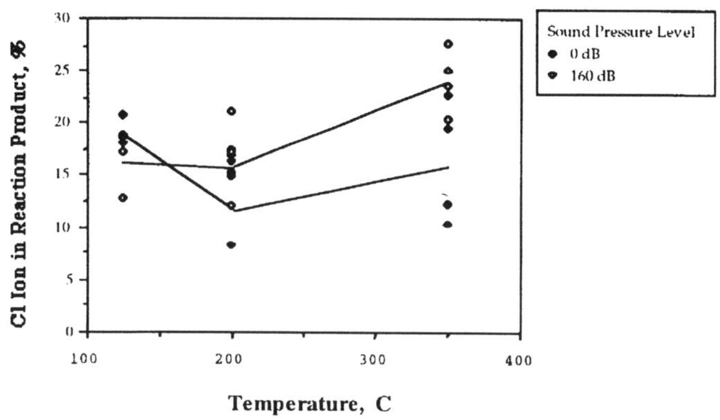

Figure 5. Reaction conversion for Small CaO fraction with and without acoustics.

Table 4 Experimental data: statistical presentation for all CaO fractions.

Large CaO	0 dB			160 dB		
Reactor Temperature °C	Average conversion %	Std Dev.	COV %	Average conversion %	Std Dev.	COV %
200	4.08	1.09	25.7	5.44	1.19	21.87
350	5.90	0.26	4.32	6.53	1.05	16.09

Medium CaO	0 dB			160 dB		
Reactor Temperature °C	Average conversion %	Std Dev.	COV %	Average conversion %	Std Dev.	COV %
125	6.68	0.69	10.27	6.36	0.44	6.97
200	5.95	0.82	13.73	11.02	1.32	12.01
350	9.62	0.30	3.07	14.37	2.58	17.95

Small CaO	0 dB			160 dB		
Reactor Temperature °C	Average conversion %	Std Dev.	COV %	Average conversion %	Std Dev.	COV %
125	22.48	5.61	24.94	16.17	3.17	19.61
200	13.62	3.65	26.77	16.86	3.23	19.16
350	16.11	5.88	36.47	24.17	3.01	12.45

sion for the Small fraction of 16% at 350°C. The greater conversion for smaller CaO particles is likely due to the greater degree of interior reaction in the smaller particles compared to the larger particle fractions. The importance of the internal reaction plays a greater role as particle size decreases [8]. These results are supported by other investigators who have reported reaction conversion of CaO and HCl to increase with increasing temperature and decreasing CaO particle size [8,9,10].

With a baseline established over a temperature and particle size range, a series of experiments were conducted at 163 dB to determine the effect of sound pressure on reaction conversion. Figures 3-5 illustrate the conversion results expressed as percent Cl ion in the reaction product for the Large, Medium and Small CaO fractions over the temperature range 125 to 350∞C. The 163 dB results are compared to the baseline results of Figure 2.

Figure 3 shows that reaction conversion in an acoustic field for the Large CaO fraction did not change from the baseline values established in Figure 2. The acoustic field had little effect on the reaction. However, for the Medium CaO fraction in Figure 4, while no change in reaction conversion is evident at 125°C, an appreciable increase has occurred at 200 and 350°C.

Figure 5 shows the conversion results for the Small CaO fraction under similar reactor conditions and SPL as described for the Large and Medium CaO fractions. A slight improvement in reaction conversion is realized only at 200 and 350°C. The reaction conversion levels at each temperature with acoustics is similar, indicating that conversion plateaus for the small CaO particles. It is likely that conversion for the Small CaO fraction is maximized due to reaction product porosity effects. The molar volume of $CaCl_2$ is 51.6 cm3/g while that of CaO is 16.9 cm^3/g, thus as the reaction proceeds, the porosity is reduced [9].

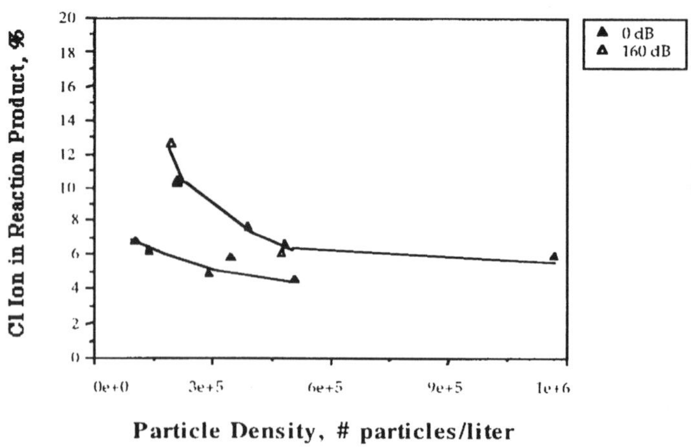

Figure 6. Reaction conversion versus reactor particle density.

Since the present experiments are at low temperatures to prevent sintering of the CaO particles, porosity reduction results from reaction progression as more $CaCl_2$ is formed.

Data scatter is quite large for the Small CaO fraction and may be linked to the variability of CaO size distribution and particle density in the reactor between experiments. Particle density refers to the number of CaO particles per liter of reactor volume (number particles/liter) during an EFR experiment. The variability is evident from the coefficient of variance (COV) presented in Table 4. Care must be taken to consistently achieve similar particle density for each experimental case in order to ensure comparison of similar conditions for both the acoustics and nonacoustics cases.

The particle density concept is illustrated in Figure 6 where the Medium CaO fraction, 200°C case with and without acoustics over a range of particle densities were intentionally run to examine the particle density effect. The figure shows that as particle density increases, conversion drops both with and without acoustics. Above 3.3x105 particles/liter conversion drops rapidly. With acoustics the conversion varies from a high of 12% and drops to about 6.5% at the highest particle density. Without acoustics, the percent of chloride ion in the reaction product

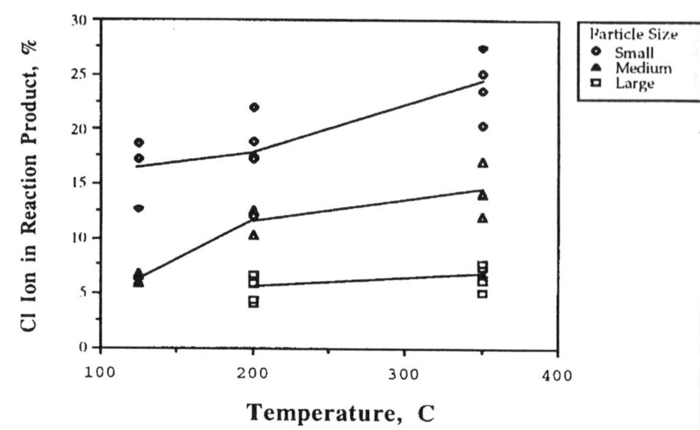

Figure 7. Reaction conversion by CaO particle size fraction with acoustics.

ranges from 4.5% to 6.5%. If the particle densities are maintained in the range 0.85x105 - 3.3x105 particles/liter, conversion is more consistent as illustrated in the 0 dB case. Comparison can be made of the no acoustics case to the case with acoustics in this particle density range. This eliminates the conversion differences between experiments due to variation in the number of CaO particles present.

Particle density for the Large CaO fraction was maintained between 0.23x105-1.42x105 particles/liter. No specific particle density range was found to be necessary for the Small CaO fraction. In other words, conversion did not vary with particle density for the Small CaO fraction. Particle density for the Small fraction ranged from 7.65x105 to 113x105 particles/liter.

Figure 7 illustrates the difference in reaction conversion due to 163 dB SPL while altering the CaO particle size fraction and reactor temperature. It is evident that as average CaO particle size decreases, reaction conversion increases. This is the trend that was evident for the cases with no sound pressure described in Figure 2. A greater conversion is achieved with the use of acoustics as was indicated above.

The data from Figures 3 - 5 are summarized in Table 4. The average conversion, standard deviation and coefficient of variation is listed at each condition of temperature and average particle size. The change in reaction conversion due to SPL is tested for significance by using a T-test to compare the difference in average value at each experimental condition.

The difference between the mean reaction conversion for the Large CaO fraction with and without acoustics is not significant. However, the Medium CaO fraction at 200 and 350∞C exhibits a significant difference between the means according to the T-test at 95% confidence. No significant difference in means is found for the Small CaO fraction at any reactor temperature. Improvement in reaction conversion is demonstrated under certain conditions of temperature and CaO fraction. The hypothesis for the improved conversion is examined next.

5.2 Change of SPL and Reaction Conversion

SPL was investigated to determine if this was a factor in creating the conversion enhancement evident from the acoustic field. Previous investigations reported values of 140 to 160 dB necessary to achieve improvements in various reaction systems under investigation.5,6 For this work, two additional levels of SPL were chosen, 140 and 150 dB. Higher sound pressure levels are not achievable with the given amplifier and compression drivers. Table 5 shows the reaction conversion for the Medium CaO fraction at 200°C.

Table 5: Reaction conversion at various SPL.

Medium Fraction, 200°C	
SPL, dB	Cl Conversion, %
0	5.95
145	6.06
152	6.04
163	10.62

The data suggests that a threshold SPL exists at 160 dB for the CaO/HCl reaction because no conversion changes are evident at 140 dB or 150 dB as compared to the base case conversion. However, conversion increases for SPL above 160 dB. Perhaps some characteristic of the flow field which results from acoustics is affecting the CaO conversion. Turbulence intensity was examined to see if a correlation exists linking the turbulence of the flow field and the conversion change observed.

5.3 Turbulence in an Acoustic Field

Measurement of SPL in the reactor section of the EFR with a pressure transducer revealed sound pressure nodes of 160 - 163 dB approximately every 30.5 cm. The pressure nodes are maximum sound pressure areas established as a result of the 600-750 Hz frequency range used in these experiments. Visually, particles injected into the reactor system can be seen mixing in these regions, then flowing to the next mixing region and so on until the particles reach the bottom of the reactor. Zones of acceleration and zones of mixing are observed throughout the length of the EFR tube due to the sound pressure nodes. Figure 8 illustrates the mixing hypothesis.

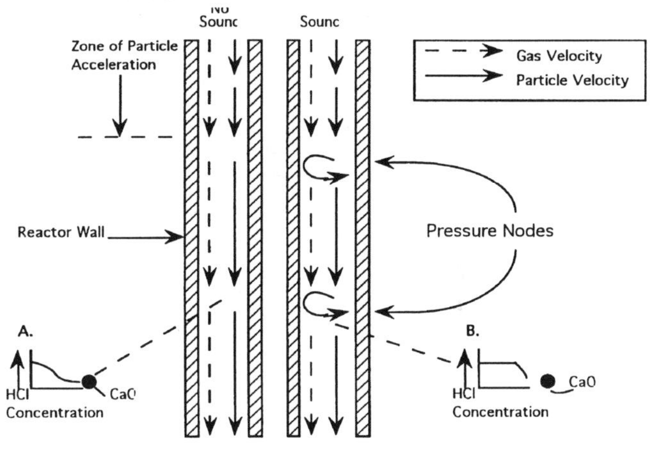

Figure 8 Mixing hypothesis

Figure 8 Mixing hypothesis

Without acoustics the CaO particles accelerate to the carrier gas velocity. With acoustics a region of maximum sound pressure (node) will possess a minimum particle and gas velocity, but a region of minimum sound pressure will possess a maximum gas and particle velocity. The flow of nitrogen carrier gas through the reactor is constant for all experiments, so average residence time should not change appreciably in the acoustic field, though regions of gas and particle deceleration and acceleration occur. It is in the alteration of gas and solids slowing and speeding that mixing seems to occur. The flow field established due to the variable acoustic pressure creates turbulence and can be characterized by the quantity turbulence intensity.

Turbulence intensity is defined as the root mean square for a velocity component referenced to a characteristic mean flow velocity [11].

$$I = \frac{\left(1/3u_i u_i\right)^{1/2}}{U_o} \tag{2}$$

where u_i is the component velocity and U_o the mean flow velocity. Turbulence intensity levels measured by laser doppler velocimetry for the downward flow component shows that turbulence intensity increases above 140 dB SPL according to measurements provided in Table 6. It is hypothesized that the increased turbulence improves the mixing of the gas phase to improve the reactant gas concentration gradient around the CaO particle.

Table 6 Turbulence intensity

SPL, dB	Turbulence Intensity, %	Standard Deviation
0	7.98	2.79
133	3.92	0.32
145	33.34	6.49
163	24.99	2.61

In steady flow, HCl concentration around a single CaO particle may steadily decrease from the bulk phase concentration to zero near the particle surface, as depicted in Figure 8A. However, in an acoustic field, the improved mixing may improve the concentration of reactant gas around the CaO particle as shown in Figure 8B. A greater HCl concentration will present more HCl to the CaO particles for reaction. However, if better mixing as illustrated through turbulence intensity was the only reason for mass transfer improvement to the CaO/HCl reaction, we would expect to see greater reaction conversion at 140 dB and not only 160 dB. Another, yet to be identified, phenomenon must be responsible for improving mass transfer to the CaO particles to improve reaction conversion. Perhaps SPL and the mixing phenomena combine to improve mass transfer to the CaO particles.

6. CONCLUSIONS

Conversion is enhanced by conducting the CaO/HC_l reaction in an acoustic field over 160 dB. The Large CaO fraction does not experience a significant improvement in reaction conversion with acoustics. The Medium and Small fractions do experience some improvement in reaction conversion with acoustics above 200 ^oC reactor temperature. The conversion improvement is statistically significant for the Medium fraction. Only a marginal conversion increase is realized for the Small fraction.

Conversion improvement is likely due to an improvement in mass transfer of HCl to the CaO particle. Turbulence intensity levels found in the resonant acoustic field increase dramatically from 3-8% to over 25-33% above 145 dB. However, conversion enhancement was not proportional to the greater mixing illustrated through the turbulence intensity measurements. Mixing in the gas phase may improve the concentration gradient of the gas to the solid particle, but further theoretical development is required to fully illustrate the mass transfer phenomenon.

Application of acoustics to any gas-solid reaction to enhance reactant mixing may be possible for the correct combination of parameters such as SPL, frequency, particle size, temperature and reaction time requirements. Practical generation of acoustic intensities up to 160 dB could be created using natural gas-fired tunable pulse combustors to develop a resonant wave inside the reaction vessel in question.

7. FUTURE WORK

Future work will involve measuring porosity for the Small CaO fraction for both reacted and unreacted samples to establish whether maximum conversion is limited by porosity reduction. A theoretical model will be proposed to account for mass transfer effects in a gas/solid reaction system to determine how mass transfer is affected by the acoustic field for the CaO/HCl reaction.

8. ACKNOWLEDGMENTS

Financial support of this work by the Institute of Paper Science and Technology and its Member Companies is gratefully acknowledged. Helpful discussions from Dr J.Y. Zhu and Dr. Jeff Empie are also gratefully acknowledged. Portions of this work will be used by J.B. for the Ph.D. degree at the Institute of Paper Science and Technology.

9. LITERATURE CITED

1. **Jozewicz, W.; Chang, J.C.S.**, "Bench-Scale Evaluation of Calcium Sorbents for Acid Gas Emission Control," *Environmental Progress*, **9** (2), pp.137-142 (Aug. 1990).

2. **Del Bueno, D., and R. Seay**, "Tighter Sulfur Dioxide Regulations Spur Better Control Technologies," *Pulp & Paper*, **63** (9), pp.190-194 (Sept. 1989).

3. **Michele, H.**, "Purification of Flue Gases by Dry Sorbents-Possibilities and Limits," *International Chemical Engineering*, **27** (2), pp.183-196 (April 1987).

4. **Muzio, L.J., and G.R. Offen**, "Assessment of Dry Sorbent Emission Control Technologies. Part I. Fundamental Processes," JAPCA, **37** (5), pp.642-654 (May 1987).

5. **Zinn, B.T., et al.**, "Application of Resonant Driving to Increase the Productivity and Thermal Efficiency of Industrial Processes," Proceedings of the International Symposium on Pulsating Combustion, August 5-8 1991.

6. **Stewart, C.R., et al.**, "Application of Pulse Combustion to Solid and Hazardous Waste Incineration," Proceedings of the International Symposium on Pulsating Combustion, August 5-8 1991.

7. **Daoudi, M., and J.K. Walters**, "A Thermogravimetric Study of the Reaction of Hydrogen Chloride Gas with Calcined

Limestone: Determination of Kinetic Parameters," *The Chemical Engineering Journal* , **47**, pp.1-9 (1991).

8. **Daoudi, M., and J.K. Walters**, "The Reaction of HCl Gas with Calcined Commercial Limestone Particles: The Effect of Particle Size," *The Chemical Engineering Journal* , **47**, pp.11-16 (1991).

9. **Mura, G., and A. Lallai**, "On the Kinetics of Dry Reaction Between CaO and Gas HCl," *Chemical Engineering Science*, **47** (9), pp.2407-2411 (1992).

10. **Weinell, C.E., et al.**, "Hydrogen Chloride Reaction with Lime and Limestone: Kinetics and Sorption Capacity," *Industrial Engineering Chemistry Research* , 31, pp.164 - 171 (1992).

11. **Panton, R.L.,** "Incompressible Flow," John Wiley & Sons, Inc., pp.712 (1984).

Operating Experience With a Full Scale Biofilter at a Hardboard Mill

Phillip J. Allen
Earth Tech, 2210 Industrial Street, P.O. Box 628, Grayline, MI 49738

Tamara S. Van Til
Earth Tech, 5555 Glenwood Hills parkway, S.E., P.O. Box 874, Grand Rapids. MI 49588

ABTco, Inc., operates a hardboard manufacturing facility located in a residential area of Alpena, Michigan. The facility was required to control odors from it's process by the Michigan Department of Environmental Quality (MDEQ). As a means to control odors, a pilot study was conducted to determine the suitability of biofiltration. We installed a full scale biofilter on a press and cooler exhaust stream which has been operating since May 1, 1996. There have been some difficulties since start up, but to date the problems appear to have been resolved.

Introduction

ABTco, Inc. entered a Consent Agreement with the state regulatory agency in 1994 for the purpose of resolving long standing odor complaints from the community and to achieve compliance with the state odor nuisance law (Rule 901) by 1998. The facility has several processes that were identified as major sources of odors. A press and cooler were the first sources at the facility to be controlled in the multi-step compliance program.

Typical technical concerns for this installation included:

- Continuous process up-time requirements.
- Condensable material which is corrosive and flammable (including linseed oil) in the process exhaust which accumulates quickly on ductwork and other surfaces.
- Freezing of the water handling equipment due to cold weather.
- The temperature of the exhaust entering the biofilter during cold weather.
- Little operating experience of add on controls on this source type.
- Possible process fires which could affect the control system.
- The highly cyclical nature of the emissions through the press and cooler cycle.

Exhaust Gas Parameters

Stack testing indicated the following exhaust gas parameters to be treated:

Parameter	*Press*	*Cooler*
Flow Rate (scfm)	30,500	17,050
Temperature (degrees F)	103	109
Moisture Content Percent	0.84	0.76
Particulate Loading (gr/dscf)	0.008	0.022
VOC* Loading (average ppm)	107	65
VOC Loading (average lb/hr)	22.1	7.5
Odor Units (ED50)**	3,377	5,196

****VOC*** = Volatile organic compounds, as carbon, as measured by the United States Environmental Protection Agency (US EPA) Reference Method 25A. Approximately 50% of the VOCs are considered condensable and/or soluble in water.

****ED50*** represents an estimate of the number of dilutions necessary to reach a concentration where 50% of the population can detect an odor.

Pilot

A pilot study was conducted for the purposes of determining the potential of biofiltration as a control for odors for this application in 1995. It was also necessary to familiarize plant personnel with the technology and establish critical operating parameters (e.g., retention time, media type, moisture requirements and wastewater production). Two pilot units (one vendor supplied and one "home made") were operated for a total of eight (8) weeks. This included two (2) weeks of initial acclimation and two weeks at 15, 30 and 60 second retention times. Pilot study results indicated that a 15 second retention time was sufficient to remove odors and VOCs from the press exhaust gas stream. The pilot study success verified that biofiltration would be an effective control for odors for this application.

Vendor Selection Process

We prepared bid specifications for odor control equipment (Regenerative Catalyst Oxidizers (RCO), Regenerative Thermal Oxidizers (RTO) and Biofilters). We had three "bona fide" bids for each technology. Biofiltration was selected based on least anticipated operating costs (Capital costs were similar for all three technologies). Items reviewed as part of the bid package for biofilters included the following:

1) Costs Associated With The Chosen System:

a) Capital Costs. The system selected had an installed cost "as bid" of \$2.3 million. This equates to \$48 scfm as treated.

b) Expected utility consumption. (*Annual Operating Costs* presented on page 6)

c) Expected amount of water blowdown. (*Annual Operating Costs* presented on page 6)

d) Life of the media and replacement cost. The selected system had a five (5) year guarantee on media life.

2) Performance Guarantees:

Based on VOC removal (odor guarantees were not available). The biofilter system has a 95% VOC removal "non methane" guarantee. The assumption was made early in the process that control of VOCs correlated with the control of odors.

3) Compressed installation schedule as required in the Consent Agreement.

4) Proposed system design:

a) Retention time/size of system (total footprint). A minimum retention time of 15 seconds was determined by the pilot study. The selected system has a 17 second retention time with three (3) out of four (4) module banks operating, and 23 second retention time with all banks operating.

b) Redundancy of critical components. The installation has redundancy in the system to ensure 99% plus up time (including media, fans, pumps, etc.). The biofilter system has four (4) banks of eight (8) media modules. Each bank can be isolated while maintaining operation of the system during periods of media changeout.

c) Airflow:

i) Upflow vs. downflow. Upflow was chosen based on concerns regarding compaction of media and moisture control. Media compaction would increase pressure drop while moisture control problems may cause channeling.

ii) Induced draft vs. forced draft. Induced draft was preferred to keep fans "clean" and to reduce maintenance and increase the life of the fans. The selected system is induced draft.

d) Materials of construction:

i) Cold weather considerations. A worst case condition of -25(F was considered for heat trace and insulation requirements.

ii) Fire protection. Fire protection is required throughout inlet ducting (water spray nozzles).

iii) Corrosive resistant materials. The system had ductwork materials of stainless steel into and including the humidifier with the remainder of the ductwork being galvanized.

iv) Bed modules. The chosen system utilizes prefabricated, precast concrete modules (32 total).

e) Media:

i) Composition. The chosen system has proprietary wood-compost media.

ii) Moisture control. The biofilter system has a timed water addition. No moisture probes were installed.

iii) Nutrient addition. The system has a timed addition of 12/12/12 fertilizer.

iv) Pressure drop monitoring. The system measures pressure drop across the entire bed system.

v) Process downtime considerations. The biofilter system has a warm building air inlet to maintain bed oxygen levels and temperature during process downtime.

f) Gas pretreatment:

i) Humidification system. The system has a spray tower humidifier which recirculates 500 gpm of water.

ii) Recirculated water treatment. The facility does have waste water treatment capability. The recirculation water system has automatic duplex filters.

iii) ***Temperature control.*** The system has steam injection to maintain inlet airstream temperature during cold weather conditions.

g) ***Interaction with plant operations (no additional staff available to operate system):***

i) ***Real-time operator interface.*** The chosen system utilizes a dedicated personal computer. This computer is programmed with an extensive graphic *Wonderware* package which is real-time interactive.

ii) ***Access.*** Ease of access to all components is critical to high uptime requirements. The bed modules have a "removable" roof system, the fans are on the ground, the water recirculation system is inside a building, and permanent access is provided for entry into the ductwork and biofilter structure.

iii) ***Manuals/training provided.***

iv) ***Ease of media changeout.*** Spare modules were purchased from the vendor which can be preloaded with media. Additional modules are available. This should provide for easy media change out. (Note: To date media has not been changed.)

INSTALLATION SCHEDULE

The state regulatory agency imposed a deadline for the system to be fully operational by May 15, 1996. In consideration of the winter conditions, the installation was accomplished in a very short time. Selection of a vendor was completed in December of 1995 and excavation of the site took place in January, 1996. Acclimation of the system required a period of over two (2) weeks, by increasing the airflow through the system incrementally by 25%. The auto control system was not operational in time to meet the deadline, so the system was operated manually for the first month. The ability to operate the system manually has been retained.

PERFORMANCE

The system overall has performed satisfactory, however, there have been a few exceptions.

1) Actual VOC removal as measured soon after acclimation was 93% non methane (calculated as propane), which is below the guaranteed level of 95%. The VOC outlet concentration was 3.74 ppm as propane. A portable Flame Ionization Detector (FID) has been used periodically to check the outlet concentration from the system. This monitoring indicates that the system continues to operate with a VOC concentration from 3.5 to 5.0 ppm as propane.

2) The system is running higher than expected on pressure drop. The system was designed for 1.5 inches W.C. of pressure drop. The actual pressure drop has varied from 1.5 to 4.7 inches W.C. and new sheaves were installed on the fans to increase the air flow. Pressure drop is measured across the entire system and moisture is not measured in the individual beds. Bed moisture control was an initial problem. Distribution hose connections were broken in three (3) beds. Once the hoses were repaired the system operated better. Currently, the pressure drop is 3.5" W.C. to 4.2 W.C.

3) The amount of water blowdown from the system is running around 15 gallons per minute to the on site wastewater treatment system which creates a concern for suspended solids. This rate is much higher than expected.

4) The fiberglass roof panels were leaking where they were connected to the prefabricated concrete. As a result, the roof had to be further sealed and caulked. This may make the roof panels more difficult to remove than expected.

5) Problems with the control system have necessitated manual operation on occasions. A few examples are as follows:

The PLC control panel was not delivered in time.

The computer interface did not work properly for the first month of operation.

Premature high pressure drop across the biobed caused the control to shutdown the system.

Faulty pressure monitors.

6) The designed blowdown rate in the recirculated water system resulted in a pH of 3.5.

7) Due to initial uncontrolled pH, the galvanized inlet duct (between the humidifier and the inlet tunnel) corroded extensively during the first three (3) months of operation. This entire section was replaced with stainless steel ducting.

8) Deterioration and erosion of the interior inlet area concrete may necessitate sand blasting and resurfacing. Stainless steel blast plates were added to the concrete in the areas of direct air blast.

9) The spare fan housing iced on the interior walls and

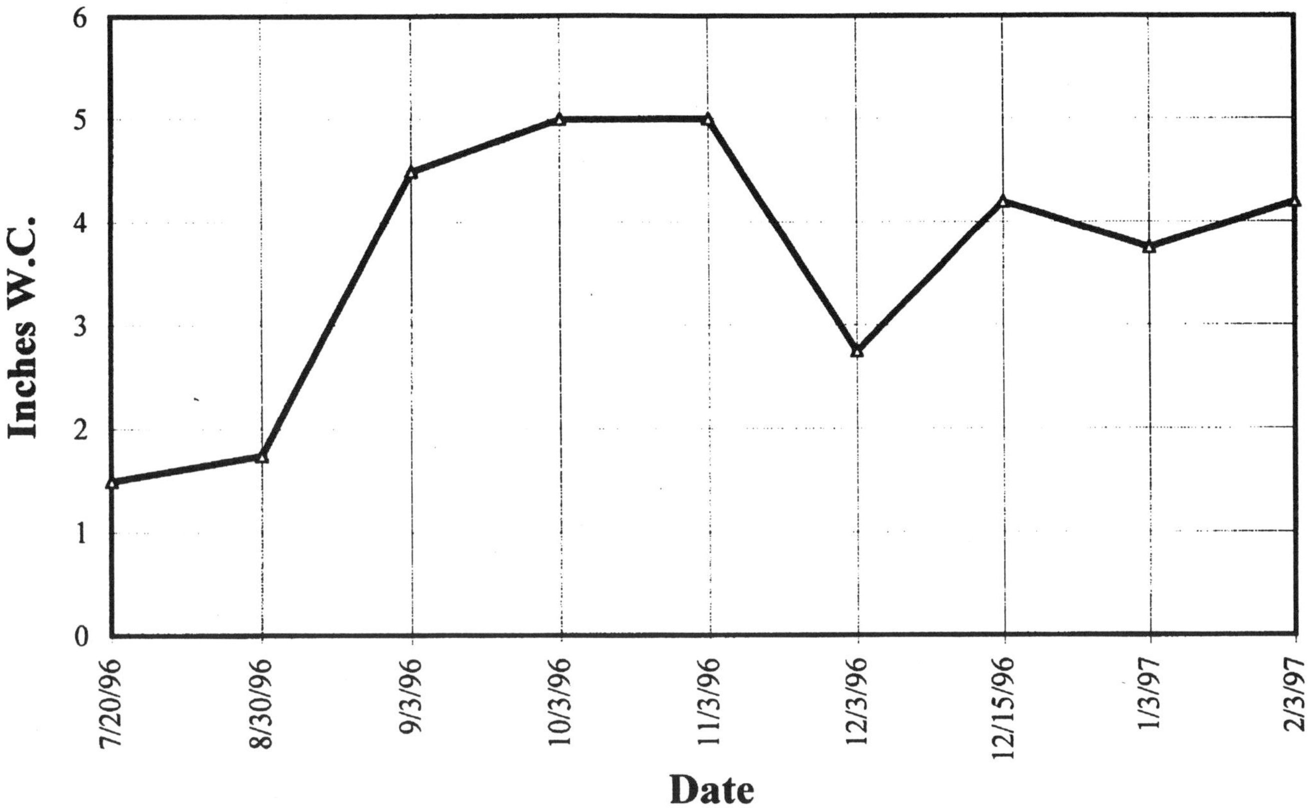

Figure 1. ABTco Inc., - Alpena, Michigan #3 Press & Cooler Biofilter Pressure Drop History.

impeller due to cold weather and heavy condensation. Current considerations of heat tracing and insulating in the housing are being discussed.

10) Approximately eighty (80) hours of unplanned press downtime has been attributed to biofilter problems in eight (8) months. This was accumulated during approximately twenty (20) occasions.

Annual Operating Costs

With regard to the above mentioned difficulties, the operating costs are slightly higher than anticipated for the system. Considering the fact that the system has not yet operated for a year. We estimate the annual operating costs to be as follows:

Water @ 50 gpm*:$63,000
Steam @ 1000 lb/hr for 6 months/year: $11,200
Fan @ 120 HP** $39,269
Sundry Electric @ 30 HP** $9,817
Bed replacement ($50,000 every 5 years): $10,000
Maintenance Spares: $20,000
Labor: $50,000
Inspection Contract (proposed): $20,000

Total Operating Costs: $224,286

* Water from blowdown and other collection points.
** Assumes 90% efficient motors.

Operation of the system has been incorporated into the current level of maintenance and production personnel. Per shift monitoring of specific check list items and sophisticated operator interface facilitate this accomplishment.

Overall System Design Evaluation

The following conclusions of the design of this particular system is the result of several months of operational experience:

Positives:

A modular system allows for maintenance on one bank of beds while maintaining operation of the remaining banks of beds.

Prefabricated concrete units loaded with media installed very easily.

Media was mixed off site and inoculated on site with micro organisms specific to the application.

Duplicity of pumps and fans reduce the amount of downtime.

Duplex strainers with automatic controls on the water recirculation and mill water systems, have minimized downtime.

Operator interface with real time interactive graphics.

Manual operation mode allows a quick restart when control problems are experienced. Manual operation allows the system to remain operational even when there are control problems.

Negatives:

There is no moisture monitoring of the media and no pressure drop measurement for individual banks. This makes troubleshooting difficult when encountering a high pressure drop across the system. It is difficult to tell if there is an air distribution problem, drying of media, or overly wet media.

The blower sizing is inadequate (may be a blower vendor problem).

The fiberglass roof limits the vacuum on the system to less than 15 inches water gauge. The intention was that the roof could be easily removed. However, because of adhesive seals and hundreds of nuts and bolts required to seal the roof, they are not as accessible as we had hoped.

The lack of pH control has caused considerable corrosion and failure of non stainless steel ducts. A pH control system has been added.

Conclusions and Recommendations

Conclusions:

This biofilter installation demonstrates the effectiveness of this technology for treating odors from the press operations in the wood products industry. It is recommended that any company considering the use of biofiltration completely evaluate the experience of the vendors and specifically their experience with the installation of full scale systems. This includes the depth of expertise in various engineering disciplines, project management and organizational skills. Provided operator interface and training should be thoroughly evaluated.

Recommendations:

Based on past experience, ABTco, Inc. has the following recommendations for companies considering biofiltration for controlling air emissions. Note that *Step 1* and *2* would be required when considering any emission control devices.

1) ***Determine Maximum Emission Scenario:***

Any control device will need to be designed to accommodate the worst case operating scenario. For a press, the level of emissions will depend on factors such as production rate, board thickness, press temperature, type and amount of added resin or binders, amount of softwood used, and the type of softwood used. Some preliminary testing may be required to determine the worst case scenario.

For biofiltration in the wood products industry, it is especially important to predict the maximum amount of terpenes emitted, since these compounds require a longer residence time. Current operations, as well as future predicted operations, will need to be examined.

2) ***Determine Exhaust Gas Parameters:***

If at all possible, this step should be done concurrently with the operating scenario defined in *Step 1*. A quantification of the exhaust gas stream to be treated will be required to obtain an accurate cost estimate from biofilter vendors and anticipated performance information. At a minimum, the following parameters should be tested:

air flow rates
moisture content
temperature
particulate loading, including size distribution
VOC loading

Speciated VOC testing may also be desired if there are considerable questions about the magnitude of terpenes or other compounds. Condensable versus non-condensable VOCs should be quantified. Individual vendors may have differing requirements as to the amount of information they need.

3) ***Determine if Pilot Testing is Needed:***

Unless a biofilter is operating on a process which is similar, it is recommended that pilot testing be done on a slipstream of the target exhaust. Vendors have pilot units available (usually at a cost), or one may be constructed. If faced with a tight schedule, be aware that pilot units may be in great demand,, and there may be a waiting period before a vendor unit is available. It is recommended that the pilot test be conducted for a period of a minimum of eight (8) weeks to allow for acclimation and testing at different residence times. The larger the pilot, the better. If you have a preferred vendor use their pilot. The pilot test may indicate if further emission testing is required.

An individual with excellent production knowledge should be assigned to oversee the pilot operation. Total knowledge of the process and equipment operation is mandatory if accurate test data and operating information are to be properly coordinated.

4) ***Closely Analyze Each Vendor Specification:***

As with any project, a request for quotation can be constructed to make the resulting bid packages as similar as possible. Regardless of how detailed the request for quotation is, there will still be considerable differences between the vendor bids. Obviously, the scope of work will need to be examined (including the amount of follow up assistance). It is suggested that at least three vendor bids be analyzed. The proposed design will require close scrutiny, with specific items of concern as follows:

upflow versus downflow
modular design
method of humidification
method of bed moisture control and detection, method of water addition
methods used to avoid air channeling at walls and in the media
information on media (overall size and distribution, geometry)
method of particulate removal, if required
control instrumentation, items monitored
guaranteed life of media (minimum 5 years recommended)
guaranteed life on concrete steel and piping (10 years recommended)
operator interface

Expect that a *Confidentiality Agreement* will need to be signed to obtain some of this information.

As with any system, the conditions of a performance guarantee will need to be closely examined.

ACKNOWLEDGEMENTS

CONTRIBUTORS TO THIS ARTICLE WERE:
Alfred J. Nadeau
Environmental/Utilities Manager
ABTco, Inc.

Reginald S. Camp
Program Director of Forest Products
EARTH TECH

Joanne M. Barthlow
Engineering/CADD Technician
EARTH TECH

Reducing VOCs From Wood Drying by Wet Line Extension

Sujit Banerjee, Wei Su
Institute of Paper Science and Technology, 500 10th St., N.W. Atlanta, GA 30318

Lawrence P. Otwell, Michael E. Hittmeier
Georgia-Pacific Corporation, P.O. Box 105605, Atlanta, GA 30348

Kenneth M. Nichols
Weyerhaeuser, Tacoma, WA 98477

INTRODUCTION

Volatile organics released from oriented strand board (OSB) and other furnishes during drying require expensive controls. The problem is especially acute for softwood, whose emissions, principally terpenes [1,2], exceed those from hardwood by an order of magnitude. Since the ratio of water:VOC released can depend upon several parameters, such as the presence of fines, options exist for lowering emissions without requiring major additional capital. The specific objectives of this study were to (a) understand the factors that promote or retard VOC emissions as the furnish dries, (b) use this knowledge to reduce VOCs during drying, and (c) to develop methods for low VOC-drying. In this paper, we will interpret some of our findings in terms of a strategy for drying pine.

EXPERIMENTAL

OSB flakes were obtained from Georgia-Pacific facilities at Grenada, MS. and Dudley, NC. Sawdust was obtained from the Weyerhaeuser Adel, GA. plant. The principal component of these furnishes was loblolly pine. Two furnaces were used in this study. A tube furnace was utilized to dry small (< 8 g.) samples. Air was metered to the furnace at 2 lpm, and the entire furnace emissions were fed to a flame ionization analyzer (FIA). This unit is used in EPA Method 25A for monitoring total gaseous non-methane emissions. A 2,750 cu in. furnace was used to dry up to 5 lbs. of furnish. Air was injected through a manifold at the base of the furnace, and a sidestream from the emissions was analyzed by the FIA. The two furnaces are referred to in this study as the small and large furnaces, respectively.

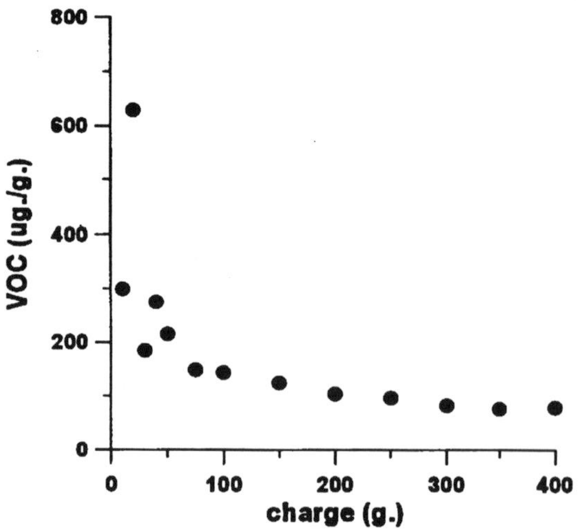

Figure 1. Dependence of VOC release on charge (wet basis).

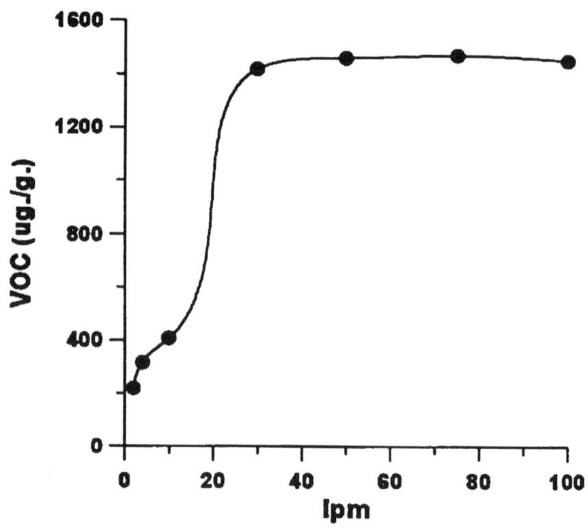

Figure 2. Dependence of emissions on flow rate (wet basis).

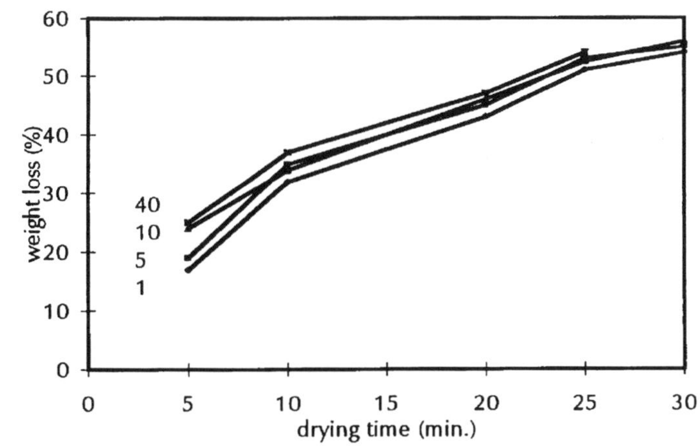

Figure 3 Effect of flow rate (1pm) on drying rate (wet basis).

Air was introduced at 4 lpm into our large furnace at 130°C, and VOCs were integrated over 1 hour. The results illustrated in Figure 1 show that smaller charges are associated with higher VOC. Except for the two highest weights in Figure 1, the furnish dried completely over the 1-hour period. Since the air flow in Figure 1 was constant, the smaller charge was associated with a higher air flow per gram of furnish. In order to isolate the effect of flow alone, the size of the charge was held constant and the air flow varied. VOCs from drying 10 g. of OSB at 130°C in our large oven for 1 hour are plotted against flow rate in Figure 2. Clearly, VOCs increase air flow, plateauing at about 30 lpm, which corresponds to 3 lpm/g. Flow rate affects the drying rate to a lesser degree as shown in Figure 3, where 50 g. of OSB were dried at flow rates of 1-50 lpm. Hence, we have the curious situation where increasing flow rate dramatically increases VOC loss, but affects the drying rate to a significant, but much smaller extent. While we are exploring the mechanism further, our working hypothesis is that the surface partially dries out at the higher flow rates. The local temperature in these micro-regions rise, and increased VOC loss occurs.

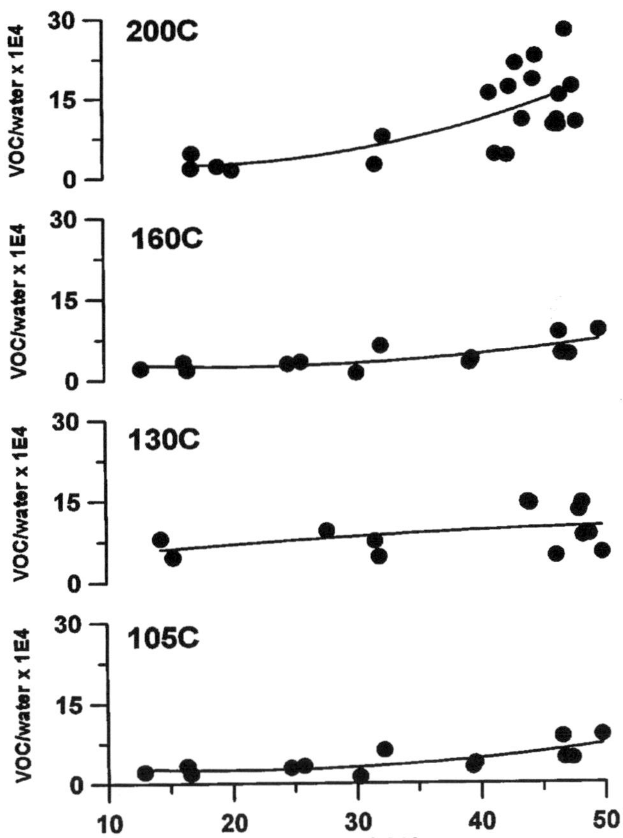

Figure 4. VOC/water ratios for OSB as a function of temperature and moisture content (initial wet basis moisture ~ 50%).

EFFECT OF TEMPERATURE ON VOC EMISSIONS

OSB flakes were homogenized by coarse-grinding in a Wiley Mill to about 1"x3". They were dried in the small furnace to various moisture end points and under various conditions of temperatures and drying periods, and the VOC /water loss determined for each charge. The results illustrated in Figure 4 show considerable scatter, but this is not unexpected given the variability in the furnish. However, it is clear that the amount of VOC released per unit of water lost is much higher at the higher temperatures, particularly after about 30% weight loss, which corresponds to the onset of the falling rate period, where transport of water to the surface becomes rate limiting [3,4]. The results combined in the 3-D plot in Figure 5 more clearly illustrate that VOC release is roughly independent of temperature up to about 30% weight loss, beyond which VOC emissions increase with temperature.

In order to further develop the temperature-VOC relationship with a more uniform furnish, we dried sawdust in our small oven at 130-180°C for 1 hour. As with OSB, two signals were

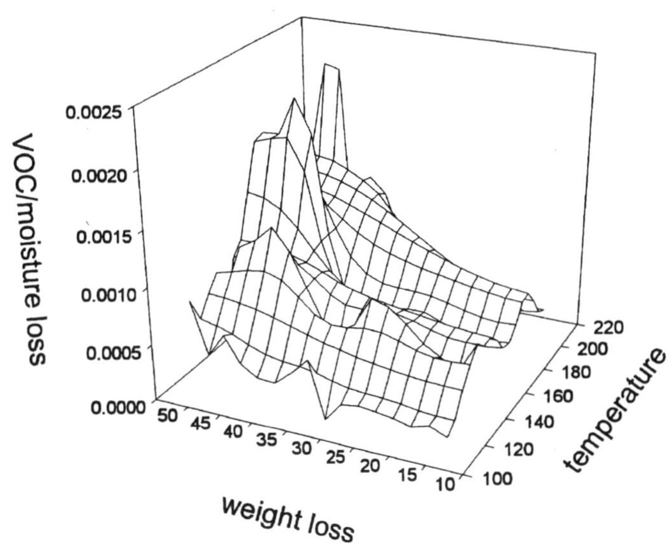

Figure 5. Relationships among VOC: water ratio, temperature and wet basis moisture content (constructed from the Figure 4 data)

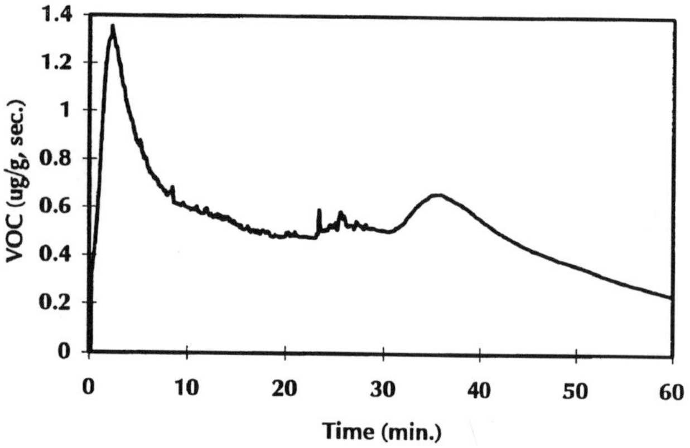

Figure 6. VOCs from sawdust at 130°C.

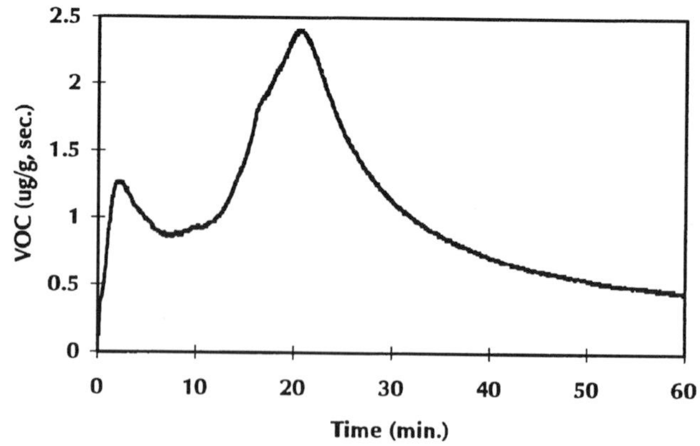

Figure 8. VOCs from sawdust at 180°C.

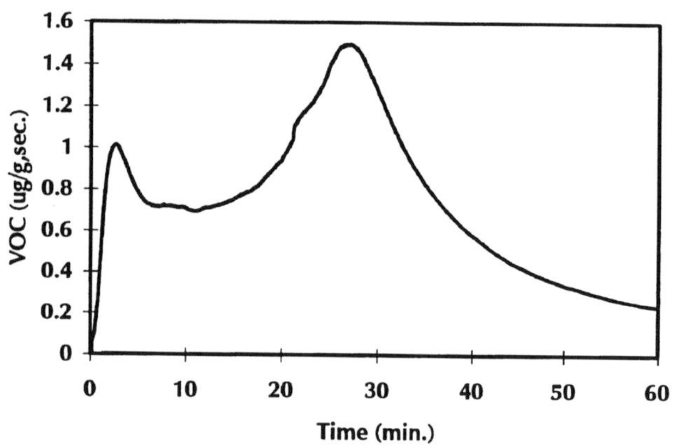

Figure 7. VOCs from sawdust at 160°C.

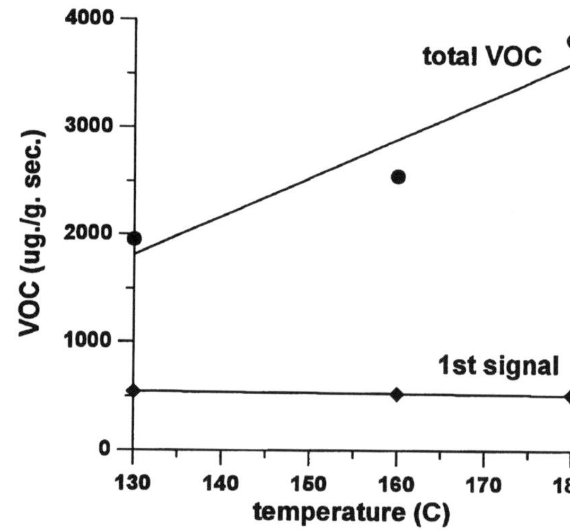

Figure 9. Dependence of VOC release on temperature for sawdust.

observed. The first emerged rapidly and corresponds to the period during which the flake temperature was below 100°C, i.e., the furnish was evaporatively cooled. The second appeared after the furnish was substantially dry and began to reach oven temperature [5]. Consider the profiles in Figures 6-8, which represent VOC signals for sawdust between 130-160°C. Note that the maximum of the first signal remains at the same level of 1-1.4 g/g. sec., but that the intensity of the second peak increases with temperature. The first signal represents VOCs released while water is still being lost; i.e. the material is evaporatively cooled to below 100°C regardless of the oven temperature. The second reflects VOCs given off after the furnish is much drier. The total VOC released over 1 hour correlates with temperature as illustrated in Figure 9. The intensity of the first peak is not temperature-dependent since the furnish temperature remains roughly constant regardless of the oven set temperature.

Figure 9 demonstrates that VOCs emerge through different mechanisms from the same furnish. One pathway is temperature dependent; the other is not. The initial signal reflects near-surface VOCs whose release is mediated by water, since the water and VOC profiles are very similar at temperatures exceeding 130°C [6]. The second peak emerges when the water is mostly gone, and its temperature dependence implicates a vapor pressure driven mechanism. As noted previously [5], VOC emissions should be reduced if the furnish is dried just to the point of emergence of the second VOC signal.

DISCUSSION

Although the mechanism through which flow rate influences VOC release is unknown, we hypothesize that dry regions develop on the surface at high flow rates. If so, then the temperature at these dry micro-regions will rise owing to the loss of evaporative cooling [5] and will promote VOC release. Support for this position comes from the temperature dependence data. The VOC rise in Figure 4 corresponds to the beginning of the falling rate regime when the surface becomes progressively less saturated, i.e., dry regions develop. The contrast in the behavior of the two signals in Figure 9 is also quite compelling. The initial signal is independent of the oven set temperature because of evaporative cooling during constant rate drying. The intensity of the second signal which is associated with late drying, increases with temperature, and is responsible for the increase in overall VOCs. Again, VOCs increase when the furnish (or portions therein) dry out.

From a practical point of view, it seems that a uniform reduction of moisture in the furnish during drying will reduce VOCs. Again, while the mechanism underlying these relationships is unknown, surface drying will impede the transport of interior moisture to the surface. Hence, evaporative cooling of the surface will be diminished, and the surface temperature will rise, at least in certain regions. While the conditions that govern VOC loss are complex, we know that a VOC surge occurs when the flake is largely dry and the flake temperature rises [5]. Thus, prolonging the onset of the temperature rise will retain the VOC in the flake and decrease the VOC:water loss ratio. We note that this is not necessarily a panacea, since the retained VOCs may well increase press vent emissions, a possibility that we plan to explore.

The optimum drying strategy should minimize temperature imbalances and moisture gradients within the flake in order to reduce "dry spots", since these regions will release a higher proportion of VOCs. Since surface dry-out will impede the movement of water to the surface, it should be avoided, or its onset delayed.

CONCLUSIONS

The principal theme that emerges from this work is that the wet line should be maintained at the surface for as long as possible to keep the furnish evenly wet during drying. Practical observations support this position. It appears that lower VOCs are associated with drying with a lower temperature conveyor system or with superheated steam, both of which tend to keep the wet line at the flake surface.

ACKNOWLEDGEMENT

This work was sponsored by a consortium comprising the Electric Power Research Institute, Southern Co., Entergy, Georgia-Pacific, and Weyerhaeuser.

LITERATURE CITED

1. **Flodin, K., and J. Andersson**, *Eur. J. Forest Pathol.*, 7, 282, (1977).
2. **Cronn, D.R., S.G. Truitt, and M.J. Campbell**, *Atmos. Environ.*, **17**, 201, (1983).
3. **Rosen, H.N.** in *Handbook of Industrial Drying*, **A.S. Mujumdar, Ed**., Dekker, NY (1995).
4. **Malte, P.C., B. Dorri, A.F. Emery, R.W. Cox, and R.J. Robertus**, *Drying Technology*, **1**, 83, 1983.
5. **Banerjee, S., M. Hutten, W. Su, L. Otwell, and L. Newton**, *Environ. Sci. Technol.*, **29**, 1135, 1995.
6. **Banerjee, S., D. Andrews-Wilberforce, and A. Fraley**, unpublished results.

Effect of Specific Edge Loading on Freeness Drop Coefficient

P.D. Bennett, V. Turt and J.M. Genco
Department of Chemical Engineering, University of Maine, Orono, ME 04469

S. Jansson and B.K. Lönnberg
Pulping Technology, Åbo Akademi University, Åbo, Finland

Refining trials were conducted in a re-circulation mode using bleached softwood and hardwood kraft pulps. In the experiments, the Canadian standard freeness was measured on the refined samples as a function of time using a variable speed double disk refiner The refining process was described using the specific edge load theory formulated by Brecht (1967). The net specific energy (E_{net}) was varied at constant values for the specific edge loading (SEL). The data for the freeness drop (ΔF) could be mathematically correlated in terms of the specific edge load (SEL) and the net refining energy (E_{net}) using the power low suggested by Leider and Nissan (1994). The freeness drop coefficient (b)

$$b = \left(\frac{-\delta F}{\delta E_{net}} \right)_{SEL} = f(SEL)$$

was found to decrease with increasing specific edge loading for both the bleached hardwood and softwood kraft pulps. The decrease in the freeness drop coefficient was thought to result from the increase in fiber hydration that occurs when refining at low SEL values.

INTRODUCTION

Refining can be defined as the mechanical treatment of pulp carried out in the presence of water usually by passing the suspension of pulp fibers through a relatively narrow gap between a rotor and a stator, both of which carry bars or knives aligned more or less across the line of flow of the stock. Refining is a critical operation in a paper mill. Commercial refiners are categorically based on geometry and are either conical or disc machines (see Figure 1). In all cases the power depends on the gas width.

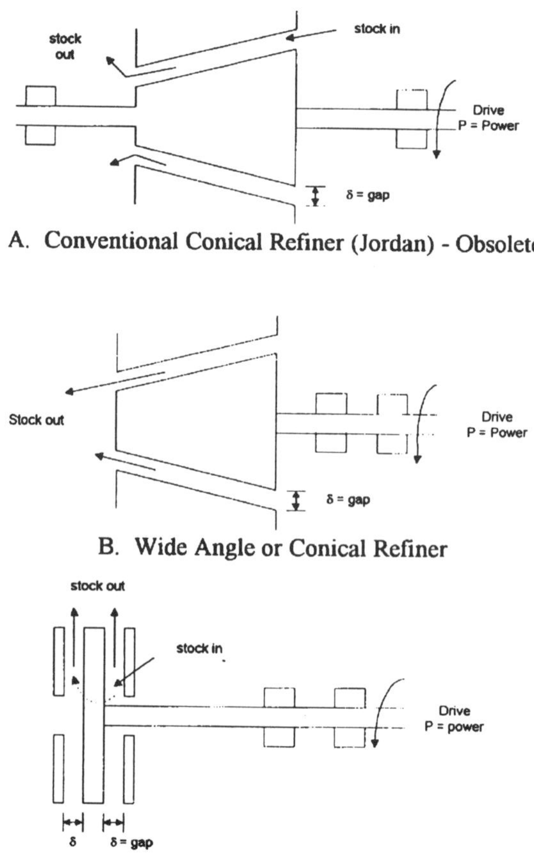

A. Conventional Conical Refiner (Jordan) - Obsolete

B. Wide Angle or Conical Refiner

C. Typical Double Disc Refiner

Figure 1. Typical Commercial Refining Machines

The primary effects of refining results in changes in the structure of the fiber and these can be categorized in the following manner:

1. Creation of new surfaces.

2. Creation of new particles, and

3. Generation of structural damage and modifications.

Figure 2 presents an illustration of the refining action on a fiber. External fibrillation means that the primary wall of the fiber has been largely removed and that the fibrils on the secondary wall have been exposed. This permits the fiber to swell. Internal fibrillation describes the breaking of internal tangential bonds in the structure of the cell wall. This makes the fiber more flexible and increases the probability of contact and bonding between the fibers. Molecular fibrillation is defined as the partial

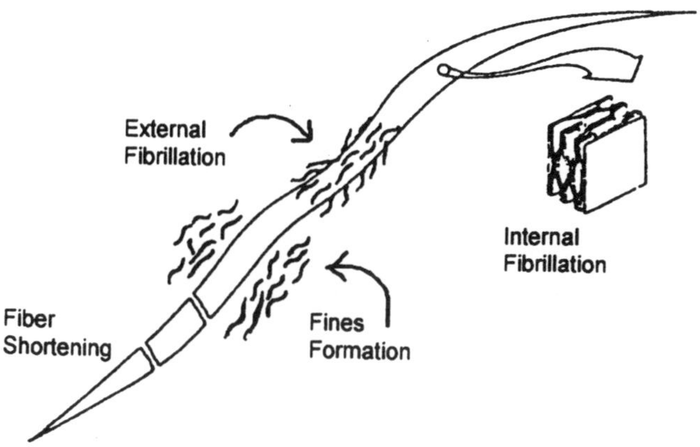

Figure 2. Illustration of the refining action on a fiber.

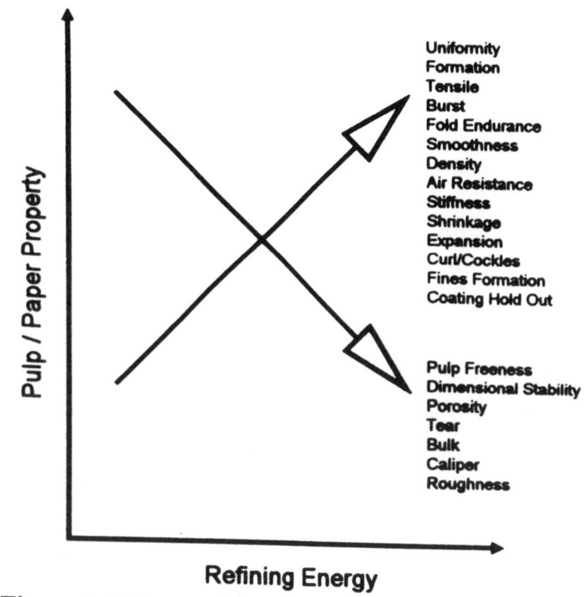

Figure 3. Effect of refining on pulp and paper properties

solubilization of the polymeric structure of the cell wall. It is characterized as a particle solution of the cell wall carbohydrates, where the molecules are still attached to the cell wall matrix from one of the ends while the other end remains in solution. Creation of a colloidal polysaccharide solution has a connection with the breakage of the hydrogen bonds on the amorphous parts of the fiber. When these bonds break, they create new hydrogen bonds with water which leads to apparent swelling in the amorphous parts. Factors that affect the electrical potential of the fiber, such as pH, consistency and ion concentration in the water, are factors which influence the swelling of the colloidal polysaccharides. The colloidal solution acts as a glue when bonds between the fibers are created. There is no direct evidence for the formation of molecular fibrillation in refining, but there are results that indicate that part of the cell wall material is being dissolved during refining. For instance, the refining of an α-cellulose fiber is nearly impossible.

Refining modifies the physical properties of pulp used in the manufacture of paper. Refining improves burst, tensile and internal strength of paper. Formation is also improved. However, refining also produces undesirable effects, such as reductions in tear strength and dimensional stability. Virtually every significant physical and optical property of pulp, hence paper, is affected by refining: see Figure 3. Another purpose of refining is to facilitate the wet end operations on the paper machine. In this application it is used to control the drainage rate on the paper machine.

Objective and Scope

The objective of this project was to determine the effect of the specific edge loading on the Freeness drop coefficient. This was done by using a variable speed refining operating in the re-circulation mode. Data are presented for refining of Northeastern softwood Kraft and Northwestern hardwood pulp samples. Specifically, the freeness drop coefficients as determined as a function of refining energy at constant values for the specific edge loading. Canadian standard freeness values were evaluated. The change in freeness was correlated with the fundamental refining variables defined in the specific edge load theory proposed by Brecht (1967)

RELATED LITERATURE

Considerable effort has been expended in quantifying the action of refining to permit refiner comparison [Danforth (1969), Stevens (1981), Lumiainen (1989), Kerekes (1990)]. The fundamentals of the refining process have been extensively studied by Hietanen (1990, 1991) and Hietanen and Ebeling (1990A, 1990B, 1983). Hartman and Higgins (1983) discuss the influence of mechanical treatment on the development of optimum sheet properties.

Early work on the beating process and its effects on fiber quality are delineated by Higgins and de Young (1967). Eklund and Lindstrom (1991) summarize the basic mechanisms of mechanical processing of fiber. Considering thermodynamics, Hietanen and Ebeling (1990A, 1990B) classify the primary effects of refining as creation of new surfaces, creation of new particles and generation of structural damage and modifications.

The effect of beating on fiber swelling and water retention value was studied by Steenberg, Sandgren, and Wahren (1960) and by Szwarcztajn and Przybysz (1975). They characterized the change in water retention value with the various pulp fractions created during beating. Giertz (1980) studied the role of fines generated during the refining process and demonstrated the important contribution that fines made to the strength of chemical pulp.

Brecht (1967), Leider and Nissan (1977) and Kerekes (1990) have proposed refining theories. Brecht's specific edge load theory is the most widely accepted theory for application to mill situations. Consequently, it was used to correlate the data in the present study.

Specific Edge Load (SEL) Theory

The power required to drive the machine depends upon its diameter (D), gap width (d) and consistency (c) of the stock being refined (Higgins, 1985).

$$P = C_P \frac{D^5 c \Omega^3}{\delta} \qquad (1)$$

where

C_P = Power coefficient
Ω = RPM
c = Consistency
δ = Gap (see Figure 1)
D = Average diameter of machine

The 'specific edge load' (SEL) is a quantity that can be used to make comparisons between competitive refining equipment. It is defined as the 'effective energy input' (joules) by the refiner motor per unit 'edge length' (m) of the bars that are used to impart energy to the pulp as it travels through the grooved of the refiner. Wultsch and Flucher (1958) found that a given pulp had similar properties when refined in an identical manner in a small laboratory refiner and in a large industrial refiner, provided both machines were equipped with tackle having bars of the same width and operated at constant specific edge load.

Brecht (1967) proposed the following equation to define the 'total bar edge length per unit of time' (L_s):

$$L_s = \Omega n_r n_s L = L_r \Omega \qquad (2)$$

where

L_s = total edge length per second, (m/sec).
Ω = revolutions per minute, (rpm).
N_r = number of rotor bars.
N_s = number of stator bars.
L = effective bar length, (m).
L_r = cutting length per revolution, (m).

The net specific energy is the net energy applied to the fibers and the fluid surrounding the fibers per unit mass of dry fiber. Thus the specific energy ($\hat{E}_{net}$) is defined by the equations:

$$\hat{E}_{net} = \frac{P - P^0}{q \rho c} \qquad (3)$$

where

$\hat{E}_{net}$ = net specific energy, kW-hr/metric ton
P = total refining power, kW
P^0 = no-load power, kW
q = flow rate, m^3/hr
ρ = stock density, kg/m^3
C = stock consistency, kg fiber/kg total mass

Since $\hat{E}_{net}$ is a direct indication of energy transferred to the pulp suspension, it is convenient to use it as a measure of the degree of refining that has been done. It gives no indication of the duration or the intensity of treatment.

The quantity L_s is used to calculate the intensity of refining or the specific edge loading (SEL):

$$SEL = \frac{P - P^0}{L_s} = \frac{P - P^0}{\Omega L_r} \qquad (4)$$

The result of the refining process should be the same as long as the SEL remains constant, even if some factors of the equation are changed. These changes can be achieved by altering certain operating variables such as rotational speed, plate tackle, or the gap between the rotor and stator plates which directly affects the total refining power applied during the process. Brecht (1967) observed identical refining results for density, tear and bursting strength at the same net specific energy consumption if the specific edge load remained constant.

Turt, Genco and Co (1994) observed that similar fiber properties were obtained provided the net refining energy and the specific edge loading were constant. The authors noted that operating the refiner at widely different rotational speeds did not change the response of the pulp to refining, provided the net specific energy ($\hat{E}_{net}$) and specific edge load (SEL) were held constant.

$$\Delta P = f(\hat{E}_{net}, SEL) \qquad (5)$$

The fiber properties could be adequately correlated in terms of a power law relationship originally suggested by Leider and Nissan (1987).

$$\Delta P = k(SEL)^a \hat{E}_{net}{}^b \tag{6}$$

where k, a, and b are coefficients that describe the pulp. Turt (1994) showed that in the recirculation mode of refining, the equation

$$F = F^0 - \frac{Qb\hat{E}_{net}{}^*}{V}t \tag{7}$$

can adequately represent the refining data for the change in freeness (F) or drainage rate for the pulp in terms of the flow rate (Q) through the refiner, the volume of the stock being refined (V) in time (t). $\hat{E}^*{}_{net}$ is the refining energy being applied to the stock being recirculated. The freeness drop coefficient was determined to be a function of the specific edge load but independent of the net refining energy.

$$b = \left(\frac{-\partial F}{\partial E_{net}}\right)_{SEL} = f(SEL) \tag{8}$$

The U. Maine data showed that the production of fines during refining can be correlated to the change in freeness. This result was taken as justification for the validity of Equation 8 for the freeness change since it was derived from the assumption where the fines content is proportional to the change in freeness of the pulp. The derivation of Equation 7 is summarized in Appendix A and forms the basis for treating the experimental data for freeness.

EXPERIMENTAL METHODS

The effect of refining on freeness was compared and correlated with the variables of the specific edge load theory. Refining experiments were performed in a 13-inch variable speed double disk refiner located in the University of Maine's Pilot Plant. The refiner is a Model 3000 Beloit Jones double disk machine similar to those used in industrial operations. The experiments were performed in a recirculation mode.

Bleached Northeastern softwood Kraft pulp samples were obtained from the S. D. Warren Paper Company (Division of SAPPI - South African Pulp and Paper Company) located in Skowhegan, Maine. Bleached Northwestern hardwood pulp samples were also obtained from Daishowa-Marubeni International Limited located in Peace River, Alberta Canada. All refining experiments were performed at 3% consistency. Details of the experimental methods are summarized in Appendix A.

The bleached softwood Kraft pulps (both Northeastern and Northwestern) were refined at specific edge load values between 0.25 and 1.25 J/m. Specific edge load values for the hardwood pulps were varied between 0.2 and 0.85 J/m.. The flow rate was set by the opening in the stock flow valve. Turt (1994) summarizes the data for the bleached Northeastern softwood Kraft pulp.

RESULTS AND DISCUSSION

Effect of SEL.

Turt (1994) correlated the fiber freeness for bleached Northwestern softwood Kraft pulp as a function of the net refining energy, E_{net}, at two different specific edge loading values (SEL=0.25 and 1.25 J/m). Three different disk speeds (600, 900, and 1200 rpm) were chosen as an independent operating variable during the refining experiments. The plate gap was varied to give the same specific edge loading. These data are summarized in Figures 4 and 5.

Figure 4 shows the change in freeness of the pulp as a function of the refining time. Data are shown for a rotational speed of 600 rpm. In the recirculation mode, the freeness decreases linearly whether the refining was being done at the low SEL or the high SEL. At the same refining time, the high SEL experiments gave pulp with lower freeness than the low SEL experiments. Alternatively, for the same freeness value, the low SEL refining requires more time because the energy input to the refiner is five times less than the high SEL refining.

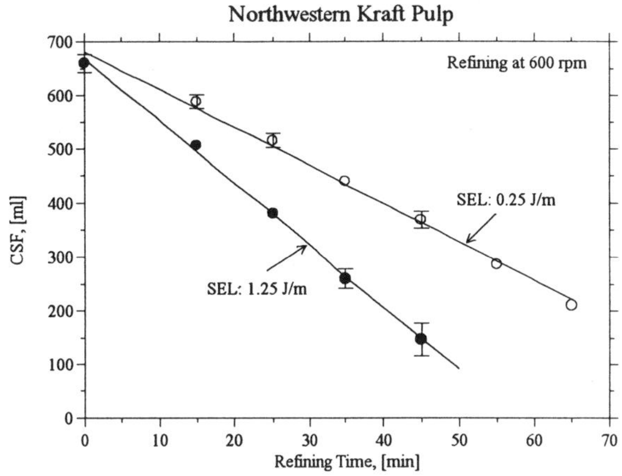

Figure 4. Canadian standard freeness versus refining time at 600 rpm for Northwestern kraft pulp. (Turt, 1994)

Figure 5 shows the change in the Canadian Standard Freeness (CSF) data plotted versus the net refining energy using refining intensity as the parameter. There is an obvious linear relationship between the CSF and the net refining energy at both specific edge loadings. The magnitude of slope of the freeness versus E_{net} curve (b value) is higher for refining at the low SEL value. At the low specific edge loading (SEL:0.25 J/m), less energy is consumed to achieve a target freeness than at the high specific edge loading. It can be considered that refining at the low SEL value causes fiber hydration and consequently the freeness decreases more rapidly for a given net refining energy (Enet).

Secondly, identical results for the freeness (CSF) were obtained when various operational disk speeds were used, as long as the specific edge load remained constant. This is the same qualitative result obtained by Brecht (1967) in his classical refining experiments defining the specific edge load theory.

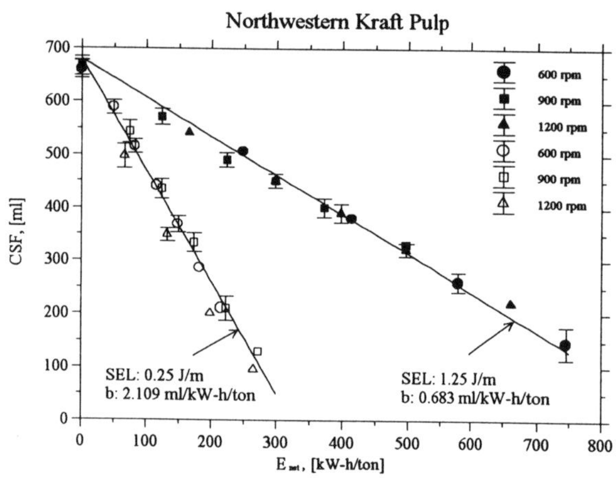

Figure 5. Canadian standard freeness versus net refining energy for Northwestern kraft pulp. (Turt, 1994)

Freeness Test.

Data for the freeness of the pulp plotted as a function of the refining time are summarized in Figures 6 and 7. The freeness decreases linearly with refining time provided the freeness is above about 200 ml. Below this value there is curvature to the plots. The early freeness drop data seen to follow the theoretical equation (Eqn.7) derived in Appendix B.

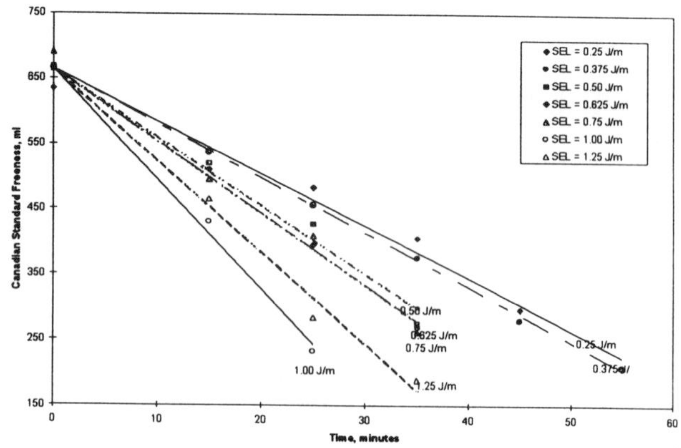

Figure 6. Canadian standard freeness versus refining time: Softwood Pulp.

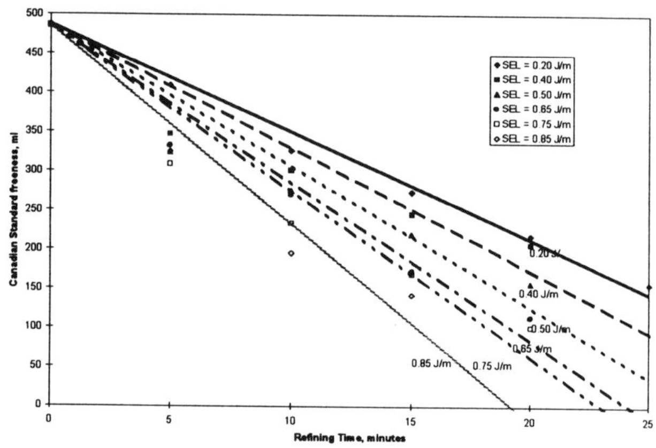

Figure 7. Canadian standard freeness versus refining time: Hardwood Pulp.

Figures 8 and 9 show the change in freeness data plotted against the net refining energy for the softwood and hardwood pulps. The data are plotted on a log-log scale according to the power law, Equation 9.

$$\Delta P = k(SEL)^a \hat{E}_{net}{}^b \quad (9)$$

Taking the logarithm of both sides,

$$\log(\Delta P) = \log k + a\log(\hat{E}_{net}) + b\log(SEL) \quad (10)$$

Since the SEL value was constant in any experiment, Equation 9 reduces to Equation 11 which is plotted in Figures 8 and 9.

$$\Delta P = k'\hat{E}^a_{net} \quad (11)$$

$$\log(\Delta P) = \log k' + a\log \hat{E}_{net} \quad (12)$$

There is a linear relationship between the freeness and the net refining energy. At the low specific edge loading, less energy is consumed to achieve a target freeness than at the high specific edge loading. Refining at the low

SEL value causes fiber hydration and the freeness decreases more rapidly for a given refining energy.

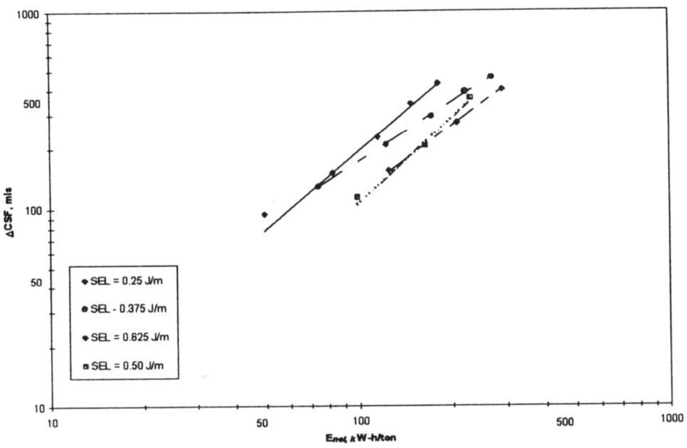

Figure 8A. Change in freeness versus refining energy: Softwood Pulp (low SEL values)

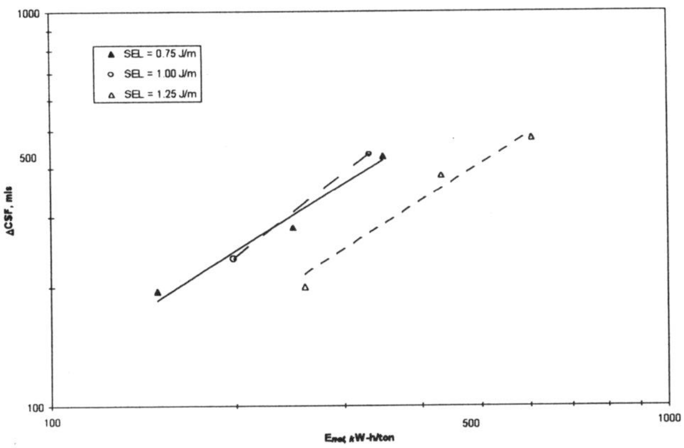

Figure 8B. Change in freeness versus refining energy: Softwood Pulp (high SEL values)

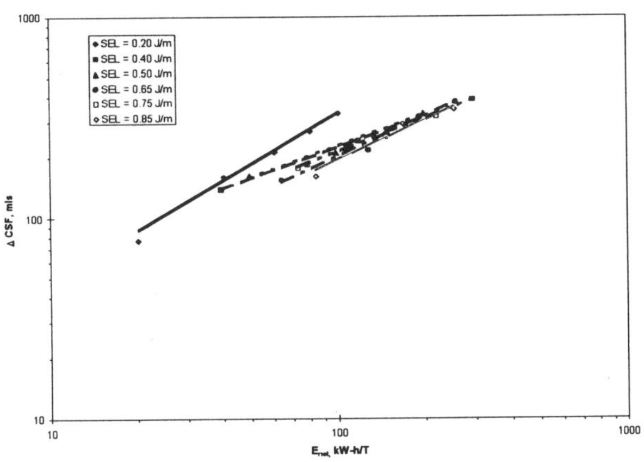

Figure 9. Change in freeness versus refining energy: Hardwood Pulp

Refining at lower specific edge load values require longer refining time to achieve a given freeness than refining with a higher specific edge load value. The magnitude of the slope is higher for the experiments with a lower specific edge load value than for the experiments with a high specific edge load.

Estimation of Freeness Drop Coefficient (b).

If the freeness of the pulp is assumed to be proportional to the amount of fines present in the pulp, a material balance will result in the relationship between the initial freeness (F^0) and the freeness (F) will result in Equation 7. The (V/Q) term is the residence time for flow through the refiner circuit. The value of (b) is obtained from the slope of the freeness versus time curve.

$$slope = \frac{bE^*_{net}}{\left(V/Q\right)} \tag{13}$$

$$b = \frac{slope \times V}{E^*_{net} \times Q} \tag{14}$$

Table 1 shows the calculated b values, and the efficiency, η, for the bleached Northwestern Kraft pulp reported by Turt (1995). Table 2 shows similar data gathered on the bleached Northeastern softwood and hardwood Kraft pulps. The b value (see Table 1) is considerably higher (2.1 ml. per kW-h/mt)) when refining at 0.25 J/m compared to that obtained (0.68 ml. per kW-h/mt) when refining at 1.25 J/m SEL. As the specific edge load increases, the refining coefficient (b) decreases indicated lower freeness drop obtained per unit of energy added to the pulp.

Table 1. Calculated b values for the Northwestern kraft pulp in the recirculation experiments (Turt,1994)

SEL J/m)	Ω (rpm)	slope (ml/min)	E_{net}* (kW-h/ton)	b value (ml/kW-h/ton)
0.25	600	7.08	19.9	2.138
0.25	900	10.0	29.6	2.026
0.25	1200	14.3	39.7	2.164
1.25	600	11.5	99.2	0.698
1.25	900	16.9	148.9	0.684
1.25	1200	22.0	198.4	0.668

Table 2. The b values and efficiency data for bleached Northeastern Softwood and hardwood pulp.

SEL, J/m	*b, kWh/T*	*h, %*
	Softwood Pulp	
0.250	2.354	73.0%
0.375	1.637	80.2%
0.500	1.566	84.4%
0.625	1.208	87.2%
0.750	0.994	89.0%
1.000	0.754	91.5%
1.250	0.599	93.1%
	Hardwood Pulp 1	
0.20	3.22	47.8
0.40	1.68	64.2
0.50	1.54	69.2
0.65	1.41	74.5
0.74	1.24	76.9
0.84	1.38	79.3

Figure 10 shows the graph of the b value as a function of the specific edge load. Data for the hardwood pulp are plotted along with data for the softwood kraft pulp.. The bleached hardwood pulp showed higher freeness drop coefficient (b) values for all of the specific edge load factors tested. The data of Figure 10 were thought to result from the greater hydration that takes place when operating at low specific edge load values (Turt, Genco and Co; 1994).

Refining Efficiency

The refining efficiency, η, is defined as the ratio of the net power ($P-P^0$) divided by the total power input to the refiner (P):

$$\eta = \frac{P - P^0}{P} = 1 - \frac{P^0}{P} \qquad (15)$$

The efficiency represents the fraction of the power to the refiner required to actually perform the mechanical action of the refiner. The data for the efficiency are shown in Figure 11. In the experimental refiner the efficiency increased with increasing SEL and was higher for the softwood pulp. No good explanation can be given for the higher efficiency for the softwood pulp. This may be related to the observation that softwood pulps can take the mechanical load applied by the refiner without clashing much better than hardwood pulps.

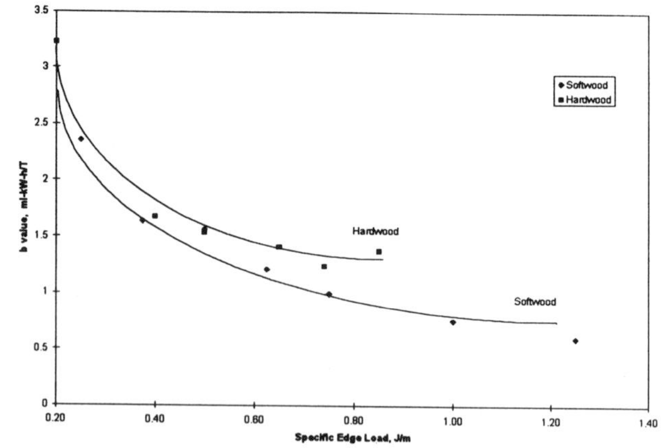

Figure 10. The relationship of b value to specific edge load

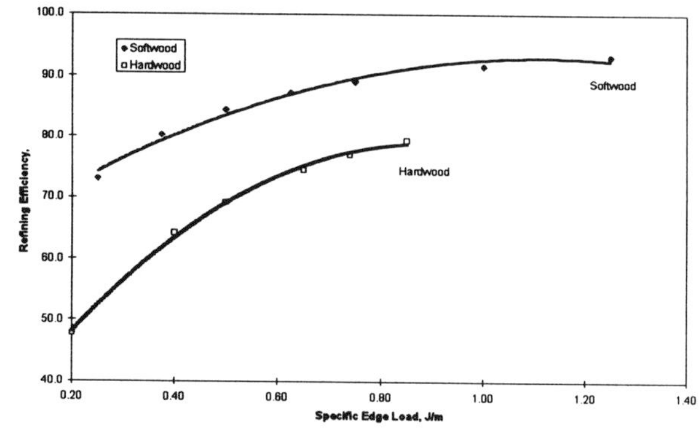

Figure 11. The efficiency versus specific edge load.

Application to the Mill

Although the current experiments were performed in the 100% re-circulation mode, they should be applicable to the mill situation. Operating in the 100% re-circulation mode would be expected to increase the width of the residence time distribution for the stock passing through the refiner compared to the once-pass mode practiced industrially. Turt (1994) showed similar pulp properties, including freeness, when refining in the once-pass and 100% re-circulation modes of operation.

Currently constant speed refining is practiced in the mill situation but suffers from a serious disadvantage. As the speed of the machine or as the grade of paper changes, this will necessitate changes in flow through the refiner. Since the rotational speed and plate pattern in the refiner are fixed, the number of specific edge loading will change. This results in variable pulp properties and poor machine performance. Often the machine operators complain of stock changes when in reality the flow through the refiner is causing the pulp properties to vary.

Variable speed refining is a potential strategy to deal with changes in flow through the refiner. To maintain approximately the same refining level, both the SEL and the E_{net} values have to remain constant. To keep both E_{net} and SEL constant, the rotational speed of the refiner, Ω, must change accordingly to account for changes in flow (Q) through the refiner. If the speed of the disk is changed the SEL value and E_{net} variables can be maintained constant when the demand for refined stock changes. Under current practice the only alternative is to change the plate pattern to vary the specific edge loading; which is impractical. Variable speed refining is more complicated than constant speed refiner. Also, a variable speed motor would be required in place of the constant speed motor.

CONCLUSIONS

When operating in the 100% re-circulation mode, the freeness drop varies linearly with time provided the freeness drop is above about 200 m. Below this value some curvature is seen with time. The freeness drop coefficient (b) decreases as the specific edge loading and the severity of the refining increases. The Canadian standard freeness could be adequately correlated in terms of the specific edge load (SEL) and the net refining energy (E_{net}) by using the power law correlation, $\Delta F = k(SEL)^a \hat{E}_{net}{}^b$. These results support the basic premise postulated by Brecht (1967) and extended by Turt, Genco and Co (1994) to include fiber properties.

Variable speed refining represents a potential solution to the problem of changing flow through the refine which causes the SEL value to vary.

APPENDIX A: EXPERIMENTAL METHODS

Hydropulping Operation

A hundred pounds of dry pulp was used for each batch of pulp tested in the experiments. At the beginning of each pulping experiment, a sufficient amount of water was added to the pulper to reach approximately 12% pulping consistency. The electrical drive of the rotor was started and the dry pulp were added to the pulper over a 5-minute time period. The pulp was dispersed in the pulper for 30 minutes and pumped over to a larger stock chest. The consistency in the stock chest was lowered to 3% for subsequent refining.

Disk Refining Experiments

Refining experiments were carried out in a refining flow loop located in Jenness Hall. The flow loop contained a 13-inch double disk refiner which was equipped with a variable speed motor. The drive on the refiner could deliver up to 112 kW of total electrical power over the operating range of 500-1750 rpm. The flow loop has a centrifugal pump to deliver the stock to the refiner and was also equipped with a flowmeter to provide various flow rates of the pulp suspension. Figure A1 shows the schematic of the flow loop.

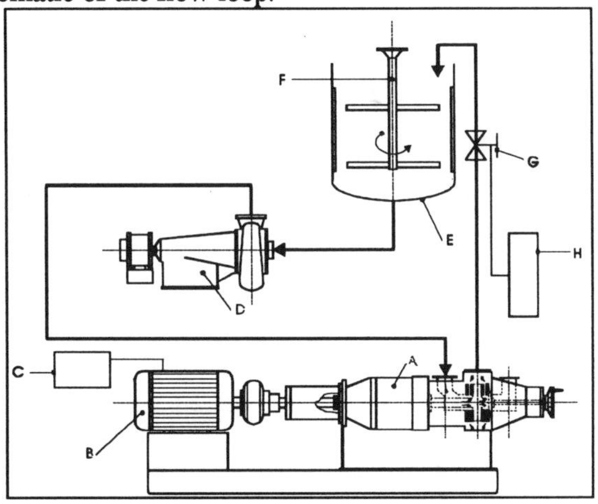

Figure A1. Recirculation refining mode.

A: 13-inch disk refiner
B: Refiner motor (112 kW)
C: Powermeter
D: Centrifugal pump
E: No. 1 machine chest
F: Agitator
G: Flow valve
H: Flow control

Figure A2 shows a representative picture of the 13-inch double disk refiner and flow configuration. Two stainless steel disk plates are mounted on the rotating head of the floating shaft. These plates are called "rotor plates". The floating shaft is situated on double row taper roller bearings to provide sufficient axial travel. One of the stationary disks is installed on a pad close to the inlet. A sliding head bearing the second stationary disk plate moves axially by a manual control and adjusts the refining gap between the rotor and stator plates. The Duo-Flo configuration permits half of the incoming stock under pressure to pass through to the eye of the second set of disks. The stock passes in parallel across both sets of disks, rejoins, and leaves the refiner through a single outlet. The data for the plate pattern is shown in Table A1.

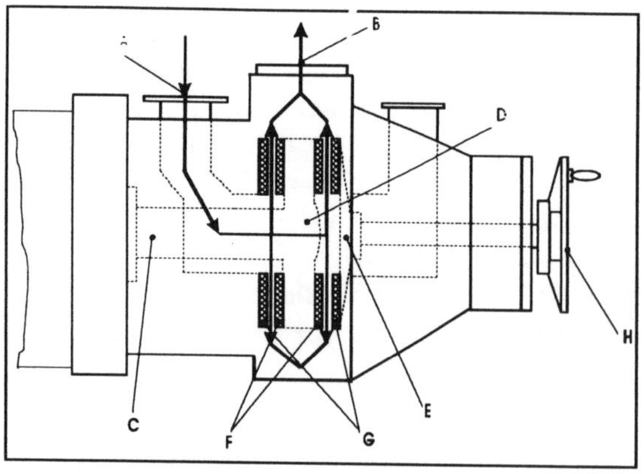

Figure A2. Duo-Flow configuration of the refiner.
A: Inlet flow
B: Outlet flow
C: Floating shaft
D: Rotating head
E: Sliding head
F: Rotor disks
G: Stationary disks
H: Manual control for setting the plate gap

Table A1. Plate Pattern for 13-inch Refiner

Plate No.	D27707128 & 129
Outer Diameter	330.2 mm
Inner Diameter	196.8 mm
Bar Width	3.175 mm
Groove Width	2.381 mm
Groove Length	73 mm
Groove Angle	24°
Cutting Length	1.62 km/rev

In the experiments, different values of specific edge loading (SEL) were studied. These values of specific edge loading were obtained by changing the power setting to the refiner. Softwood data was obtained at a rotational speed of 600 rpm. The hardwood experiments were conducted at 1000 rpm. A stroboscopic tachometer was used to obtain the actual rpm value. The flow rate was set by opening and closing a stock flow valve in the flow loop. The stock flow rate was measured by pumping the stock from one tank to another and measuring the level of the stock in the first tank.

The pulp consistency was checked before refining by taking samples from three different locations in the tank. It was adjusted to approximately 3% for each experiment.

Table A2 shows the experimental conditions used in these experiments.

The no-load power was determined for the stock with the refiner plates backed off at the corresponding rotational speed. The total refining power required to achieve a given value of the specific edge loading (SEL) was estimated from Eqn. A1.

$$SEL = \frac{(P - P^0)}{\Omega L_r} \tag{A1}$$

This was accomplished by varying the power (P) at a given rotational speed (Ω). The no-load power is represented by the symbol (P^0). Since the refining was performed in a recirculation mode the specific energy for the stock in the tank was estimated from the equation

$$E_{net} = \frac{(P - P^0)t}{V\rho c} = \frac{(P - P^0)t}{(Qt')\rho c} \tag{A2}$$

$$E_{net} = \left[\frac{(P - P^0)}{(Q\rho c)}\right]\frac{t}{t'} \tag{A3}$$

$$E_{net} = \frac{E^*_{net}t}{t'} \tag{A4}$$

where (Q) is the flow through the refiner, (ρ) is the density of the stock, (c) is the stock consistency, (t) is the time and (t') is) the residence time for stock flow through the tank having stock volume (V).

Before the refining power was set, unrefined samples were collected from various locations in the stock tank. When the total power was applied, the experiment was started. Samples from three different locations in the stock tank were collected. Blended samples were taken every 5 to 10 minutes during the experiment. Samples were collected and stored (750 ml each).

Analysis of the Samples

Canadian Standard Freeness. Canadian standard freeness or stock drainage rate was measured according to TAPPI test method T-227. Each sample of pulp suspension was first diluted to 0.3% consistency and 1000 ml of diluted stock was measured in a 1 liter graduated cylinder. After pouring the stock into the freeness chamber, the top lid and the air-cock were closed, and the bottom lid opened. After waiting for 5 seconds from the time that the addition of stock was completed, the air-cock was opened to start the flow. The volume discharged from the side orifice was recorded when the side discharge ceased. Each sample was tested three times and the average freeness and standard error from the average calculated.

The change in Canadian Standard Freeness is defined as:

$$\Delta CSF = (CSF^{0}) - (CSF) \quad \text{(A5)}$$

where
(CFS°) =Freeness of the unrefined pulp
(CFS) =Freeness of the refined pulp

APPENDIX B: DERIVATION OF THE EQUATION FOR THE FREENESS CHANGE IN THE RECIRCULATION REFINING EXPERIMENTS

Figure B1. shows a typical refining system with recirculation. If the freeness change of the pulp is assumed to be proportional to the amount of fines particles present, then:

$$w = mF + e \quad \text{(B1)}$$

$$w' = mF' + e$$

where

w = fines fraction of the pulp entering the refiner (fraction)

w' = fines fraction of the pulp after the refiner (fraction)

F = freeness of the pulp entering the refiner (ml)

F' = freeness of the pulp after the refiner (ml)

m = slope (1/ml)

e = intercept (fraction)

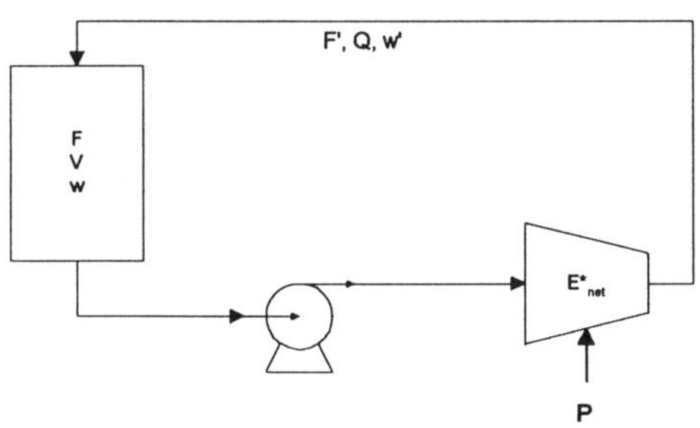

Figure B1. Refining in the recirculation mode.

The following assumptions are made for writing a mass balance for the fines fraction around the tank:

1. Constant pulp flow rate.

2. The tank has an effective stirrer such that the fines content (w) in the tank is uniform throughout the tank and that the pulp stream leaving the tank has the same fines content as that in the tank.

3. The stock density remains constant during the process.

4. The stock volume in the tank remains constant.

$$Q\rho w' = Q\rho w + V\rho \frac{dw}{dt} \quad \text{(B2)}$$

where Q = flow rate (m3/min.)

ρ = stock density (kg/m3)

V = pulp volume (m3)

Using Eq. B.1, the change in fines can be related to the changes in freeness:

$$dw = mdF \quad \text{(B3)}$$

If Eq. B1 and Eq. B3 are substituted into Eq. B2:

$$QF' = QF + V\frac{dF}{dt} \quad \text{(B4)}$$

Equation B4 is the differential equation for the freeness change taking place in the recirculation refining system.

If it is further assumed that the freeness after the refiner (F') is a function of the net energy (E_{net}) put into the pulp in the refiner and the specific edge loading (SEL), the freeness change across the refiner is :

$$F' = f(E_{net}, SEL)$$

$$dF' = \left(\frac{\partial F'}{\partial E_{net}}\right)_{SEL} dE_{net} + \left(\frac{\partial F'}{\partial SEL}\right)_{E_{net}} dSEL \quad \text{(B5)}$$

We can define:

$$b = -\left(\frac{\partial F'}{\partial E_{net}}\right) SEL \quad \text{(B6)}$$

$$a = \left(\frac{\partial f'}{\partial SEL}\right)_{e_{NET}} \quad \text{(B7)}$$

Under the assumption of refining at constant specific edge loading (SEL), Eq. A5 becomes:

$$dF' = \left(\frac{\partial F'}{\partial E_{net}}\right)_{SEL} dE_{net} = -bdE_{net} \quad \text{(B8)}$$

where b is the negative slope of the curve of the freeness against net refining energy. The coefficient b is the reduction in freeness per unit energy input to the refiner. From Figure 4.1, b is a function of the specific edge loading (SEL). Assuming that b does not depend on E_{net}, Eq. B8 can be integrated across the refiner to give:

$$\int_{F}^{F'} dF' = -b\int_{0}^{E^{*}_{net}} dE_{net} \qquad \text{(B9)}$$

where E^{*}_{net} is the net specific energy input to the pulp as it passes once through the refiner.

$$F' = F - bE^{*}_{net} \qquad \text{(B10)}$$

If Eq. B10 is substituted into Eq. B4:

$$Q\left[F - bE^{*}_{net}\right] = QF + V\frac{dF}{dt} \qquad \text{(B11)}$$

$$V\frac{dF}{dt} = -QbE^{*}_{net} \qquad \text{(B12)}$$

$$\int_{F^0}^{F} dF = \left[-\frac{Qb}{V}E^{*}_{net}\right]\int_{0}^{t} dt \qquad \text{(B13)}$$

$$F = F^0 - \frac{QbE^{*}_{net}}{V}t = F^0 - \frac{bE^{*}_{net}}{\left(V/Q\right)}t \qquad \text{(B14)}$$

where F = freeness leaving the tank at time t (ml)

F^0 = initial freeness (ml)

Q = flow rate (m^3/min.)

b = slope of freeness drop curve per unit energy input to the refiner (ml/kW-h/ton)

E^{*}_{net} = net specific energy input to the pulp as it passes once through the refiner (kW-/ton)

V = volume of the tank (m3)

t = refining time (min.)

Equation B14 is the equation for the freeness change in the recirculation mode. Equation B14 can be rewritten in terms of the one-pass time (t_{pass}). Since the one-pass time is related to the fluid volume (V) in the system and the flow rate (Q):

$$F = F^0 - bE^{*}_{net}\left(t/t_{pass}\right) \qquad \text{(B15)}$$

REFERENCES

Brecht, W., "A method for the comparative evaluation of bar equipped beating devices," *TAPPI*, **50**(8): 40A-44A (Aug. 1967)

Danforth, D.W., "Stock preparation: theory/practice," *Southern Pulp and Paper Manufacture*, **32**(7): 52-53 (1969).

Eklund, D. and T. Lindstrom, *Paper Chemistry: An Introduction*, DT PAPER SCIENCE PUBLICATIONS, Grankulla, Finland, pp. 48-52 (1991).

Giertz, H.W., "Fundamental Concepts of refining," In *International Symposium on Refining*, Madison, WI, pp. 324-330 (Sept. 16-18, 1980).

Hartman, R.R. and B.G. Higgins, "Mechanical Treatment of Pulp Fibers for Sheet Property Development," In *International Paper Physics Conf.*, Cape Cod, MA, p 41-50 (Sept. 18-24, 1983).

Hietanen, S., "Development of a Fundamentally New Refining Method," *Paperi ja Puu*, **72**(10): 952-960 (1990).

Hietanen, S., "Effect of Control Variables and Length of Refining Zone on the Refining Result with a New Pilot Refiner," *Paperi ja Puu*, **73**(1): 52-61 (1991).

Hietanen, S. and K. Ebeling, "Fundamental Aspects of the refining Process," *Paperi ja Puu*, **72**(2): 158-170 (1990).

Hietanen, S. and K. Ebeling, "A New Hypothesis for the Mechanics of Refining," *Paperi ja Puu*, **71**(2): 172-179 (1990).

Hietanen, S., and K. Ebeling, "Homogeneity in Refining Action," In *International Paper Physics Conf.*, Cape Cod, MA, pp 27-39 (Sept. 18-24, 1983).

Higgins, B. "Description of Model REFNRI" contained in MAPPS Model Documentation Manual, Release 2.0, Institute of Paper Science and technology, Atlanta, Georgia (August, 1985).

Higgins, H.G. and D. deYong, "The Beating Process: Primary Effects and Their Influence on Pulp and Paper Properties," In *The Formation and Structure of Paper*, F.Bolam, Ed., TECH. SECT. BRIT. PAPE RAND BOARD MAKERS ASSN., **Vol.2**, pp. 651-695 (1962).

Kerekes, R., "Characterization of Pulp refiners by a C-Factor," *Nordic Pulp and Paper Research Journal*, **1**: 3-8 (1990).

Leider, P. and A.H. Nissan, "Understanding the disk refiner," *TAPPI*, **60**(10): 85-90 (1977).

Lumiainen, J., "A New Approach to the Critical Factors Affecting Refining Intensity and Refining Results in Low-Consistency Refining," *Sunds Defibrator*, Jylha Oy Valkeakoski, Finland, 11:18 (1989).

Lumiainen, J., "Refining of Secondary Fibers," Sunds Defibrator, 3rd Annual International Pulping Conference, Finland.

MacDonald, G.R., and J.N Franklin, "*Pulp and Paper Manufacture: Papermaking and Paperboard Making*," **Vol. 3**, 2nd ed., pp. 131-183 (1970).

Steenberg, B., B. Sandgren, and D. Wahren, "Studies on Pulp Crill. Part 1: Suspended Fibrils in Paper Pulp Fines," *Svensk Paperstidning*, **63**(12): 395-397 (1960).

Stevens, W.V., "Refiner analysis offers improved systems efficiency, horsepower usage," *Pulp & Paper*, pp. 64-67 (March 1981).

Szwarcztajn, E. and K. Przybysz, "Investigations of water retention value of beaten cellulose fibers," *Cellulose Chemistry and Technology*, **9**:597-607 (1975).

Turt. V., "Effect of Refining on Fiber Properties," M.S. Thesis, University of Maine (1994).

Turt, V., Genco, J. and Co, A., "Effect of Refining on Fiber Properties," 1994 Tappi Engineering Conference, Chicago, Illinois (1994).

Index

A

Acoustic enhancement 116

Alkali

charges, optimization of 20

profiling 1

B

Biofilter, full scale 124

Black liquor

char, characterization of 67

sprays, distribution of mass flows in 58

Bleaching

extraction, lignin removal during 52

of colored broke 96

sequence, organoslov-based 42

Boilers, kraft recovery 77

C

Calcium oxide, dry HCl scrubbing with 116

Cartons, re-pulping of 104

Carryover deposition, study of 77

Chemicals, using non-chlorine 104

Colored broke, bleaching of 96

D

Deinking, novel approaches to 87

Delignification

fir wood 7

oxygen. 20, 26

shredded spruce 7

Dry HCl scrubbing 116

E

Elemental chlorine free (ECF) processes 96

F

Fir wood, delignfication of 7

Freeness drop coefficient 134

G

Gassification, black liquor char 67

H

Hardboard mill 124

Hardwoods southern 26

K

Kraft

pulping 1

recovery boiler(s) 77, 110

L

Laser-printed paper 87

Lignin removal, limiting 52

Liquor analysis 1

M

Mass

flows, distribution of 58

spectrometry 67

Mill, hardboard 124

Molecular beam mass spectrometry 67

N

Novel

approaches to deinking 87

extraction stage, study of 42

Non-chlorine chemicals, using 104

O

Organoslav-based bleaching sequence 42

Oxygen

delignification. 20, 26

optimization of 20

sodium carbonate and 7

P

Paper, xerographic and laser-printed 87

Physico-chemical processes 52

Poly-coated cartons 104

R

Real-time liquor analysis 1

Repulping of cartons 104

S

Shredded spruce, delignification of 7

Southern hardwoods 26

Specific edge loading, effect of 134

Spectrometry, mass 67

T

Thermogravimetry, using 67

Totally chlorine free (TCF) processes 96

V

Volatile organic compounds (VOCs), reducing 130

W

Wet-line extension, wood drying by 130

Wet-strength cartons 104

Wood drying 130

X

Xerographic paper 87